Artificial Intelligence and Animal Ecology: A Review

Editors

Lidia Ghosh
RCC Institute of Information Technology
Kolkata, India

Amiyangshu De
Indian Institute of Technology Kharagpur
West Bengal, India

CRC Press
Taylor & Francis Group
Boca Raton London New York

CRC Press is an imprint of the
Taylor & Francis Group, an **informa** business
A SCIENCE PUBLISHERS BOOK

First edition published 2026
by CRC Press
2385 NW Executive Center Drive, Suite 320, Boca Raton FL 33431

and by CRC Press
4 Park Square, Milton Park, Abingdon, Oxon, OX14 4RN

CRC Press is an imprint of Taylor & Francis Group, LLC

Library of Congress Cataloging-in-Publication Data (applied for)

ISBN: 978-1-032-78234-8 (hbk)
ISBN: 978-1-032-79573-7 (pbk)
ISBN: 978-1-003-49278-8 (ebk)

DOI: 10.1201/9781003492788

Typeset in Times New Roman
by Prime Publishing Services

Preface

The advent of artificial intelligence (AI) has catalyzed rapid technological progress, driving advancements in complex problem-solving, optimization, and predictive modeling. In the realm of animal ecology, AI has emerged as a transformative tool, enabling detailed analyses of behavioral ecology, habitat utilization, conspecific and heterospecific interactions, and species' adaptive mechanisms in response to environmental variability. AI's computational frameworks are intrinsically linked to natural and ecological processes, drawing from the Darwinian principle of survival of the fittest. This interconnectedness reflects a dual approach: a bottom-up perspective, where ecological systems inspire AI development, and a top-down perspective, where AI deepens our understanding of ecological complexities and aids in the conservation of biodiversity.

Artificial Intelligence and Animal Ecology: A Review, delves into this reciprocal relationship, illustrating how AI models are shaped by biological phenomena and, conversely, how AI enhances ecological research. The book spans diverse interdisciplinary domains, highlighting bio-inspired optimization methods—such as swarm intelligence, evolutionary computation, and predator-prey dynamics—alongside AI-driven ecological modeling and conservation strategies. Techniques like Genetic Algorithms, Particle Swarm Optimization, and Ant Colony Optimization replicate natural processes to improve ecological forecasting, species distribution modeling, and conservation planning.

Contributions from eminent researchers and ecologists provide critical insights into AI's evolution through ecological principles and its application in solving real-world environmental challenges. The chapters explore AI's role in replicating natural behaviors, interpreting species communication networks, understanding interspecific dynamics within fragmented habitats, and addressing ecological stressors in rapidly shifting environments. Additionally, the book examines advanced conservation technologies, predictive ecological models, and resilience strategies. It also projects future trends, positioning AI as a pivotal force in ecological science—offering innovative pathways for biodiversity preservation, ecosystem management, and sustaining ecological equilibrium amid accelerating global environmental change.

We hope this book serves as a valuable resource for researchers, conservationists, and policymakers alike, fostering further innovations in AI-driven ecological research. We extend our sincere gratitude to all contributors, reviewers, and institutions that have supported this endeavour.

Lidia Ghosh
Amiyangshu De

Contents

Preface iii

1. Bio-Inspired Evolutionary and Swarm Optimization Algorithms 1
   Lidia Ghosh and Antara Ghosh
2. When Optimization Techniques are Inspired from Marine Ecology and Terrestrial Animal Behaviors 18
   Amiyangshu De and Manob Jyoti Saikia
3. AI Advancements in Understanding Animal Behavioral Paradigms 44
   Antara Ghosh and Dipti Debnath
4. Vocal Communications in the Animal Kingdom 65
   Naureen Afrose, Rideb Chakraborty, Pratibha Bhowmick and Mithun Bhowmick
5. Predicting Interactions Among Species in Ecological Networks: A Roadmap to Address Eltonian Shortfall 82
   Subhajit Roy, Sayantani Saha and Ayan Mondal
6. Interactions Among Species in Ecological Communications 122
   Chiraz Yemmen and Sami Fattouch
7. AI Tools in Navigating the Nexus: Understanding Behavioral Effects of Altered Animal Ecology on Developmental Disorders 144
   M. Sasirekhamani, J.A. Johnson and G. Chelladurai
8. AI Applications in Addressing and Studying Developmental Challenges 168
   Anoosha Tahir, Aneela Ashraf, Madiha Fatima, Saba Zafar, Naunain Mehmood and Hira Muqaddas
9. Importance of AI-Tech in Managing a Protected Area: Considering Forest Health Dynamics 193
   Amlan Jyoti Kalita , Jyotishman Deka, Chayanika Devi and Pankaj Kalita

10. Exploring the Role of Artificial Intelligence in Microbial Identification, Environmental Distribution, and Ecological Interactions 221
Abrar Hussain, Areej Alvi, Shahana Rasheed Qureshi, Sara Parveen, Um-E-Habiba-U-Nisa and Syed Abid Ali

11. Future Directions and Emerging Technologies in AI for Animal Ecology 244
Nik Fadzly and Zi-Han Yeo

Index 267

1 | Bio-Inspired Evolutionary and Swarm Optimization Algorithms

Lidia Ghosh* **and Antara Ghosh**

Department of Computer Application, RCC Institute of Information Technology, Kolkata, West Bengal, India

This chapter provides a comprehensive overview of bio-inspired optimization algorithms, focusing on evolutionary and swarm intelligence techniques that draw on natural processes. By mimicking behaviors observed in biological entities—such as the survival strategies of animal groups, evolutionary adaptations, and swarm dynamics—these algorithms offer robust solutions for complex optimization problems across various domains. Core algorithms discussed include Genetic Algorithms (GAs), Differential Evolution (DE), Particle Swarm Optimization (PSO), Firefly Algorithm (FA), and others, each representing unique strategies to balance exploration and exploitation within the search space. Additionally, the chapter explores recent applications of these algorithms in fields such as engineering, healthcare, and finance, highlighting their adaptability and efficiency in solving real-world optimization challenges.

Introduction

Artificial intelligence (AI) has become a cornerstone in tackling a vast array of scientific and engineering problems, from information processing to intricate optimization challenges. Over the years, a range of techniques, including genetic algorithms, neural networks, evolutionary algorithms, and fuzzy logic, has evolved to meet the demands of these complex tasks (Fathi & Parian, 2021; Unal & Basiftchi, 2022; Kouhalvandie et al., 2022; Al-Qaysi et al., 2023). These intelligent optimization methods have proven invaluable across domains such as engineering, science, medicine, and satellite technology, especially in anomaly detection and system fault management. The driving inspiration for these advancements often stems from the natural world, particularly

*Corresponding author: lidiaghosh21@gmail.com

the balanced processes observed in ecosystems, commonly referred to as "ecological equilibrium".

The collaborative behaviors observed in various species, like ant colonies, bird flocks, and bee swarms, have significantly influenced algorithm development in fields such as computer science (Sharma et al., 2022). By emulating these interactions, researchers have created swarm intelligence algorithms that exhibit properties such as adaptability, scalability, and self-organization—qualities that enhance performance in problem-solving applications (Kaswan et al., 2023). These metaheuristics mimic animal and insect strategies, allowing for the efficient search of optimal solutions across numerous fields without requiring gradient information (Turgut et al., 2023).

Optimization, in essence, is the pursuit of an ideal solution, though this can be impractical for complex problems that would require exhaustive searches (Wu et al., 2022). Instead, metaheuristic algorithms solve such issues by imitating group behaviors observed in nature. For instance, algorithms like Artificial Bee Colony (ABC), Particle Swarm Optimization (PSO), and Grey Wolf Optimizer (GWO) mimic the behaviors of bees, fish, birds, and wolves, respectively (Hackett, 2020, Darvishpoor et al., 2023). The foundation of these optimization algorithms is rooted in natural evolution, with significant contributions from Charles Darwin and John Holland, who pioneered Genetic Algorithms (GAs) in the 1960s, utilizing concepts such as selection, crossover, and mutation. The 1980s saw the advent of swarm intelligence techniques inspired by the behavior of ants, bees, and birds. Ant Colony Optimization (ACO), introduced by Marco Dorigo in the 1990s and modeled on ant foraging, alongside Kennedy and Eberhart's Particle Swarm Optimization, which simulates bird and fish social behaviors, exemplify this approach.

The synergy of swarm intelligence with evolutionary algorithms has produced robust optimization techniques that harness the strengths of both methods (Li, Ke, 2021; Chong et al., 2021; Rojas et al., 2022). These bio-inspired algorithms support a wide spectrum of applications, from optimizing construction parameters and aerodynamic designs to coordinating robotics and telecommunications networks (Tang et al., 2021; Suchi et al., 2023; Sahu et al., 2022). In healthcare, they assist in tasks like molecular docking and image segmentation, while in finance, they are used for portfolio management. They also contribute to sustainable development by optimizing renewable energy systems and improving neural network architectures, demonstrating versatility in solving large, nonlinear, and constraint-laden problems (Roni et al., 2022; Berditchevskaia et al., 2022; Ünal et al., 2022).

Evolutionary Algorithms

Genetic Algorithms

The natural world has long inspired technological innovation, and Genetic Algorithms (GAs) are a prime example, drawing on principles of natural selection and genetics to solve complex optimization problems (Liu et al., 2023). GAs use a 'population' of candidate solutions that undergo processes akin to genetic recombination and mutation, gradually evolving towards optimal solutions based on a fitness function that reflects the problem's objectives (Omidvar et al., 2021; Jiang et al., 2024). Higher-fitness solutions are more likely to be reproduced, aligning with Darwin's "survival of the fittest" (D'Angelo & Palmieri, 2021). GAs introduce randomness while using historical data from previous generations to guide searches, distinguishing them from simple random methods. Key elements include the population of candidate solutions, chromosomes (representing solutions), genes (solution elements), and alleles (gene values), as shown in Figure 1. The algorithm iterates through processes of fitness evaluation, selection, reproduction, and adaptation to evolve solutions (Crespo-Herrara et al., 2021).

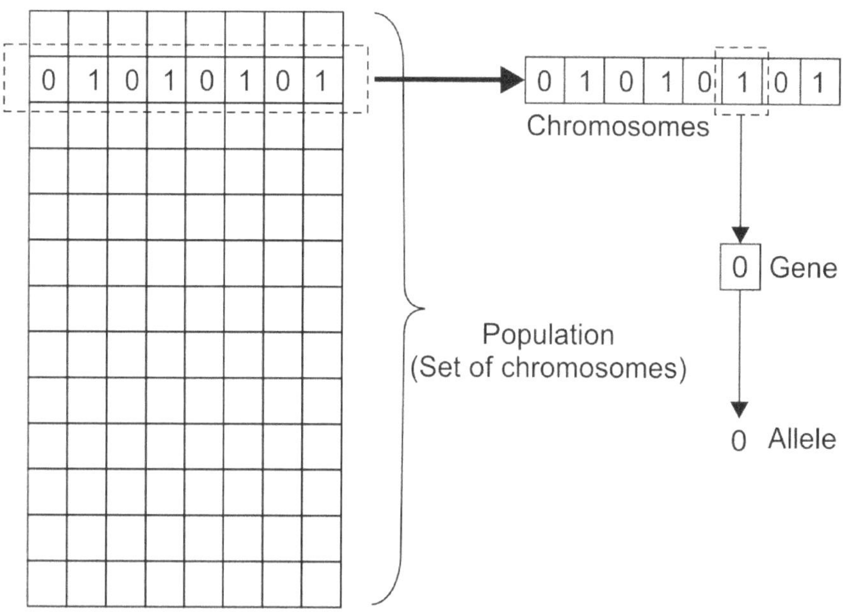

Figure 1: Example of genetic algorithm.

In computational applications, GAs represent solutions as genotypes (encoded forms) that may differ from phenotypes (real-world representations) (Haghrah et al., 2021; Banzhaf, & Bakurov, 2024). For instance, in the 0/1 Knapsack Problem (Setzer, et al., 2020), the phenotype solution is a chosen set of items, while the genotype represents this as a binary sequence. Through selection, crossover, and mutation, GAs achieve a balance between diversity and convergence. **Selection** is critical, with methods like **Fitness Proportionate Selection** (e.g., Roulette Wheel) and **Tournament Selection** ensuring that fitter candidates are more likely to reproduce (Al Hijjawi & Awajan, 2024). **Rank Selection** bases selection on rank rather than fitness, preserving diversity, while **Random Selection** is less commonly used due to lack of direction.

Crossover combines parental genes to create offspring. Techniques include:

- **One-Point Crossover:** Swaps genes after a random crossover point, balancing simplicity and efficiency.
- **Multi-Point Crossover:** Allows more diversity but can disrupt beneficial gene sequences.
- **Uniform Crossover:** Randomly selects genes from either parent, preserving diversity but risking disruption of effective gene combinations.
- **Whole Arithmetic Recombination:** Useful for real-valued chromosomes, averaging parental genes.
- **Order Crossover (OX1):** Preserves sequence order for scheduling or routing problems.

Mutation introduces diversity, with methods like:

- **Bit Flip Mutation:** Flips bits in binary chromosomes, simple but limited to binary cases.
- **Random Resetting:** Assigns random values within the gene range, good for integer encodings.
- **Swap Mutation:** Swaps gene positions to vary sequence order, particularly for permutations.
- **Scramble Mutation:** Shuffles gene subsets to explore new solutions.
- **Inversion Mutation:** Reverses gene order within a subset, useful for problems needing sequence integrity.

Through these mechanisms, GAs effectively navigate complex search spaces, adapting solutions iteratively for enhanced optimization.

Differential Evolution

Differential Evolution (DE) is a nature-inspired optimization algorithm (Sheta et al., 2020), drawing on principles from animal ecology rather than specific animal behaviors. DE's mechanisms reflect the adaptive dynamics seen in ecosystems, where populations evolve in response to environmental pressures.

1. **Population-based Search (Ecosystem Analogy):** DE maintains a population of candidate solutions throughout the optimization process, akin to a group of species adapting within an ecosystem.
2. **Mutation (Ecological Adaptation):** DE modifies individuals by adding the scaled difference between two random solutions to another. This mirrors genetic variation in ecosystems, where mutation aids adaptation to environmental changes.
3. **Crossover (Combination of Genetic Traits):** In DE, solutions recombine similarly to gene exchange in nature, which increases genetic diversity and enhances resilience to changing environments.

4. **Selection (Survival of the Fittest):** DE selects individuals based on performance, akin to Darwin's "survival of the fittest", where the best-suited organisms are more likely to survive and reproduce.
5. **Exploration and Exploitation (Ecosystem Dynamics):** DE balances exploration (searching new areas) and exploitation (refining existing solutions). This mirrors how species expand into new niches or optimize behaviors to thrive in their environments.
6. **Diversity in Population (Biodiversity):** Maintaining diversity in DE helps avoid premature convergence and enhances robustness, much like biodiversity strengthens ecosystems against disturbances.
7. **Self-organizing Behavior (Emergence):** DE's iterative process of self-organization resembles ecosystems, where populations adapt and balance interactions within their environment.

While DE does not imitate specific animal behaviors, its principles—population dynamics, adaptation, variation, and selection—are rooted in ecological processes, showing how nature-inspired algorithms can effectively tackle complex optimization challenges.

Swarm Intelligence Algorithms

Particle Swarm Optimization

Particle Swarm Optimization (PSO) (Gad, 2022) draws inspiration from the collective behavior of organisms like birds and fish working together toward a shared objective. In PSO, a swarm of particles, each representing a potential solution, explores the solution space to identify the optimal result (Passaro et al., 2008; Kiranyaz et al., 2009). Each particle updates its position based on its personal best-known position (pbest) and the global best position (gbest) achieved by the swarm, allowing for iterative convergence towards optimal solutions (Jain et al., 2022). This collaborative approach makes PSO particularly effective for diverse optimization problems.

Unlike GAs, where competition drives selection, PSO relies on cooperation, with successful particles influencing their neighbors (Li et al., 2017). Particles continuously share information, adjusting their search paths based on personal and collective successes, thereby moving closer to the global optimum. PSO has been successfully applied in fields like agriculture, finance, geology, and climate science. In some cases, particles also consider a local best (lbest) when interacting within a subset of neighbors, further refining the search process (Rahmani et al., 2012; Turgut & Turgut, 2020; Hu et al., 2020). A flowchart in Figure 2a outlines the organized execution steps for a clearer understanding of the PSO process.

Firefly Algorithm

Fireflies are small, nocturnal insects known for their ability to produce bioluminescent light, which is generated chemically in the lower abdomen and

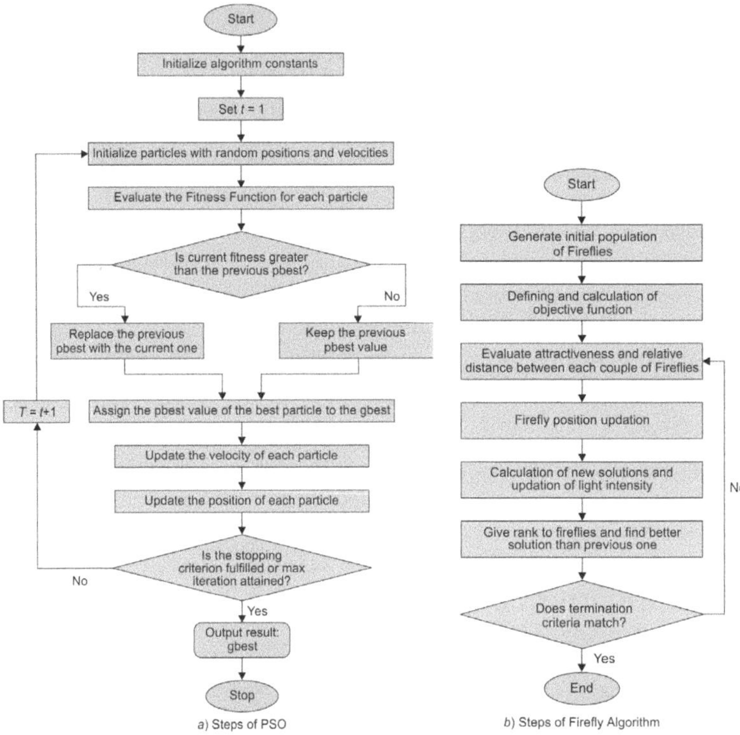

Figure 2: Flowcharts explaining the various Swarm Optimization Algorithms.

lacks infrared or ultraviolet frequencies (Carlson & Copeland, 1985). Fireflies use this light primarily for mating, predation, and as a defense mechanism (Evon, 2020). Their flashing patterns inspired the Firefly Algorithm (FA), a metaheuristic optimization method based on attraction behavior. The FA relies on three core principles:

1. Fireflies are unisexual, allowing any firefly to attract another.
2. A firefly's brightness (linked to the optimization objective) determines its attractiveness, with brighter fireflies attracting others more strongly. This attraction decreases with distance.
3. Fireflies with the same brightness move randomly, aiding exploration.

The FA's attraction mechanism leads fireflies to cluster around local optima, while random walks help explore new solutions. This balance enables FA to efficiently solve both global and local optimization problems, including constrained and NP-hard challenges, often outperforming traditional algorithms with parallel processing capabilities.

FA has been applied across numerous domains due to its versatility and effectiveness. In engineering, it has optimized structural designs like pressure tanks, beams, and springs, consistently outperforming algorithms like PSO, DE, and SA. In clustering, FA is effective for data grouping in data mining, pattern detection, and customer segmentation. In routing, it has been used to find optimal paths in transportation and communication networks, reducing delays in logistics. FA also supports industrial scheduling, improving resource utilization and meeting deadlines, while in image processing, it enhances segmentation quality and reduces image size. In healthcare, FA aids in image denoising, tumor detection, and optimizing drug delivery systems.

Ant Colony Optimization

In the 1950s, French entomologist Pierre-Paul Grassé introduced 'stigmergy', a form of indirect communication where insects modify their environment, leaving cues that guide others (Abdolrasol et al., 2021). This concept explains collective behaviors like those seen in ant colonies, where ants deposit pheromones along paths to food, prompting others to follow the most marked routes. This process optimizes resource gathering through positive feedback and autocatalysis, as demonstrated by Deneubourg et al. in the "double bridge experiment". In this study, ants initially chose between two equal-length bridges randomly, but over time, the bridge with more pheromones became the preferred path, showing their capacity for optimizing path selection (Heylighen, 2011; Way & Khoo, 1992; Chen et al., 2021).

In ant colony optimization (ACO), this behavior is applied to solve problems like the traveling salesman problem, where ants (simulated agents) move across a graph that represents cities, with each edge representing a route (Dorigo et al., 2003; Ariyasingha and Fernando, 2015). The traveling salesman's problem has a set of cities which are given and all the distances between them are known. The goal is to determine the minimum distance that would be needed to pass through each city exactly once and come back to the starting city. More precisely, the problem is to identify a Hamilton cycle that is of minimum cost in a fully connected graph. In ACO, the problem is tackled by simulating a number of artificial ants moving on a graph that encodes the problem itself: in it, each vertex indicates a city and each edge stands for a relationship between two cities and means a route that can be used. Every edge is coupled with a variable referred to as pheromone collected and then read and modified by ants (Montgomery, 2005). ACO is a metaheuristic algorithm and it is used iteratively (Dorgio & Stuzle, 2003). At each of the iterations, a number of artificial ants are taken into consideration. Each of them constructs a solution by proceeding from vertex to vertex on the graph with the provision that a vertex, which the current girl has not visited in her walk, must be visited. Indeed, at every step of the solution construction, an ant chooses the following vertex to be visited according to a stochastic mechanism that is biased by the pheromone. In the case of the vertex i, the next conduct is chosen randomly with the new holes that have never been visited before (as shown in the Figure 3a). Notably, if j has not been visited before, then the minimum distance within j is equal to the first element of the Distance list. chosen with probability that depends on the pheromone related to edge (i, j), which must be proportional to the given pheromone. At the end of an iteration, depending on the quality of the solutions that ants have built, the values of the pheromones are changed with the intention that in the following iterations the ants will lean toward building better solutions.

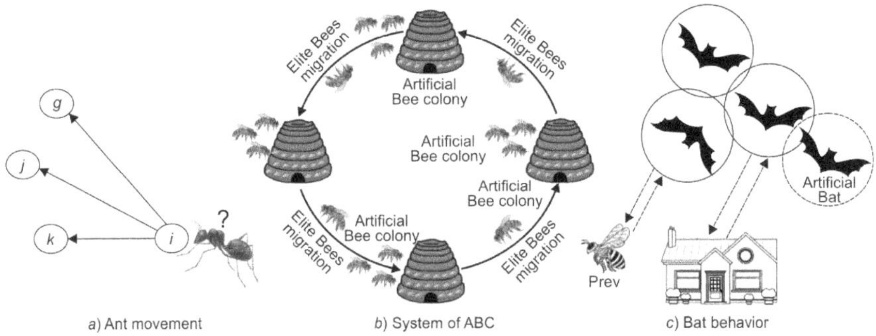

Figure 3: Various swarm optimization algorithms inspired by animal behavior.

Artificial Bee Colony

The Artificial Bee Colony (ABC) algorithm (Karaboga et al., 2011) is inspired by the complex, efficient behaviors observed in natural bee swarms, particularly their foraging process (Figure 3b). Bees in a swarm are divided into employed bees, onlooker bees, and scout bees, each playing a specific role in gathering food and communicating its location, quality, and distance to the rest of the hive (Tasgetiren et al., 2013). Employed bees explore and exploit food sources, then share their findings through specific dances. Onlooker bees select food sources based on these dances, while scout bees search for new sources when existing ones are depleted (Beekman & Lew, 2008).

Honey bees use two primary dances to convey food source information: the round dance, indicating nearby sources, and the more detailed waggle dance, which communicates direction and distance relative to the sun. This form of communication optimizes the foraging process and exemplifies self-organization and division of labor (Gardner et al., 2008; Rohrseitz & Tantz, 1999). The ABC algorithm replicates this behavior as a metaheuristic for optimization. It begins by generating initial food sources (solutions), which the employed bees evaluate and communicate to onlookers. Scout bees search for new sources if needed, iterating this process until the best solution is found. Enhancements to the basic ABC, such as the Interactive ABC (IABC) and ABCgBest variants, improve convergence and solution diversity through modified selection and control parameters, outperforming other heuristic methods in some cases (Srinivasan, 2011; Awadallah et al., 2019).

ABC has been widely applied across domains, including function optimization, job scheduling, and the Traveling Salesman Problem (TSP). In engineering, it optimizes assembly line balancing and parameter tuning in machine learning models. ABC has also proven effective in neural network training, image processing, medical applications like tumor detection, and ECG signal denoising. In wireless sensor networks, it manages energy and data transmission, while in finance, it supports portfolio optimization (Karboga & Oztu, 2011). Through these applications and continuous improvements, the ABC algorithm—rooted in the natural foraging behaviors of honey bees—has become a versatile and powerful tool in computational intelligence.

Bat Algorithm

The Bat Algorithm (BA) is a bio-inspired optimization method that mimics the echolocation behavior of bats (Russ, 2021). In nature, bats use bio-sonar to navigate and hunt by emitting ultrasonic waves that bounce off objects; the returning echoes provide spatial information to help them avoid obstacles. BA replicates this process, treating each virtual bat as a potential solution that 'navigates' the search space using echolocation (shown in Figure 3c). The algorithm uses parameters such as loudness and pulse frequency to balance exploration and exploitation, similar to how bats modulate these factors in response to their environment. Higher loudness values facilitate broader exploration, while lower values focus on refining known solutions, increasing the chances of finding a global optimum.

The algorithm updates the frequency, velocity, and position of each bat at each time step using the following equations:

$$f_i = f_{min} + (f_{max} - f_{min}) \cdot \beta \tag{1}$$

$$v_{it} = v_{i(t-1)} + (x_{i(t-1)} - x^*) \cdot f_i \tag{2}$$

$$x_{it} = x_{i(t-1)} + v_{it} \tag{3}$$

Here, $\beta \in [0,1]$ is a random number, and f_i represents the frequency for each bat. The best global solution x^* is determined by comparing all bats' positions at each iteration. A new solution is generated based on the best solution when a random number exceeds the pulse emission rate, as shown in equation (4):

$$X_{new} = X_{old} + \epsilon A_t \tag{4}$$

where $\epsilon \in [-1,1]$ is a random number, and A_t represents the average loudness of all bats at the current iteration. Loudness A_i and pulse emission rate r_i are updated as follows:

$$A_{i(t+1)} = \alpha A_{it}, \tag{5}$$

and

$$r_{i(t+1)} = r_{i0}\,[1 - \exp\exp(\gamma t)] \tag{6}$$

where α and γ are constants. The algorithm continues iterating until a termination condition is met.

BA is widely applied due to its flexibility and effectiveness. In feature selection, it optimizes classification models by reducing features while preserving accuracy. In image processing, it aids in thresholding, compression, and enhancement, improving quality and reducing file size. BA is also effective in control engineering for tuning parameters to maintain system stability. In speech and signal processing, it reduces noise and enhances quality, and in image segmentation, it determines optimal intensity thresholds, making BA a versatile tool across multiple fields.

Nature-Inspired Algorithms

Cuckoo Search Algorithm

The Cuckoo Search Algorithm (CSA) is a metaheuristic inspired by the parasitic breeding behavior of certain cuckoo species (Figure 4a). These birds lay their eggs in the nests of other species, sometimes removing host eggs to improve their own offspring's chances of survival. By mimicking host eggs and hatching earlier, cuckoo chicks often push out the host's eggs, thus ensuring their own survival and enhancing reproductive success. CSA captures this strategy in its algorithmic structure, where 'nests' represent potential solutions, and new solutions (eggs) replace poorer ones based on a probability factor.

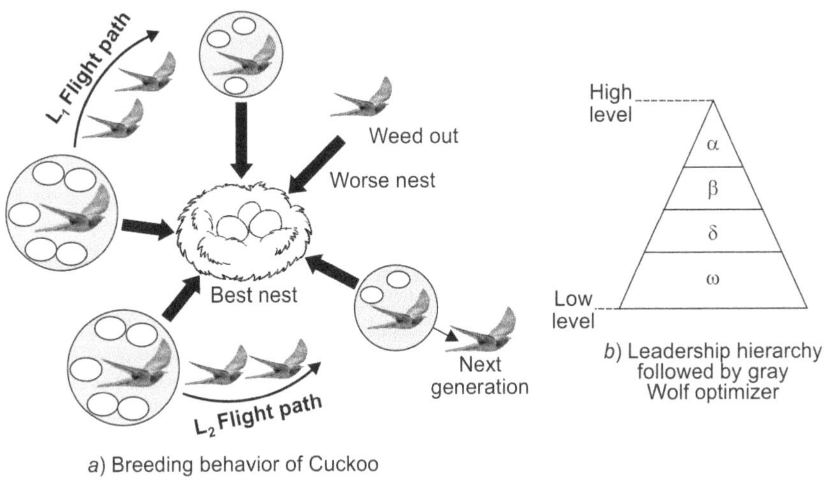

Figure 4: Cuckoo search algorithm and gray wolf optimizer algorithm inspired by animal behvior.

CSA incorporates the Levy flight mechanism, a natural foraging pattern, which enhances its ability to explore large search spaces effectively (Singh, 2021). This blend of behaviors creates a strong balance between exploration and exploitation, making CSA well-suited for diverse optimization tasks. Applications include neural networks, where it optimizes weights and biases, and support vector machines for parameter tuning. CSA is also used in wireless sensor networks for node deployment and clustering, and in image processing for tasks like enhancement and compression. In engineering, it improves traffic flow, tunnel construction, water distribution, and aerodynamics (Bunde, 2018; Ouallane et al., 2022).

In CSA, each cuckoo represents a candidate solution, while nests symbolize the search space. Levy flights (paths L1 and L2) guide cuckoos through the search space, helping them locate the "best nest" (optimal solution). Poor solutions are discarded, and better solutions form the next generation. This iterative process continues until the algorithm converges toward the optimal solution, balancing exploration and exploitation to identify high-quality results.

Gray Wolf Optimizer

Social Hierarchy and Hunting Behavior

The gray wolf (Canis lupus) is an apex predator that lives in packs, typically consisting of 5–12 members (El-Kenawy et al., 2020), organized in a strict social hierarchy essential for survival and hunting efficiency (Figure 4b). At the top are the alpha wolves, a dominant male and female, responsible for making key decisions on hunting and daily activities (Hajihosseini & Hutcherson, 2021). Alphas lead by example and command respect, although they may occasionally mirror other pack members' actions to maintain group harmony (Dubey et al., 2021, Nakra et al., 2024).

Supporting the alphas are the beta wolves, who assist in decision-making and uphold the pack's structure (Ghasemi et al., 2021). In the alpha's absence, a beta may take on leadership. Omega wolves occupy the lowest rank, often diffusing pack tensions by assuming submissive roles, and sometimes nurturing younger or injured pack members (Entrikin, 2023). Wolves that are neither alphas, betas, nor omegas are classified as deltas; they act as community leaders, guards, and hunters, maintaining pack security and resources. The hunting behavior of gray wolves also reflects their complex social structure and is characterized by several phases:

a. Tracking, chasing, and approaching the prey.
b. Stalking and encircling the prey until it can no longer escape.
c. Coordinated attacks to subdue the prey.

Algorithmic Framework

The GWO algorithm models these social dynamics within the wolf pack. In this framework, the optimal solution is represented by the Alpha wolf (α), the second-best solution by the Beta wolf (β), and the third-best solution by the Delta wolf (δ). The remaining solutions correspond to Omega wolves (ω).

a. Encircling the Prey:

$$\vec{D} = |\vec{C} \cdot \vec{X}_p(t) - \vec{X}_p(t)| \tag{7}$$

$$\vec{X}(t+1) = \vec{X}_p(t) - \vec{A} \cdot \vec{D} \tag{8}$$

Here, t indicates the current iteration, $\vec{A}$ and $\vec{C}$ are coefficient vectors, $\vec{X}_p$ is the position vector of the prey, and $\vec{X}$ indicates the position vector of a gray wolf. The following equations describe the position updates of the wolves:

$$\vec{X}_1(t+1) = X(t) - \vec{A}_1 \cdot \vec{D} \tag{9}$$

$$\vec{A} = 2\vec{a}\vec{r}_1 - \vec{a} \text{ and } \vec{C} = 2\vec{r}_2 \tag{10}$$

The components of decrease linearly from 2 to 0 over the course of iterations, and $\vec{r}_1$, $\vec{r}_2$ are random vectors in [0,1]. Omega wolves adjust their positions based on the positions of the alpha, beta, and delta wolves, benefiting from their superior knowledge of potential prey locations:

$$\vec{D}_\alpha = |\vec{C}_1 \cdot \vec{X}_\alpha(t) - \vec{X}(t)|, \tag{11}$$

$$\vec{D}_\beta = |\vec{C}_2 \cdot \vec{X}_\beta(t) - \vec{X}(t)|, \tag{12}$$

$$\vec{D}_\gamma = |\vec{C}_3 \cdot \vec{X}_\gamma(t) - \vec{X}(t)| \tag{13}$$

Updating positions for the wolves is performed as follows:

$$\vec{X}_1(t+1) = \vec{X}_\alpha(t) - \vec{A}_1 \cdot \vec{D}_\alpha, \tag{14}$$

$$\vec{X}_2(t+1) = \vec{X}_\beta(t) - \vec{A}_2 \cdot \vec{D}_\beta, \tag{15}$$

$$\vec{X}_3(t+1) = \vec{X}_\gamma(t) - \vec{A}_3 \cdot \vec{D}_\gamma \tag{16}$$

Finally, the overall position is determined by:

$$\vec{X}(t+1) = \frac{\vec{X}_1 + \vec{X}_2 + \vec{X}_3}{3} \tag{17}$$

b. Attacking Prey (Exploitation):
To simulate the gray wolves' attacking behavior, the value of $\vec{a}$ should be less. The coefficient $\vec{A}$ varies randomly within the interval [–2a,2a], with $|A| < 1$ leading to an assault on the target (exploitation).

c. Searching for Prey (Exploration):
When exploring for prey, gray wolves may deviate from their current targets in pursuit of more suitable options. The component $\vec{C}$ in the GWO encourages exploration, with a random value ranging from [0,2]. A value of $C > 1$ emphasizes attack, while $C < 1$ de-emphasizes it.

GWO has been widely applied across various fields due to its flexibility and effectiveness in solving complex optimization problems. In engineering design, it has been used for optimizing operations and tuning controllers with superior results compared to other algorithms. It has also been applied in scheduling, particularly in solving the unit commitment (UC) problem, demonstrating its efficiency across different system scales. In robotic path planning, GWO has been integrated with other algorithms for enhanced localization, while in power dispatch problems, it has outperformed several algorithms in finding optimal solutions. Additionally, GWO has been utilized for clustering tasks and in financial applications like bankruptcy prediction, where it evolves advanced models for higher accuracy and performance.

Conclusion

Bio-inspired optimization algorithms have demonstrated exceptional adaptability and effectiveness in addressing a diverse array of optimization problems. By leveraging principles from evolution, swarm behavior, and ecological interactions, these algorithms have achieved notable success across industries, from engineering and robotics to healthcare and finance. The collaborative, adaptive characteristics of swarm intelligence models, coupled with the selection-driven mechanisms of evolutionary algorithms, contribute to their broad applicability and continued relevance. Future advancements may focus on hybrid models and improved convergence rates, fostering innovations in optimization methodologies. The continuous development of these algorithms, inspired by nature, promises significant contributions to complex problem-solving in both theoretical and practical contexts.

References

Abdolrasol, Maher G.M., S.M. Suhail Hussain, Taha Selim Ustun, Mahidur R. Sarker, Mahammad A. Hannan, Ramizi Mohamed et al. (2021). Artificial neural networks based optimization techniques: A review. *Electronics,* 10(21): 2689.

Al Hijjawi, Bushra, and Arafat Awajan. (2024). Genetic algorithms: Theory, genetic operators, solutions, and applications. *Evolutionary Intelligence,* 17(3): 1245–1256.

Al-Qaysi, Ahmed Mazin Majid, Altug Bozkurt, and Yavuz Ates. (2023). Load forecasting based on genetic algorithm – Artificial neural network-adaptive neuro-fuzzy inference systems: A case study in Iraq. *Energies,* 16(6): 2919.

Ariyasingha, I.D.I.D., and T.G.I. Fernando. (2015). Performance analysis of the multi-objective ant colony optimization algorithms for the traveling salesman problem. *Swarm and Evolutionary Computation,* 23: 11–26.

Awadallah, Mohammed A., Mohammed Azmi Al-Betar, Asaju La'aro Bolaji, Emad Mahmoud Alsukhni, and Hassan Al-Zoubi. (2019). Natural selection methods for artificial bee colonies with new versions of onlooker bees. *Soft Computing,* 23(15): 6455–6494.

Banzhaf, Wolfgang, and Illya Bakurov. (2024). On the nature of the phenotype in tree genetic programming. In: *Proceedings of the Genetic and Evolutionary Computation Conference*, pp. 868–877. Melbourne VIC Australia.

Beekman, Madeleine, and Jie Bin Lew. (2008). Foraging in honeybees—When does it pay to dance? *Behavioral Ecology,* 19(2): 255–261.

Berditchevskaia, Aleks, Eirini Maliaraki, and Konstantinos Stathoulopoulos. (2022). A descriptive analysis of collective intelligence publications since 2000, and the emerging influence of artificial intelligence. *Collective Intelligence,* 1(1): 26339137221107924.

Bunde, Armin, Jürgen Caro, Jörg Kärger, and Gero Vogl (Eds). *Diffusive Spreading in Nature, Technology and Society*. Cham: Springer International Publishing.

Carlson, Albert D., and Jonathan Copeland. (1985). Flash communication in fireflies. *The Quarterly Review of Biology,* 60(4): 415–436.

Chen, Yongxin, Tryphon T. Georgiou, and Michele Pavon. (2021). Stochastic control liaisons: Richard Sinkhorn meets Gaspard Monge on a Schrodinger bridge. *Siam Review,* 63(2): 249–313.

Chong, Hue Yee, Hwa Jen Yap, Shing Chiang Tan, Keem Siah Yap, and Shen Yuong Wong. (2021). Advances of metaheuristic algorithms in training neural networks for industrial applications. *Soft Computing,* 25(16): 11209–11233.

D'Angelo, Gianni, and Francesco Palmieri. (2021). GGA: A modified genetic algorithm with gradient-based local search for solving constrained optimization problems. *Information Sciences,* 547: 136–162.

Darvishpoor, Shahin, Amirsalar Darvishpour, Mario Escarcega, and Mostafa Hassanalian. (2023). Nature-inspired algorithms from oceans to space: A comprehensive review of heuristic and meta-heuristic optimization algorithms and their potential applications in drones. *Drones,* 7(7): 427.

Dorigo, Marco, and Thomas Stützle. (2003). The ant colony optimization metaheuristic: Algorithms, applications, and advances. *Handbook of Metaheuristics,* 250–285.

Dubey, Mahima, Vijay Kumar, Manjit Kaur, and Thanh-Phong Dao. (2021). A systematic review on harmony search algorithm: Theory, literature, and applications. *Mathematical Problems in Engineering,* 2021(1): 5594267.

El-Kenawy, El-Sayed M., Marwa Metwally Eid, Mohamed Saber, and Abdelhameed Ibrahim. (2020). MbGWO-SFS: Modified binary grey wolf optimizer based on stochastic fractal search for feature selection. *IEEE Access,* 8: 107635–107649.

Entrikin, Kelsey. (2023). Predatory seduction: Scenting as a catalyst for power hierarchy in Omegaverse fanfiction. *Transformative Works and Cultures,* 40.

Evon, Carrie Rose. (2020). Messy Postcolonialism in Joss Whedon's Firefly-Serenity Verse. *Slayage,* 18(2): 34–63.

Fathi, Milad, and Jafar Amiri Parian. (2021): Intelligent MPPT for photovoltaic panels using a novel fuzzy logic and artificial neural networks based on evolutionary algorithms. *Energy Reports*, 7: 1338–1348.

Gad, Ahmed G. (2022). Particle swarm optimization algorithm and its applications: A systematic review. *Archives of Computational Methods in Engineering,* 29(5): 2531–2561.

Gardner, Kathryn E., Thomas D. Seeley, and Nicholas W. Calderone. (2008). Do honeybees have two discrete dances to advertise food sources? *Animal Behaviour,* 75(4): 1291–1300.

Ghasemi, Mohsen, Karamollah Bagherifard, Hamid Parvin, Samad Nejatian, and Kim-Hung Pho. (202). Multi-objective whale optimization algorithm and multi-objective gray wolf optimizer for solving next release problem with developing fairness and uncertainty quality indicators. *Applied Intelligence,* 51(8): 5358–5387.

Hackett, Paul M.W. (2020). *The Complexity of Bird Behaviour*. Springer. https://doi.org/10.1007/978-3-030-12192-1.

Haghrah, A., M.A. Nekoui, Morteza Nazari-Heris, and Behnam Mohammadi-Ivatloo. (2021). An improved real-coded genetic algorithm with random walk based mutation for solving combined heat and power economic dispatch. *Journal of Ambient Intelligence and Humanized Computing,* 12(2021): 8561–8584.

Hajihosseini, Azadeh, and Cendri A. Hutcherson. (2021). Alpha oscillations and event-related potentials reflect distinct dynamics of attribute construction and evidence accumulation in dietary decision making. *Elife,* 10: e60874.

Heylighen, Francis. (2011). *Stigmergy as a Generic Mechanism for Coordination: Definition, Varieties, and Aspects*. Available from: http:/pespmc1. vub. ac. be/Papers/Stigmergy-WorkingPaper. pdf (accessed 04.2014)

Hu, Wang, Yu Zhang, Junjie Hu, Yan Qi, and Guoming Lu. (2020). A fast particle swarm optimization algorithm by refining the global best solution. In: *Proceedings of the 2020 2nd Asia Pacific Information Technology Conference*, pp. 128–135.

Jain, Meetu, Vibha Saihjpal, Narinder Singh, and Satya Bir Singh. (2022). An overview of variants and advancements of the PSO algorithm. *Applied Sciences,* 12(17): 8392.

Jiang, Feifeng, Jun Ma, Christopher John Webster, Alain JF Chiaradia, Yulun Zhou, Zhan Zhao, and Xiaohu Zhang. (2024). Generative urban design: A systematic review on problem formulation, design generation, and decision-making. *Progress in Planning,* 180: 100795.

Karaboga, Dervis, and Celal Ozturk. (2011). A novel clustering approach: Artificial Bee Colony (ABC) algorithm. *Applied Soft Computing,* 11(1): 652–657.

Kaswan, Kuldeep Singh, Jagjit Singh Dhatterwal, and Avadhesh Kumar. (2023). *Swarm Intelligence: An Approach from Natural to Artificial*. John Wiley & Sons.

Kiranyaz, Serkan, Turker Ince, Alper Yildirim, and Moncef Gabbouj. (2009). Fractional particle swarm optimization in multidimensional search space. *IEEE Transactions on Systems, Man, and Cybernetics, Part B (Cybernetics)*, 40(2): 298–319.

Kouhalvandie, Lida, Ibraheem Shayea, Serdar Ozoguz, and Hafizal Mohamad. (2022). Overview of evolutionary algorithms and neural networks for modern mobile communication. *Transactions on Emerging Telecommunications Technologies,* 33(9): e4579.

Li, Ke. (2021). Decomposition multi-objective evolutionary optimization: From state-of-the-art to future opportunities. arXiv preprint arXiv:2108.09588.

Li, Lingling, Licheng Jiao, Jiaqi Zhao, Ronghua Shang, and Maoguo Gong. (2017). Quantum-behaved discrete multi-objective particle swarm optimization for complex network clustering. *Pattern Recognition,* 63: 1–14.

Liu, Songbai, Qiuzhen Lin, Jianqiang Li, and Kay Chen Tan. (2023). A survey on learnable evolutionary algorithms for scalable multiobjective optimization. *IEEE Transactions on Evolutionary Computation,* 27(6): 1941–1961.

Luo, Zhaohui, Wei Luo, Junhang Xie, Jian Xu, and Longyan Wang. (2022). A new three-dimensional wake model for the real wind farm layout optimization. *Energy Exploration & Exploitation,* 40(2): 701–723.

Montgomery, Erin James. (2005). *Solution Biases and Pheromone Representation and Selection in Ant Colony Optimisation*. PhD diss., Bond University.

Nakra, Abhilasha, and Manoj Duhan. (2024). Nature-inspired harmony search algorithm: Latest developments and variants. *IUP Journal of Computer Sciences,* 18(1).

Omidvar, Mohammad Nabi, Xiaodong Li, and Xin Yao. (2021). A review of population-based metaheuristics for large-scale black-box global optimization—Part I. *IEEE Transactions on Evolutionary Computation,* 26(5): 802–822.

Ouallane, Asma Ait, Assia Bakali, Ayoub Bahnasse, Said Broumi, and Mohamed Talea. (2022). Fusion of engineering insights and emerging trends: Intelligent urban traffic management system. *Information Fusion,* 88: 218–248.

Passaro, Alessandro, and Antonina Starita. (2008). Particle swarm optimization for multimodal functions: A clustering approach. *Journal of Artificial Evolution and Applications,* 2008: 1–15

Rahmani, Rasoul, Mohd Fauzi Othman, Rubiyah Yusof, and Marzuki Khalid. (2012). Solving economic dispatch problems using particle swarm optimization by an evolutionary technique for initializing particles. *Journal of Theoretical and Applied Information Technology,* 46(2): 526–536.

Rohrseitz, K., and J. Tantz. (1999). Honey bee dance communication: Waggle run direction coded in antennal contacts? *Journal of Comparative Physiology A,* 184: 463–470.

Rojas, Juan Miguel Lujano, Rodolfo Dufo Lopez, and Jose Antonio Dominguez Navarro. (2022). *Genetic Optimization Techniques for Sizing and Management of Modern Power Systems*. Elsevier.

Russ, Jon (Ed.). (2021). *Bat Calls of Britain and Europe: A Guide to Species Identification.* Pelagic Publishing Ltd.

Sahu, Adarsh, Jyotika Mishra, and Namrata Kushwaha. (2022). Artificial intelligence (AI) in drugs and pharmaceuticals. *Combinatorial Chemistry & High Throughput Screening,* 25(11): 1818–1837.

Setzer, Thomas, and Sebastian M. Blanc. (2020). Empirical orthogonal constraint generation for Multidimensional 0/1 Knapsack Problems. *European Journal of Operational Research,* 282(1): 58–70.

Sharma, Abhishek, Abhinav Sharma, Jitendra Kumar Pandey, and Mangey Ram. (2022). *Swarm Intelligence: Foundation, Principles, and Engineering Applications*. CRC Press.

Sheta, Alaa, Hossam Faris, Malik Braik, and Seyedali Mirjalili. (2020). Nature-inspired metaheuristics search algorithms for solving the economic load dispatch problem of power systems: A comparison study. *Applied Nature-inspired Computing: Algorithms and Case Studies*: 199–230.

Singh, Himanshu, Sethu Venkata Raghavendra Kommuri, Anil Kumar, and Varun Bajaj. (2021). A new technique for guided filter based image denoising using modified cuckoo search optimization. *Expert Systems with Applications,* 176: 114884.

Srinivasan, Mandyam V. (2011). Honeybees as a model for the study of visually guided flight, navigation, and biologically inspired robotics. *Physiological Reviews,* 91(2): 413–460.

Suchi, Burak, Fakhre Ali, and Ian K. Jennions. (2023). Understanding the role of sensor optimization in complex systems. *Sensors,* 23(18): 7819.

Tang, Jun, Gang Liu, and Qingtao Pan. (2021). A review on representative swarm intelligence algorithms for solving optimization problems: Applications and trends. *IEEE/CAA Journal of Automatica Sinica,* 8(10): 1627–1643.

Tasgetiren, M. Fatih, Quan-Ke Pan, P.N. Suganthan, and Adalet Oner. (2013). A discrete artificial bee colony algorithm for the no-idle permutation flowshop scheduling problem with the total tardiness criterion. *Applied Mathematical Modelling,* 37(10–11): 6758–6779.

Turgut, Mert Sinan, and Oguz Emrah Turgut. (2020). Differential evolution based global best algorithm: An efficient optimizer for solving constrained and unconstrained optimization problems. *SN Applied Sciences,* 2(4): 600.

Turgut, Oguz Emrah, Mert Sinan Turgut, and Erhan Kırtepe. (2023). A systematic review of the emerging metaheuristic algorithms on solving complex optimization problems. *Neural Computing and Applications,* 35(19): 14275–14378.

Ünal, Hamit Taner, and Fatih Başçiftçi. (2022). Evolutionary design of neural network architectures: A review of three decades of research. *Artificial Intelligence Review,* 55(3): 1723–1802.

Way, M.J., and K.C. Khoo. (1992). Role of ants in pest management. *Annual Review of Entomology,* 37(1): 479–503.

Wu, Xindong, Xingquan Zhu, and Minghui Wu. (2022). The evolution of search: Three computing paradigms. *ACM Transactions on Management Information Systems (TMIS)*, 13(2): 1–20.

2 | When Optimization Techniques are Inspired from Marine Ecology and Terrestrial Animal Behaviors

Amiyangshu De[1*] and Manob Jyoti Saikia[2]

[1] Advanced Technology Development Center, Indian Institute of Technology Kharagpur, Kharagpur, India

[2] Electrical and Computer Engineering Department, University of Memphis, Memphis TN 38152, USA

Nature-inspired optimization algorithms have emerged as robust tools for solving complex, multidimensional problems across diverse domains. This chapter explores a spectrum of bio-inspired algorithms rooted in the adaptive and cooperative behaviors of marine, terrestrial, and insect species. Marine ecology inspires algorithms such as the Whale Optimization Algorithm (WOA) and Dolphin Echolocation Algorithm (DEA), which leverage unique behaviors like bubble-net feeding and echolocation for effective optimization. Similarly, terrestrial animal behaviors underpin strategies like the Elephant Herding Optimization (EHO) and Lion Optimization Algorithm (LOA), emulating social dynamics and cooperative hunting. Insect-inspired algorithms, such as the Dragonfly Algorithm (DA) and Moth-Flame Optimization (MFO), mimic swarm intelligence and navigation patterns to address non-linear and dynamic problems. The chapter provides a comparative analysis of these algorithms in terms of exploration, exploitation, scalability, and adaptability in dynamic environments. It also highlights their applications in engineering, artificial intelligence, telecommunications, and environmental management. The discussion extends to hybrid models and future integration with machine learning frameworks, emphasizing the potential for innovation in addressing real-world challenges through biologically inspired computation.

*Corresponding author: amiyangshude88@gmail.com

Introduction

Nature-inspired optimization algorithms (Yang, 2020; Odili et al., 2018; Mandal, 2023) have garnered considerable attention due to their capacity to address complex problems that often challenge traditional methods. Within this broad category, algorithms modeled after the behaviors of marine and terrestrial animals have exhibited remarkable efficiency and versatility across fields such as engineering, economics, and artificial intelligence (AI). These bio-inspired algorithms harness the adaptive, cooperative strategies observed in animals to find optimal or near-optimal solutions within multidimensional spaces.

The appeal of bio-inspired optimization techniques (Johnvictor, 2022; LaTorre et al., 2021) lies in their ability to emulate natural behaviors like hunting, foraging, and social cooperation. These algorithms are frequently compared with conventional optimization methods such as linear programming, gradient descent, and brute-force search (LaTorre et al., 2021). However, bio-inspired techniques often provide more robust solutions for non-linear, multimodal, and dynamic problems. The historical evolution of these algorithms suggests vast potential for innovation by continuing to draw inspiration from the natural world. The major surprising fact is that the majority (more than ~70%) of the planet earth surface are water-bodies; hence, marine ecology has come out as a thriving resource for bioinspired AI techniques.

One prominent example is the Whale Optimization Algorithm (WOA) (Rama et al., 2020), inspired by the hunting techniques of humpback whales, particularly their bubble-net feeding behavior. This algorithm simulates the spiral motion whales use to encircle prey, and it has been widely applied in image processing, feature selection, and network design. Various WOA variants have been developed to improve convergence speed and accuracy. Similarly, the Dolphin Echolocation Algorithm (DEA) (Buchanan et al., 2021) mimics dolphins' ability to use echolocation—emitting sound waves and processing their reflections—to locate objects underwater. Its applications extend from signal processing to the optimization of engineering design parameters, with algorithmic variations enhancing performance across different constraints. The Sperm Whale Algorithm (SWA) (Chambault et al., 2021) is another marine-inspired approach, drawing on the deep-diving and social behaviors of sperm whales. It is primarily used for tasks like resource allocation and environmental monitoring, leveraging the whales' structured group dynamics for solution exploration.

Ecological niches from the terrestrial realm, Elephant Herding Optimization (EHO) (Li et al., 2020) models the tight-knit family units and herding behavior of elephants, particularly the matriarch-led social structure. The algorithm simulates individuals' movements within a herd to explore and exploit the search space effectively. EHO variants are applied to solve intricate engineering complexities and improve multi-objective optimization. In a similar vein, Lion Optimization Algorithm (LOA) (Hussain et al., 2022) draws on the unique social structures and coordinated hunting tactics of lion prides, emulating both cooperative and competitive behaviors. LOA has been applied in areas like energy management and scheduling, with improvements focused on enhancing global search capabilities. The Giraffe Kicking Optimization Algorithm (GKO) (Menaka, 2023), though less widely known, is inspired by giraffes' powerful defensive kicks. Its unique approach has been applied in specific engineering contexts, providing alternative strategies for global optimization. Likewise, the Wolf Search Algorithm (Sangwan & Bhatia, 2020) replicates the pack-hunting behavior of wolves, simulating how they communicate and strategize to locate prey. This algorithm has been successfully employed in network optimization and resource management, with enhancements aimed at improving convergence speed and balancing exploration and exploitation.

Insect-inspired optimizations include the Dragonfly Algorithm (DA) (Elkorany et al., 2022), which is based on the swarming behavior of dragonflies, and displays collective movement patterns during hunting or migration. This algorithm has been applied in data clustering, robotic path planning, and other fields requiring dynamic adaptability. It excels at balancing exploration and exploitation in large, complex search spaces. Moth-Flame Optimization (MFO) (Hou et al., 2022), inspired by the navigation of moths using moonlight, simulates their spiral movements toward light sources. It has proven useful in real-world optimization tasks such as feature selection and machine learning model training, with numerous variants focusing on improving convergence accuracy and stability. Additionally, the Salp Swarm Algorithm (SSA) (Tubishat et al., 2021), based on the chain-like swarming behavior of salps, explores the search space effectively. Its applications span medical imaging, control system design, and optimization in uncertain environments, with ongoing development of variants to better handle large-scale optimization problems.

Marine Ecology-inspired Algorithms

Whale Optimization Algorithm

The WOA is a nature-inspired meta-heuristic algorithm that models the unique hunting behavior of humpback whales, particularly their bubble-net feeding technique (as shown in Figure 1a) (Brodzicki et al., 2021). This hunting technique, primarily observed in humpback whales, involves the creation of a spiral bubble net to trap schools of fish or krill. The whales swim in circles around the prey, releasing air bubbles from their blowholes to create a cylindrical net of bubbles. This disorients and traps the prey in a shrinking circle, making it easier for the whales to lunge upward and capture them.

The WOA translates this intelligent and complex hunting behavior into an optimization process aimed at finding optimal solutions to problems (Ibrahim et al., 2024). The algorithm is built around two key phases—exploitation (encircling and spiral attack) and exploration (search for new prey)—to balance local and global searches for solutions.

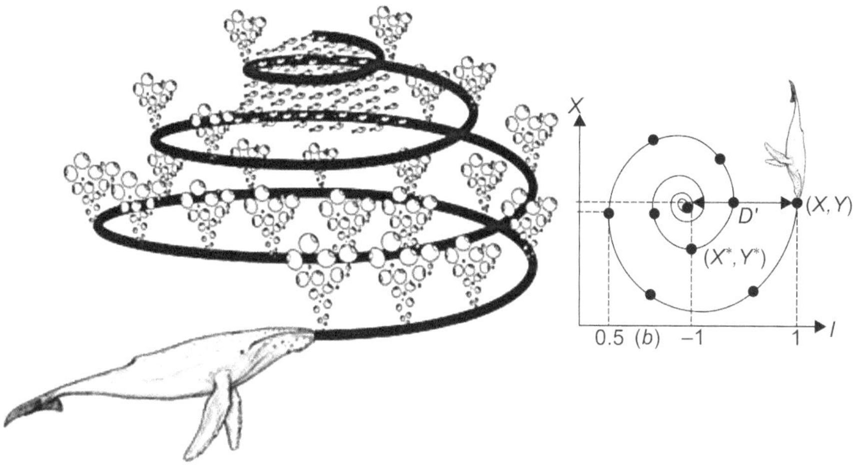

Figure 1: (a) Bubble-net feeding technique of whale and (b) spiral movement.

Encircling Prey (Exploitation)

In WOA, the prey (representing the optimal or near-optimal solution to a problem) is encircled by the whale (search agent). The best-known position of a whale (optimal solution found so far) is constantly updated as the algorithm runs. Other whales move towards this best solution, adjusting their positions iteratively. This mimics the behavior of a whale tightening its bubble net to encircle its prey (Cirion, 2021). Mathematically, this is achieved through the following equations:

$$\vec{D} = |\vec{C} \cdot \vec{X}^*(t) - \vec{X}(t)|, \tag{1}$$

$$\vec{X}(t+1) = \vec{X}^*(t) - \vec{A} \cdot \vec{D} \tag{2}$$

Here,

- $\vec{X}^*(t)$ represents the best-known position (optimal solution),
- $\vec{X}(t)$ represents the current position,
- $\vec{A}$ and $\vec{C}$ are coefficient vectors that control the whale's movement,
- $\vec{D}$ is the distance between the current position and the best solution.

$\vec{A}$ is gradually reduced over the iterations to simulate the tightening of the bubble net, which brings the search agents closer to the best solution.

Spiral Attack (Exploitation)

The spiral attack mimics the upward spiraling movement of the whale as it moves toward the surface to capture prey trapped within the bubble net, as depicted in Figure 1b. In WOA, this spiral movement is integrated to ensure a diverse approach to refining the search around the best solution.

The mathematical representation of this spiral movement is:

$$\vec{X}(t+1) = \vec{D'} \cdot e^{bl} \cos\cos(2\pi l) + \vec{X}^*(t), \tag{3}$$

where:

- D' is the distance between the whale's current position and the prey,
- b is a constant defining the shape of the spiral,
- l is a random number between −1 and 1,
- $\vec{X}^*(t)$ is the best-known position.

This logarithmic spiral movement allows the algorithm to search around the best solution in a non-linear fashion, improving the chances of finding the true global optimum.

Exploration (Global Search)

While exploitation focuses on refining solutions near the best-known solution, exploration allows WOA to search for potentially better solutions elsewhere in the search space. This is critical in optimization algorithms to avoid premature convergence on local optima. The exploration phase is mathematically modeled by selecting a random whale instead of the best one to update the positions of the other whales (Jin et al., 2021).

The position update in the exploration phase is given by:

$$\vec{D} = |\vec{C} \cdot \vec{X}_{rand}(t) - \vec{X}(t)| \tag{4}$$

$$\vec{X}(t+1) = \vec{X}_{rand}(t) - \vec{A}. \vec{D}\ , \tag{5}$$

where $\vec{X}_{rand}(t)$ is the randomly chosen position of a whale.

The coefficient vector $\vec{A}$ plays a key role in controlling the balance between exploration and exploitation. It decreases linearly over time, allowing the algorithm to shift from exploration in the early stages to exploitation in the later stages. When $|A|>1$, the whales move towards random positions (exploration), and when $|A|<1$, the whales move towards the best solution (exploitation).

Hybrid of Exploitation and Exploration

In WOA, exploitation and exploration are combined in a probabilistic manner to ensure the algorithm searches both locally and globally. At each iteration, there is a 50% chance that the algorithm will either update the position using the spiral movement (exploitation) or perform an encircling mechanism (exploration). This hybrid approach ensures a robust search for the global optimum (Kang et al., 2022)].

Key Advantages of WOA

1. **Balance of Exploration and Exploitation:** WOA efficiently balances the exploration of new solutions and the exploitation of known good solutions.

This helps in avoiding local optima while improving the convergence rate towards the global optimum.

2. **Simple Yet Effective:** The algorithm is relatively simple to implement yet powerful enough to solve complex optimization problems.
3. **Multidimensional Applicability:** The WOA can be extended to *n*-dimensional search spaces, making it applicable to a wide range of optimization problems across different fields.

The WOA has been applied in diverse fields, such as engineering design for structural optimization and control tuning, machine learning for feature selection and classification, image processing for segmentation, and robotics for path planning (Gharehchopogh & Gholizadeh, 2019). It is also used in communication networks for bandwidth allocation, energy systems for load dispatch, finance for portfolio optimization, and bioinformatics for gene selection and protein folding (Meraihi et al., 2022).

Dolphin Echolocation Algorithm

Echolocation Principles

The DEA is a meta-heuristic optimization method inspired by the natural echolocation behavior of dolphins. Like other meta-heuristic algorithms, DEA consists of a group of search agents that explore the feasible solution space using randomization and predefined rules, which are often based on natural phenomena. In DEA, dolphins use sonar—a specialized sound wave—known as a 'click' to detect and identify objects in their environment (Gracic et al., 2024), as described in Figure 2. Upon emitting the sound wave, dolphins analyze the reflected echo to estimate the distance, size, and position of an object. This

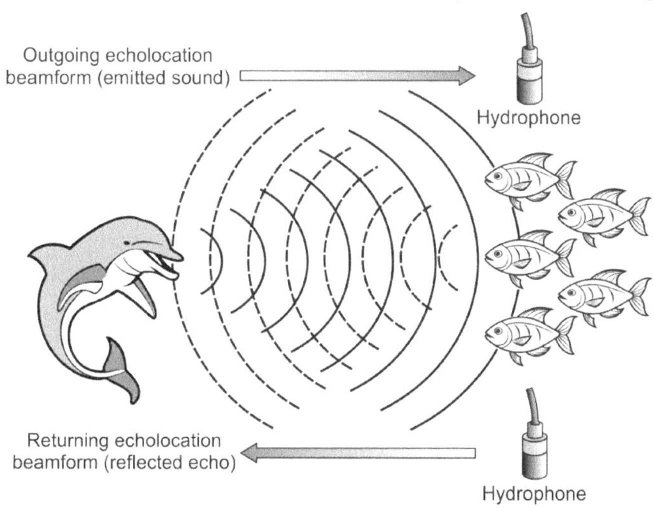

Figure 2: Natural echolocation behavior of dolphin.

process is then repeated in rapid succession, allowing the dolphin to refine its understanding of the object's location and characteristics.

In the context of optimization, DEA mirrors this echolocation process by simulating how dolphins emit sound waves, receive feedback (echoes), and adjust their position to locate the optimal solution (Kipnis et al., 2022). Each search agent in the DEA represents a dolphin that 'clicks' to evaluate a solution in the problem space. The reflected echo provides information about the quality of the solution, enabling the algorithm to adjust the dolphin's position iteratively in search of better solutions. The key element here is the use of feedback to fine-tune the search, analogous to how dolphins lock onto and track a target using echolocation.

DEA also incorporates randomization to avoid getting stuck in local optima, and it uses sound wave propagation models to explore new solutions. As dolphins adjust the frequency and intensity of their sonar to study objects more closely, DEA adapts its parameters during the optimization process to enhance convergence and search efficiency (Su et al., 2022). Through iterative adjustments, the algorithm gradually hones in on the global optimum, just as dolphins zero in on their prey. Variants of DEA, such as Adaptive Echolocation Mechanism and Multi-objective DEA, have been introduced to improve the algorithm's performance. For example, in the Adaptive Echolocation Mechanism, parameters like signal frequency and amplitude are adjusted dynamically to enhance convergence. In multi-objective problems, DEA is tailored to handle multiple conflicting objectives by modifying the standard approach to generate a set of optimal solutions, rather than a single solution. The scheme of DEA is listed below:

Algorithm 1: Dolphin Echolocation Algorithm

1. **Initialize Search Agents:** Set up a population of agents (dolphins) with random positions in the solution space.
2. **Emit Sonar Clicks (Evaluate Solutions):** Each agent emits a sonar click to evaluate the quality (fitness) of the current solution.
3. **Receive Echo (Feedback):** Agents receive feedback in the form of echoes, which reflect the quality of their current position in the solution space.
4. **Update Position of Agents:** Based on the intensity of the echo (feedback), agents adjust their position to move closer to better solutions.
5. **Adjust Sonar Frequency and Amplitude:** Dynamically modify the frequency and amplitude of the sonar clicks to fine-tune the search process, improving exploration or exploitation as needed.
6. **Check Termination Condition:** Evaluate whether the stopping criteria are met (e.g., a maximum number of iterations or a satisfactory fitness level).
    - If the termination condition is met, proceed to Step 7.
    - If not, repeat from Step 2.
7. **Output Optimal Solution:** Once the stopping criteria are satisfied, the best solution found is output as the optimal solution.

Sperm Whale Algorithm

The Sperm Whale Algorithm (SWA) is an optimization method inspired by the behavior of sperm whales, particularly their deep-sea hunting and surface breathing cycles (Sadayappan et al., 2023). The algorithm uses these behaviors to explore and exploit a solution space effectively. Here's how optimization is performed in SWA:

Algorithm 2: Sperm Whale Algorithm

1. **Initialization:** A population of sperm whales (agents) is initialized with random positions in the search space, representing potential solutions.
2. **Upward and Downward Movement:** Each whale undergoes two phases in its search cycle:
   - *Surface Breathing (Exploration):* Whales ascend to the surface, representing exploration of the search space (Das, 2023). During this phase, whales move to new random positions, encouraging global search and exploration of diverse regions.
   - *Deep Dive (Exploitation):* Whales dive deep to hunt for food (squids), symbolizing the exploitation phase. Here, the whales refine their search by moving toward better solutions in the deeper parts of the search space.
3. **Mirroring Mechanism:** For each whale, a mirror image of its position is created in the search space. This mirror image provides an additional solution candidate, and the quality of the original and mirror solutions are compared (Darvishpoor et al., 2023). However, only the most promising local solution is kept, enhancing convergence by focusing on the best candidates.
4. **Best and Worst Individual Comparison:** The positions of the worst-performing whales are updated using information from the best-performing whales. The worst solutions are replaced by new positions, which are influenced by the positions of the best whales, ensuring that the population evolves toward better solutions.
5. **Optimization Loop:** The whale population repeatedly performs the surface and deep dive cycles, adjusting positions based on feedback from the objective function and improving the quality of solutions over time.
6. **Termination Condition:** The optimization process continues until a termination condition is met (Wu, et al., 2022), such as reaching a maximum number of iterations or achieving a satisfactory solution.

This process allows the SWA to balance between global search (exploration) and local search (exploitation), making it effective in optimizing complex problems with multiple objectives or constraints. Variants of SWA, such as the Improved Sperm Whale Algorithm (ISWA) and Hybrid Sperm Whale Algorithm (HSWA), have been developed to enhance its performance by improving convergence speed and incorporating features from other optimization techniques.

Terrestrial Animal-inspired Algorithms

Elephant Herding Optimization

Elephant Herding Optimization (EHO) is inspired by the social herding behavior of elephants (Figure 3), particularly how elephant clans are organized and led by a matriarch. The optimization process in EHO mimics these social dynamics through specific mechanisms that update the position of each elephant in a population of clans (Li & Wang, 2022).

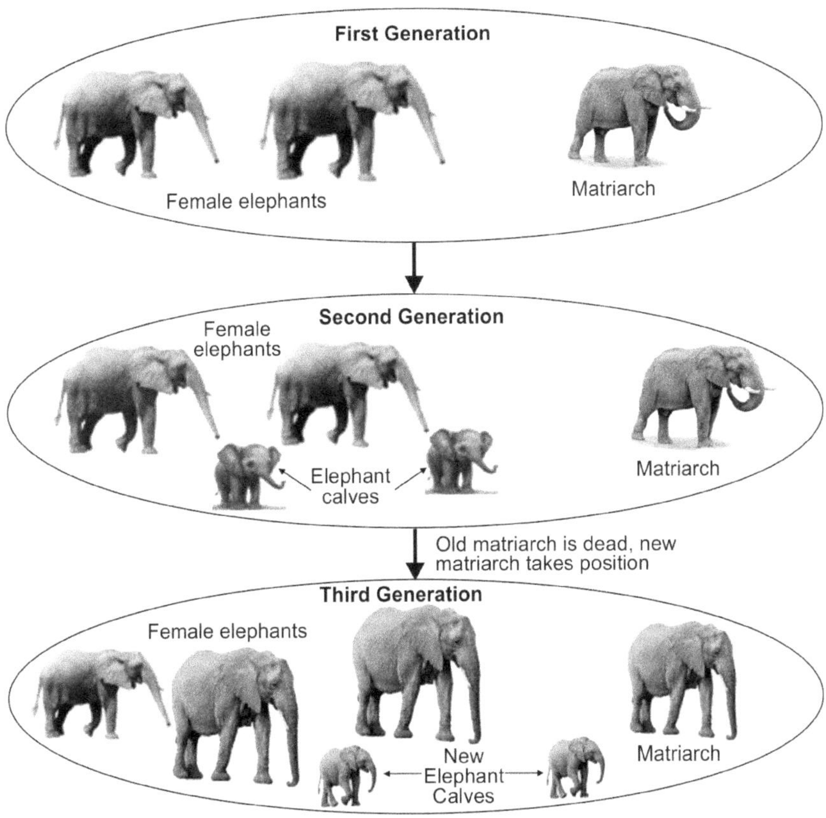

Figure 3: EHO inspired by social herding behavior of elephants.

1. **Herding Behavior:** Elephants in the wild live in groups called clans, which are predominantly led by a female matriarch. The female elephants prefer to stay with their relatives, while the male elephants eventually leave the group to live independently. This natural separation of male elephants is reflected in the optimization process, where some elephants are separated from the clan and live far away.

In the EHO algorithm, assumptions include:

- The size of elephant clans remains constant.
- Male elephants leave the clan at a certain point to live separately.

2. **Clan-updating Operator:** The clan-updating operator simulates the leadership of the matriarch and the social influence she has on the other elephants in the clan. Each elephant's new position is updated as a function of the best-performing elephant (matriarch) in the clan. The position of elephant j in clan c_i is updated using the following formula:

$$X_{new,c_{i,j}} = X_{c_{i,j}} + a \times X_{best,c_i} - X_{c_{i,j}} \times r \tag{6}$$

where,

- $X_{new,c_{i,j}}$ is the new position of elephant j in clan c_i.
- $X_{c_{i,j}}$ is the current position of elephant j in the same clan.
- $X_{best,c_{i,}}$ is the position of the matriarch (best-performing elephant in clan c_i).
- $a\in [0,1]$ is a scaling factor.
- $r\in [0,1]$ is a random value to introduce variability.

3. **Center of Clan:** In addition to following the matriarch, elephants are also influenced by the center of the clan. The center is calculated as the average position of all elephants in the clan, representing the collective behavior. The position of the center in the d-th dimension is calculated as:

$$X_{center,c_{i,d}} = \frac{1}{n_{c_i}} \times \sum_{j=1}^{n_{c_i}} X_{c_{i,,j,d}} \tag{7}$$

where,

- $X_{new,c_{i,j}}$ is the center of clan c_i in the d-th dimension.
- n_{c_i} is the number of elephants in the clan.
- $X_{c_{i,,j,d}}$ is the d-th dimension of elephant j in the clan.

Elephants move toward the center as follows:

$$X_{new,c_{i,j}} = \beta \times X_{center,c_i} \tag{8}$$

where, $\beta\in[0,1]$ controls the influence of the center on the elephant's new position.

Thus, EHO effectively balances exploration (by separating male elephants and using randomization) and exploitation (by following the matriarch and moving toward the clan center). Through these mechanisms, EHO adapts to solve various continuous and discrete optimization problems.

Lion Optimization Algorithm

The **Lion Optimization Algorithm (LOA)** is inspired by the social behavior, hunting, and territorial dynamics of lions. The key steps in optimization through LOA can be described as follows:

1. **Social Structure and Initialization:** The algorithm mimics the lion's social structure, dividing the lion population into two categories: **nomads** and **residents** (organized into prides). A percentage of lions are designated as **nomads** and the rest as **resident lions**, who form prides randomly. Each pride consists of females and males in specific proportions (LeFlore, 2022). Resident males dominate a pride, while nomads roam independently in search of better territories. The **nomad lions** wander freely through the search space, attempting to explore new areas and improve the quality of their solutions.
2. **Territory and Position Representation:** Each lion's position represents a potential solution, and its territory refers to its best-known solution found so far, akin to the best position in previous iterations. The optimization goal is to update these positions iteratively to find better solutions. The pride's territory is defined as the collection of the best positions held by the members of the pride, and each member strives to improve its position based on local information (its past best solution) and interactions with others.
3. **Hunting Process:** Female lions, known for their group hunting abilities, play a significant role in the optimization process (Nhalungo, 2024). In each pride, a certain number of female lions go hunting, exploring the search space more effectively. They surround the prey (the best solution) and attempt to improve upon it. While some females actively hunt, others migrate within the territory, contributing to exploration and exploitation.
4. **Territorial Defense and Male Competition:** The **male lions** defend their pride's territory. Occasionally, nomad males challenge resident males for dominance. If a strong nomad male defeats a resident male, it takes over the pride, pushing out the weaker male. This process helps ensure that stronger solutions replace weaker ones, driving the optimization process forward. In LOA, this corresponds to replacing suboptimal solutions with better ones.
5. **Mating and Lion Pride Dynamics:** Mating takes place between resident males and females, potentially producing new offspring (solutions) that inherit qualities from their parents (Allen, 2022), similar to genetic crossover in other evolutionary algorithms. As young males mature, they are forced to leave the pride and become nomads, leading to increased exploration of the search space.
6. **Nomad Exploration:** Nomads (both males and females) continually roam through the search space, seeking better regions (solutions). Their primary role is to explore areas outside the pride's current territory. If a nomad finds a better solution, it might challenge a resident lion, displacing it and improving the population's overall quality.
7. **Death and Survival of Lions:** Weaker lions (suboptimal solutions) may be eliminated due to competition, famine, or other pressures. This ensures that only the strongest solutions survive, improving the overall quality of the population over time.
8. **Stopping Condition:** The process of pride formation, hunting, male competition, nomadic exploration, and mating continues iteratively until a

predefined stopping condition (such as a maximum number of iterations for convergence criteria) is met.

By modeling the social and territorial behaviors of lions, LOA effectively balances exploration (via nomads) and exploitation (via resident prides), allowing it to search the solution space for optimal or near-optimal solutions.

Giraffe Kicking Optimization Algorithm

Giraffe Kicking Optimization (GKO) is based on the behavior of mother giraffes, specifically the way they kick. This kicking action serves as a metaphor for the algorithm's ability to maintain a balance between **exploration** and **exploitation** during optimization processes (Bridle, 2022). The biological background of the algorithm is inspired by the kicking action of mother giraffes, which is used to protect their young from predators. This action is mapped into an optimization technique that 'kicks' nodes in a Vehicular Adhoc Network (VANET) to wake up the minimum number of sensor nodes necessary for network efficiency, thus conserving energy and prolonging network lifespan (Lee, 2021).

The optimization process involves the following key aspects:

1. **Exploitation vs Exploration:** The kicking action is modeled to strike a balance between refining known solutions (exploitation) and searching for new solutions (exploration). This ensures that the optimization process doesn't overly focus on local optima and instead continues to explore better global solutions.
2. **Fitness Function:** GKO utilizes a multi-fitness function based on factors like residual node energy, intra-cluster distance, and the degree of sensor nodes within a network. This helps in determining which sensor nodes should be activated or serve as cluster heads.
3. **Clustering:** GKO is combined with the C-mean clustering algorithm to group sensor nodes and designate cluster heads for efficient data transmission. This reduces unnecessary node activations and improves throughput.
4. **Energy Efficiency:** The algorithm minimizes energy consumption by ensuring only the necessary number of sensor nodes are awake, optimizing network resources and prolonging the overall network lifetime.

In short, GKO uses the biological metaphor of a giraffe's kicking behavior to manage sensor networks by enhancing network throughput, reducing energy consumption, and prolonging the life of the network (Prakash et al., 2022). This is achieved through intelligent node activation and efficient routing, striking a balance between exploration and exploitation. The flowchart of GKO is shown in Figure 4.

Wolf Search Algorithm

The Wolf Search Algorithm (WSA) is a bio-inspired optimization algorithm modeled after the hunting behaviors and social interactions of wolves. The

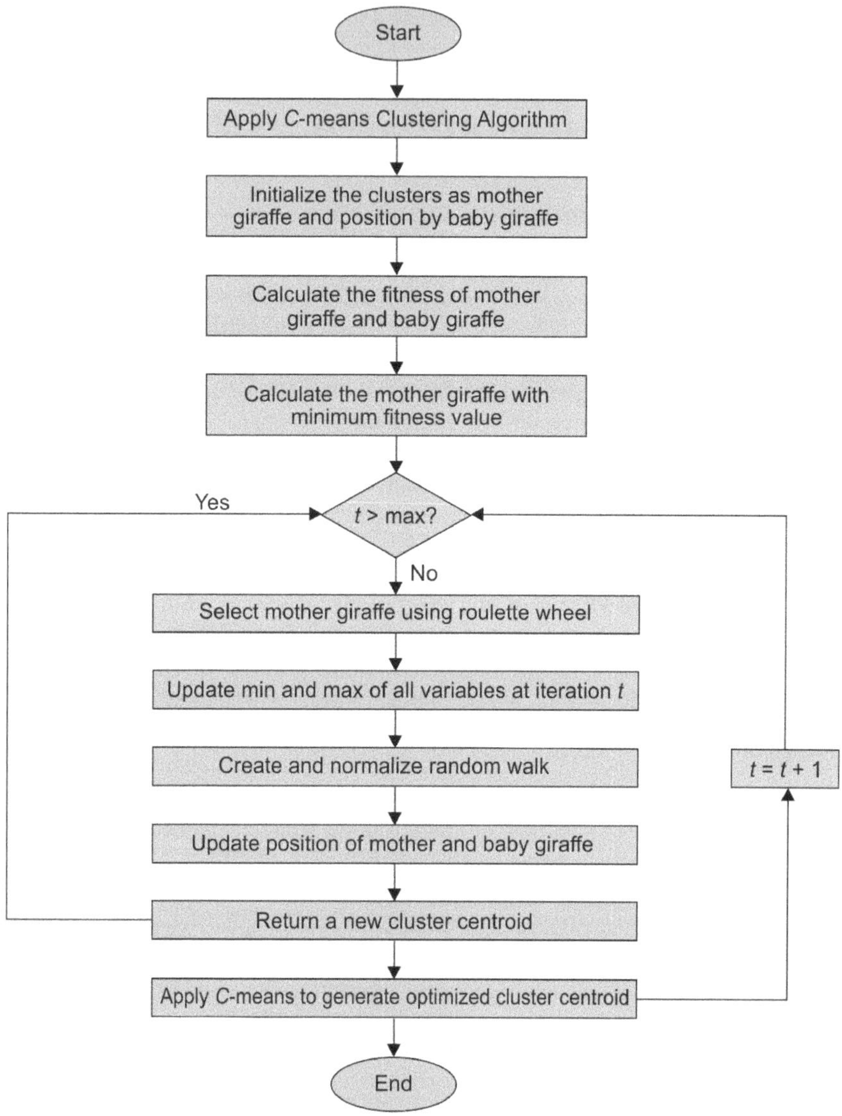

Figure 4: Flowchart of giraffe kicking optimization algorithm.

key characteristics of WSA are its blend of individual search efforts and semi-cooperative behavior among wolves (Dragoi & Dafinescu, 2021), which makes it distinct from other swarm intelligence-based methods like Particle Swarm Optimization (PSO) or Firefly Algorithm.

Key Features of WSA

Individual Search and Self-organization: Each wolf in the WSA operates independently, performing localized random searches based on its own traits, and does not engage in long-range communication with others. This differs from algorithms where communication is central, such as PSO, where agents follow the leader. The wolves can only merge and share information if a neighboring wolf is in a better position (i.e., a better solution).

Visual Range and Flocking Motion: Wolves maintain awareness of their environment within a defined visual range, allowing them to detect prey (the global optimum), peers (other wolves), or threats. They tend to move based on this sensory input, but if no significant targets are detected, they exhibit random movement (Brownian motion) to explore new areas.

No Central Leader: Unlike many bio-inspired algorithms that rely on a centralized leader (as in PSO or Firefly), WSA distributes the search responsibility across all wolves. Each wolf acts as an independent leader, moving toward the best solution it can find, thereby exploring multiple regions of the search space simultaneously (Wolf, 2022).

Threat Evasion and Escaping Local Optima: Wolves are cautious hunters and are designed to evade threats in their environment. If a wolf encounters a predator (modeled as being trapped in a local optimum), it performs a large jump away from the current position to explore new regions of the solution space. This mechanism helps prevent the algorithm from getting stuck in local optima.

Memory and Long-distance Sensing: Wolves in nature are known for their ability to remember locations, track prey over long distances, and mark their territory. Similarly, the WSA incorporates memory mechanisms that allow wolves to retain information about past search areas and optimal solutions. This feature aids the exploration of the search space without revisiting unfruitful regions.

Semi-Cooperative Behavior: Although wolves hunt in packs, their coordination is loose, and they make individual decisions. This behavior is mirrored in WSA, where wolves may benefit from their peers' positions but are not tightly bound to follow one another. This balance between cooperation and individuality ensures efficient search space exploration and exploitation. The flowchart is given in Figure 5.

Search Strategy in WSA

- **Local Search:** Each wolf searches its local region based on its current position and traits, continuously updating its location as it seeks to improve its standing.
- **Flocking and Swarming:** Wolves can merge with others in their visual range if another wolf offers a superior position, but they avoid rigid flocking behaviors seen in other swarm-based algorithms. Each wolf remains somewhat independent in its movements.
- **Brownian Motion (BM):** When no prey or threat is detected, wolves move randomly to explore new areas of the solution space.

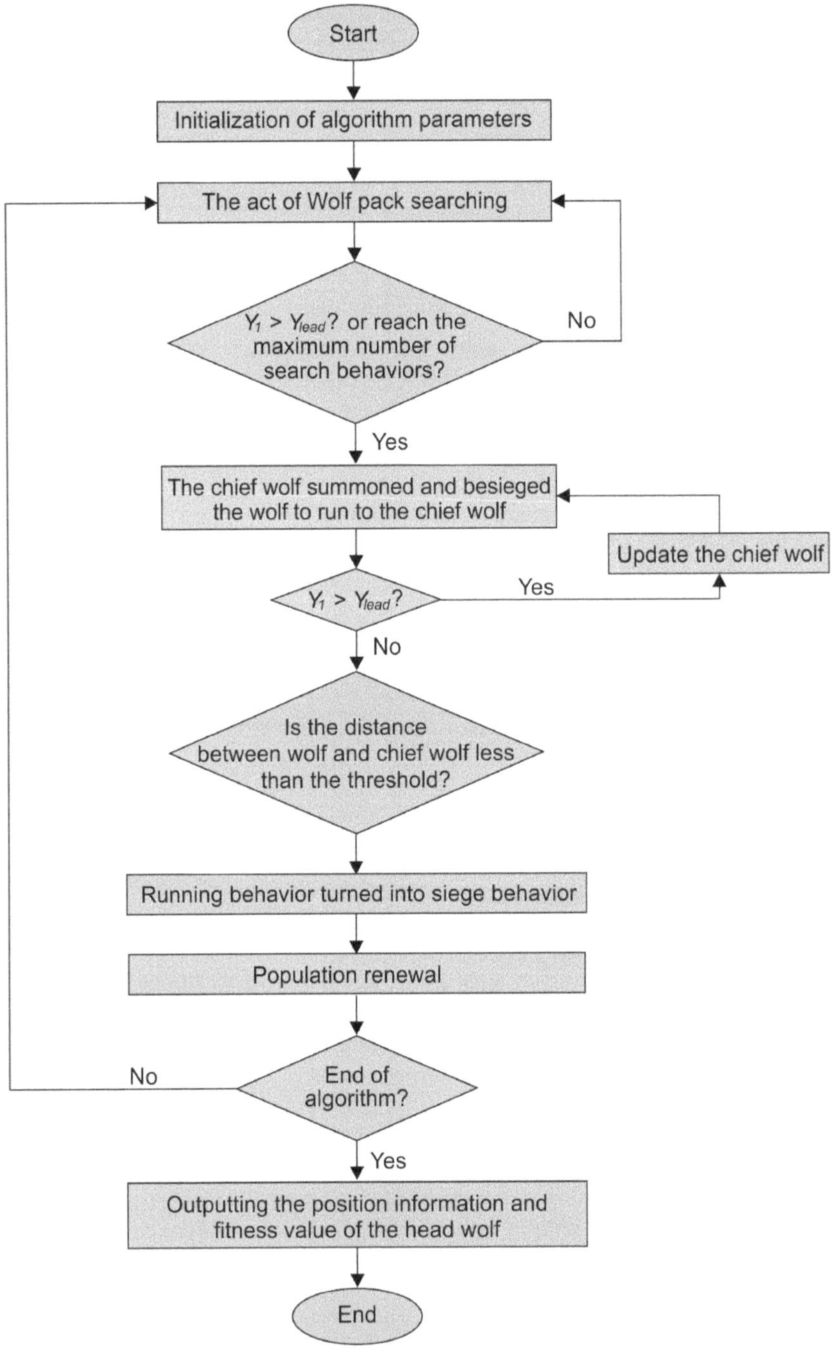

Figure 5: Flowchart of wolf search algorithm.

- **Threat Response:** If a wolf encounters a 'threat' (analogous to a local optimum or worse-performing solution), it makes a long-distance move to escape the trap, allowing it to continue exploring more promising areas.

WSA offers a unique optimization approach that blends independent, local searches with semi-cooperative swarming behavior. By allowing wolves to act as independent search leaders and incorporating mechanisms to escape local optima, WSA can effectively explore and exploit solution spaces without being constrained by rigid communication protocols or leadership structures seen in the other algorithm (Dong, 2022).

Insect-inspired Algorithms

Dragonfly Algorithm

The DA is inspired by the swarming behavior of dragonflies in nature, particularly their dynamic and static swarming movements, which serve as the basis for the two main optimization processes: **exploration** and **exploitation** (Alshinwan et al., 2021). Exploration helps dragonflies (and the algorithm) search for new areas in the solution space, while exploitation focuses on refining the current solutions to reach the global optimum.

Dragonflies exhibit two key types of swarming:

1. **Static Swarming:** This occurs when dragonflies group together in one area, often for feeding. It is primarily associated with **exploitation** because the focus is on refining and improving the solutions around a local optima.
2. **Dynamic Swarming:** Dragonflies move over larger areas in search of food or avoiding predators. This phase is associated with **exploration**, helping the algorithm search broadly across the solution space to discover new potential solutions.

In DA, these behaviors are simulated using attraction and repulsion forces, as well as alignment mechanisms:

- **Attraction:** Ensures dragonflies (solutions) are drawn toward the best candidate solutions.
- **Repulsion:** Helps them avoid poor solutions or obstacles.
- **Alignment:** Ensures that solutions move coherently, maintaining a balance between exploration and exploitation.

Figure 6 illustrates the behavior of dragonflies in the DA through five key mechanisms: **separation**, **alignment**, **cohesion**, **attraction towards food**, and **distraction from enemies**. Let's break them down:

(a) **Separation:** This is the tendency of each dragonfly to avoid collisions with its neighbors by maintaining a minimum distance from nearby individuals. It

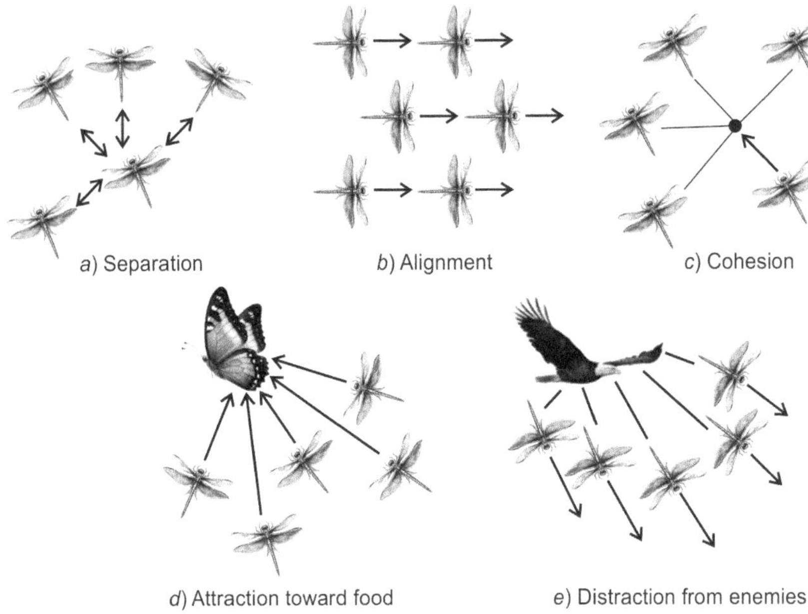

Figure 6: Steps of dragonfly algorithm.

helps in preventing overcrowding and ensures a spread-out exploration of the search space.

(b) Alignment: In this step, the dragonflies adjust their velocities based on the average heading of neighboring individuals. It ensures that they move cohesively in a unified direction, guiding the swarm collectively during the search process.

(c) Cohesion: Cohesion represents the tendency of dragonflies to move toward the centre of their neighbors (Emambocus et al., 2022). This helps the dragonflies (solutions) cluster together, which is crucial for exploitation and focusing on promising regions in the search space.

(d) Attraction towards food: Dragonflies are attracted to potential food sources, represented here by the butterfly. This mechanism guides dragonflies toward the best solutions found so far, contributing to the exploitation of the solution space.

(e) **Distraction from enemies:** This step shows the dragonflies avoiding predators (represented by the bird). In terms of the algorithm, this corresponds to repulsion from bad solutions or traps, ensuring that the dragonflies move away from poor solutions.

These behaviors are combined dynamically in the algorithm to balance between exploration and exploitation, ensuring an efficient search for optimal solutions. DA is effective because of its ability to balance exploration and exploitation, making it suitable for solving a wide range of optimization problems, including scheduling, feature selection, image processing, and economic dispatch. Its simplicity and versatility have also led to the development of multiple variants like Binary DA, Multi-objective DA, and hybrid models that further enhance its performance for specific tasks.

Moth Flame Optimization

The Moth Flame Optimization (MFO) algorithm is a population-based algorithm inspired by the navigation method used by moths in nature called "transverse orientation". Moths maintain a fixed angle with the moon's light as they fly in a straight path (Storms et al., 2022). However, in artificial light, they tend to spiral around it, as they mistakenly treat it like the moon. This biological behavior is mimicked in the MFO algorithm to explore a solution space effectively.

Optimization Process in MFO

1. **Initialization and Population:** MFO begins by generating a population of moths, each representing a potential solution in the problem's search space. Each moth's position corresponds to a candidate solution, and the objective function evaluates how good the solution is (i.e., the moth's fitness).
2. **Flame Generation:** The flames are considered the best solutions found so far by the moths. As the optimization process progresses, these flames represent the elite solutions towards which the moths are attracted. This process allows the algorithm to perform a global search by scouting for better solutions.
3. **Exploration and Exploitation:** MFO balances exploration (searching new areas of the solution space) and exploitation (refining the current best solutions). The position of moths is updated by calculating a spiral-shaped path towards the flames, simulating the moths' natural flight pattern around light sources.
    - **Exploration:** During early stages, moths explore the search space widely to find promising regions.
    - **Exploitation:** As the optimization progresses, moths converge towards the flames, focusing on exploiting the best solutions.
4. **Spiral Movement Mechanism:** The key feature of MFO is the spiral path formula (Figure 7), where each moth moves closer to a flame in a logarithmic spiral. This movement is mathematically modeled to guide moths in spiraling towards the flames, reducing the distance between the moth and its respective flame over time.
5. **Adaptation of Flame Number:** The number of flames decreases over iterations to ensure convergence. In the initial stages, the algorithm allows more flexibility by maintaining a higher number of flames. As the algorithm proceeds, this number is reduced to help focus on refining the top solutions, thereby improving exploitation.

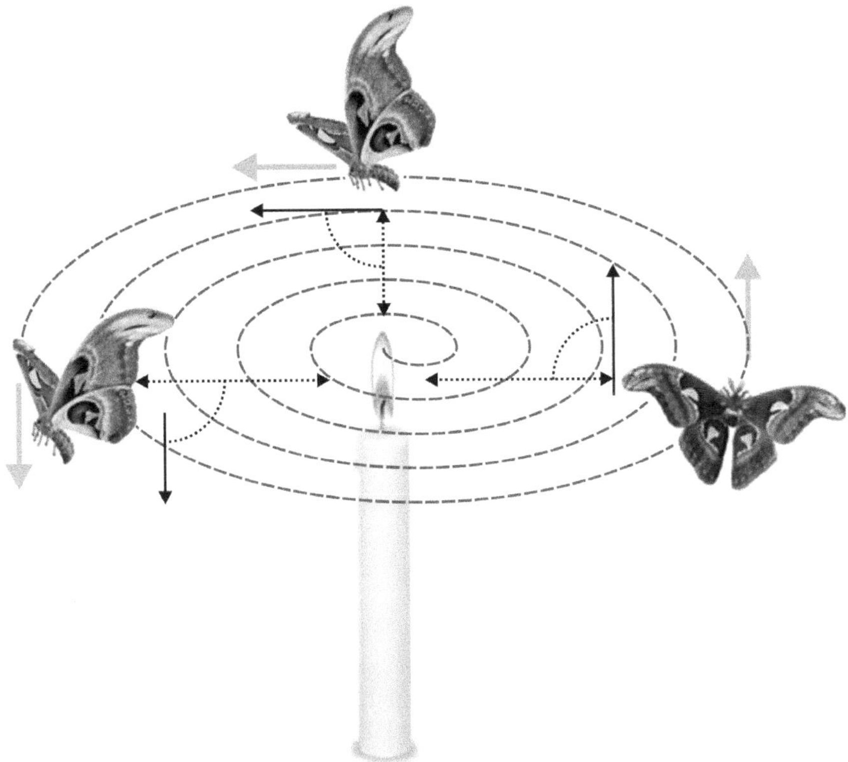

Figure 7: Moth flame optimization algorithm inspired by navigation method used by moths.

6. **Update of Flames**: Flames are updated at each iteration, meaning that as moths find better solutions, they replace the flames (previous best solutions). This dynamic flame adjustment ensures that the algorithm consistently refines the quality of the best solutions.

The strengths of MFO algorithm are stated below:

- **Global Search Capability:** MFO uses a powerful local and global search strategy through spiral movements, ensuring both diversity (exploration) and convergence (exploitation) (Sahoo et al., 2023).
- **Versatility:** MFO is simple and flexible, and it can be applied to a wide range of optimization problems, such as scheduling, parameter estimation, image processing, machine learning, and more.

The MFO algorithm and its variants are applicable across various domains, including engineering design, energy systems, telecommunications, healthcare, and cybersecurity, due to its ability to handle complex optimization problems efficiently.

Salp Swarm Algorithm

The optimization in the SSA is inspired by the swarming behavior of salps in nature, specifically their coordinated movement in a chain formation (Figure 8). This biological behavior is modeled mathematically to solve optimization problems by balancing exploration (global search) and exploitation (local search).

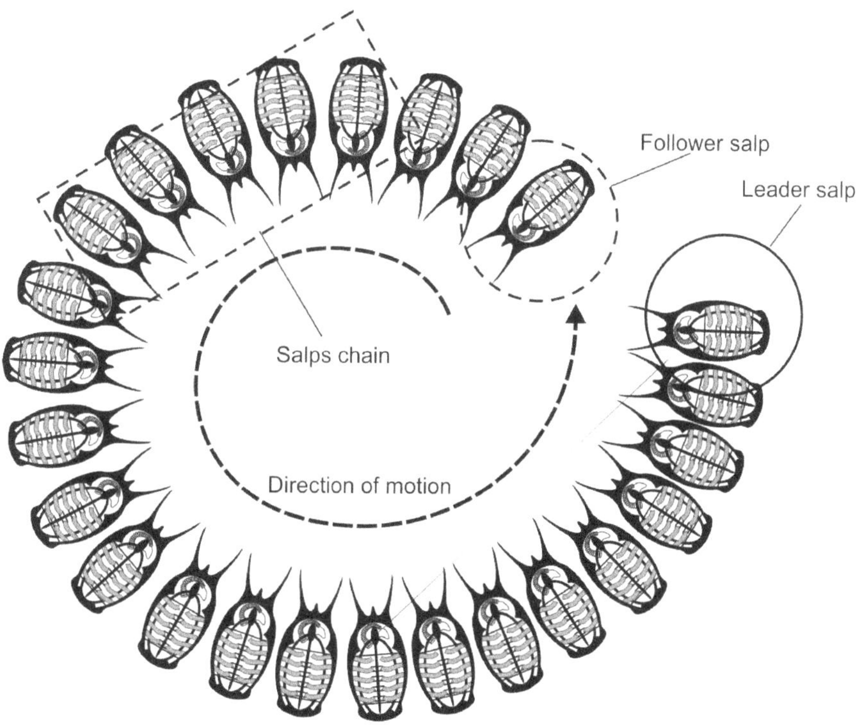

Figure 8: Salp swarm algorithm inspired by the swarming behavior of salps.

Key Elements of SSA Optimization

1. **Swarming Behavior and Chain Formation:** Salps move in interconnected chains, and this coordinated movement improves their swimming efficiency and feeding rates. Similarly, SSA models a population of salps where:
   - The first salp in the chain acts as a leader, guiding the direction of movement, akin to an **exploration** phase where global search of the solution space is conducted.
   - The following salps in the chain update their positions based on the movement of the leader, representing the **exploitation** phase, where solutions are fine-tuned locally by following the leader's trajectory.

2. **Mathematical Modeling:** The position of each salp in the population represents a candidate solution to the optimization problem. The leader updates its position based on its distance from the global best solution, using a formula that mimics the pulsating swimming of salps (Darvishpoor et al., 2023). The followers then adjust their positions relative to the leader to ensure a harmonious movement toward the global optimum. The mathematical equations governing this process ensure a balance between exploration and exploitation:
    - **Leader update:** The leader searches the space by making significant jumps toward promising areas.
    - **Follower update:** The following salps move incrementally, based on their predecessors, to exploit regions close to the best solutions found by the leader.
3. **Filter Feeding Mechanism:** Salps' filter-feeding behavior, where they pass water through their bodies to collect plankton, symbolizes how the algorithm evaluates solutions. As the salps move, they "capture" better solutions through coordinated efforts. The SSA optimizes by continuously refining the population of solutions, similar to how salps adjust their filter-feeding while swimming through nutrient-rich waters.
4. **Coordination and Efficiency:** The energy-efficient pulsating motion of salps, where they contract and relax their gelatinous bodies, is mirrored in the SSA by a dynamic adjustment of search intensity. When the leader detects better solutions, the followers can coordinate their movement toward that direction. This coordination is key for efficiently exploring the solution space without wasting computational resources.
5. **Exploration and Exploitation Balance:** In SSA, the **exploration** is ensured by the leader's random and large steps in the solution space (Romeh & Mirjalili, 2023), while the **exploitation** is managed by the followers who refine their positions relative to the leader. This chain formation is crucial for maintaining a balance between exploring new areas of the search space and exploiting known good areas to optimize solutions.

By leveraging the natural behaviors of salps, SSA efficiently explores large solution spaces while fine-tuning local areas, making it suitable for solving a wide variety of complex optimization problems.

A Comparative Assessment of Optimization Techniques

Table 1 compares three different categories of bio-inspired algorithm categories—Marine animal-inspired, Terrestrial animal-inspired, and Insect-inspired—across three criteria: exploration vs. exploitation, flexibility in dynamic environments, and scalability. This will highlight the similarities and uniqueness of such optimization techniques in solving real-life problems.

Table 1: Comparative analysis

Criteria	Marine Animal-inspired Algorithms	Terrestrial Animal-inspired Algorithms	Insect-inspired Algorithms
1. Exploration vs. Exploitation	Generally, have an inclination toward exploration	Provide both value propositions (Kasaragodu, 2023)	Strengths in exploration while at the same time may require some fine-tuning in exploitation (Lahari & Janamala, 2024)
2. Flexibility in dynamic environments	Usually, better suited for these environments	Of moderate flexibility, and suitable for many environments but often more complex models needed (Rafeeq et al., 2021)	Flexible for many environments with similar characteristics but likely to involve the use of behavioral models
3. Scalability	Some of them may not be very scalable in a number of contexts because of their principle of broad sampling (Gharehchopogh & Gholizadeh, 2019)	Moderately scalable	Highly scalable, particularly those grounded on swarm intelligence

Discussion and Future Directions

The optimization techniques evolved from mimicking ecological behaviors of living organisms from marine and terrestrial behaviors has unveiled a rich tapestry of algorithms capable of addressing complex, real-world challenges across various fields. These bio-inspired algorithms draw upon the efficiency and adaptability observed in natural systems, offering novel ways to approach optimization problems. Techniques such as the WOA and DEA have demonstrated their potential in solving problems in fields such as engineering, data science, and AI. Similarly, terrestrial-inspired algorithms, such as the LOA and GFO, have shown promise in addressing multi-objective and constrained optimization problems. By mimicking behavioral patterns of species in the natural world, these algorithms not only expand the toolbox for solving complex problems but also open avenues for more sustainable and adaptive approaches in optimization. The integration of marine and terrestrial behaviors into computational paradigms represents a burgeoning field that will likely continue to evolve as more species behaviors are studied and translated into algorithmic form.

Several promising avenues remain open for future exploration. Firstly, hybrid algorithms that combine marine and terrestrial behavior inspirations could enhance the performance of individual techniques. For example, blending strategies from both environments might result in algorithms that are more robust and flexible in addressing dynamic and real-time optimization challenges. Additionally, there is significant scope for integrating these algorithms with machine learning frameworks, particularly in areas such as deep learning optimization and reinforcement learning. The adaptability of natural systems could offer more efficient learning processes for intelligent systems, leading to breakthroughs in artificial intelligence applications. Moreover, the exploration of novel behaviors from less-studied species, both marine and terrestrial, could yield even more powerful optimization strategies. As new biological insights emerge, they could be directly translated into computational models, fostering a continuous cycle of innovation. Finally, further empirical studies on the application of these bio-inspired algorithms in real-world problems, such as climate modeling, healthcare, and resource management, will be crucial. Such applications will not only validate the effectiveness of these algorithms but also inspire further refinements and adaptations to meet the ever-evolving demands of complex systems in the natural and technological world.

References

Allen, C. (2022). *Importance of Older Males: Influence of Age Structure on Social Dynamics between Males in An Elephant Bull Area*. PhD thesis. University of Exeter.

Alshinwan, Mohammad, Laith Abualigah, Mohammad Shehab, Mohamed Abd Elaziz, Ahmad M. Khasawneh, Hamzeh Alabool, and Husam Al Hamad. (2021). Dragonfly algorithm: A comprehensive survey of its results, variants, and applications. *Multimedia Tools and Applications,* 80: 14979–15016.

Bridle, James. (2022). *Ways of Being: Animals, Plants, Machines: The Search for a Planetary Intelligence*. Penguin UK.

Brodzicki, Andrzej, Michał Piekarski, and Joanna Jaworek-Korjakowska. (2021). The whale optimization algorithm approach for deep neural networks. *Sensors,* 21(23): 8003.

Buchanan, Caleb, Ying Bi, Bing Xue, Ross Vennell, Simon Childerhouse, Matthew K. Pine et al. (2021). Deep convolutional neural networks for detecting dolphin echolocation clicks. In: *2021 36th International Conference on Image and Vision Computing New Zealand (IVCNZ),* pp. 1–6. IEEE.

Chambault, Philippine, Sabrina Fossette, Mads Peter Heide-Jørgensen, Daniel Jouannet, and Michel Vély. (2021). Predicting seasonal movements and distribution of the sperm whale using machine learning algorithms. *Ecology and Evolution,* 11(3): 1432–1445.

Cirino, Erica. (2021). *Thicker than Water: The Quest for Solutions to the Plastic Crisis*. Island Press.

Darvishpoor, Shahin, Amirsalar Darvishpour, Mario Escarcega, and Mostafa Hassanalian. (2023). Nature-inspired algorithms from oceans to space: A comprehensive review of heuristic and meta-heuristic optimization algorithms and their potential applications in drones. *Drones*, 7(7): 427.

Das, Raju. (2023). *Potential Contribution of Carbon Sequestration in the North Eastern Part of the Bay of Bengal Bangladesh.* PhD diss. Chattogram Veterinary & Animal Sciences University, Khulshi, Chattogram.

Dong, Lin, Xianfeng Yuan, Bingshuo Yan, Yong Song, Qingyang Xu, and Xiongyan Yang. (2022). An improved grey wolf optimization with multi-strategy ensemble for robot path planning. *Sensors,* 22(18): 6843.

Dragoi, Elena Niculina, and Vlad Dafinescu. (2021). Review of metaheuristics inspired from the animal kingdom. *Mathematics,* 9(18): 2335.

Elkorany, Ahmed S., Mohamed Marey, Khaled M. Almustafa, and Zeinab F. Elsharkawy. (2022). Breast cancer diagnosis using support vector machines optimized by whale optimization and dragonfly algorithms. *IEEE Access,* 10: 69688–69699.

Emambocus, Bibi Aamirah Shafaa, Muhammed Basheer Jasser, Angela Amphawan, and Ali Wagdy Mohamed. (2022). An optimized discrete dragonfly algorithm tackling the low exploitation problem for solving tsp. *Mathematics,* 10(19): 3647.

Gharehchopogh, Farhad Soleimanian, and Hojjat Gholizadeh. (2019). A comprehensive survey: Whale Optimization Algorithm and its applications. *Swarm and Evolutionary Computation,* 48): 1–24.

Gracic, Mak, Guy Gubnisky, and Roee Diamant. (2024). Review of Cetacean's click detection algorithms. arXiv preprint arXiv:2402.04735.

Hou, Guolian, Linjuan Gong, Bo Hu, Huilin Su, Ting Huang, Congzhi Huang et al. (2022). Application of fast adaptive moth-flame optimization in flexible operation modeling for supercritical units. *Energy,* 239: 121843.

Hussain, Irshad, Ibrar Ullah, Wajid Ali, Ghulam Muhammad, and Zulfiqar Ali. (2022). Exploiting lion optimization algorithm for sustainable energy management system in industrial applications. *Sustainable Energy Technologies and Assessments,* 52: 102237.

Ibrahim, Abdelhameed, El-Sayed M. El-kenawy, Nima Khodadadi, Marwa M. Eid, and Abdelaziz A. Abdelhamid. (2024). Guided whale optimization algorithm (guided WOA) with its application. In: *Handbook of Whale Optimization Algorithm*, pp. 243–251. Academic Press.

Jin, Qibing, Zhonghua Xu, and Wu Cai. (2021). An improved whale optimization algorithm with random evolution and special reinforcement dual-operation strategy collaboration. *Symmetry,* 13(2): 238.

Johnvictor, Anita Christaline, Vaishali Durgamahanthi, Ramya Meghana Pariti Venkata, and Nishtha Jethi. (2022). Critical review of bio-inspired optimization techniques. *Wiley Interdisciplinary Reviews: Computational Statistics,* 14(1): e1528.

Kang, Yan, Xuekun Yang, Bin Pu, Xiaokang Wang, Haining Wang, Yulong Xu, and Puming Wang. (2022). HWOA: An intelligent hybrid whale optimization algorithm for multi-objective task selection strategy in edge cloud computing system. *World Wide Web,* 25(5): 2265–2295.

Kasaragodu Anebagilu, Prajna. (2023). *Bio-inspired Optimization in Integrated River Basin Management.* Leibniz Universitat Hannover..

Kipnis, Dror. (2022). *Interference-Tolerant Approaches for Underwater Acoustic Detection of Marine Fauna.* PhD diss. University of Haifa (Israel).

Lahari, Pappu Soundarya, and Varaprasad Janamala. (2024). Hunter–Prey Optimization Algorithm: A review. *Journal of Electrical Systems and Information Technology,* 11(1): 19.

LaTorre, Antonio, Daniel Molina, Eneko Osaba, Javier Poyatos, Javier Del Ser, and Francisco Herrera. (2021). A prescription of methodological guidelines for comparing bio-inspired optimization algorithms. *Swarm and Evolutionary Computation,* 67: 100973.

Lee, Michael Sutton. (2021). *A Novel Intersection-based Clustering Scheme for VANET.* The University of Alabama.

LeFlore, Eric G. (2022). *Pride in Our Prides: Mitigating Human-lion Conflict in the Okavango Delta, Botswana.* Cambridge University Press.

Li, Wei, and Gai-Ge Wang. (2022). Elephant herding optimization using dynamic topology and biogeography-based optimization based on learning for numerical optimization. *Engineering with Computers,* 38(Suppl 2): 1585–1613.

Li, Wei, Gai-Ge Wang, and Amir H. Alavi. (2020). Learning-based elephant herding optimization algorithm for solving numerical optimization problems. *Knowledge-Based Systems,* 195: 105675.

Mandal, Pawan Kumar. (2023). A review of classical methods and Nature-Inspired Algorithms (NIAs) for optimization problems. *Results in Control and Optimization,* 2023: 100315.

Menaka, G., Srinivas Aluvala, R. Kumar, Zainab Abed Almoussawi, and Muntather Almusawi. (2023). A teaching and learning based optimization algorithm for efficient routing in vehicular ad hoc networks. In: *2023 International Conference on Ambient Intelligence, Knowledge Informatics, and Industrial Electronics* (*AIKIIE*), pp. 1–5. IEEE.

Meraihi, Yassine, Asma Benmessaoud Gabis, Amar Ramdane-Cherif, and Dalila Acheli. (2022). Advances in coyote optimization algorithm: Variants and applications. In: *Advances in Computational Intelligence and Communication: Selected Papers from the 2nd EAI International Conference on Computational Intelligence and Communications* (*CICom 2021*), pp. 99–113. Cham: Springer International Publishing.

Nhalungo, Sofia. (2024). *The Effects of Sport Hunting on Impala (Aepyceros melampus melampus) Population Demography and Behaviour in the Niassa Special Reserve.* PhD diss. Universidade Eduardo Mondlane.

Odili, Julius Beneoluchi, Ahmad Noraziah, Radzi Ambar, and Mohd Helmy Abd Wahab. (2018). A critical review of major nature-inspired optimization algorithms. *The Eurasia Proceedings of Science Technology Engineering and Mathematics,* 2: 376–394.

Prakash, P. Suman, Dwaram Kavitha, and P. Chenna Reddy. (2022). Energy and congestion-aware load balanced multi-path routing for wireless sensor networks in ambient environments. *Computer Communications*, 195: 217–226.

Rafeeq, Mohammed, Siti Fauziah Toha, Salmiah Ahmad, and Mohd Asyraf Razib. (2021). Locomotion strategies for amphibious robots: A review. *IEEE Access,* 9: 26323–26342.

Rana, Nadim, Muhammad Shafie Abd Latiff, Shafi'I. Muhammad Abdulhamid, and Haruna Chiroma. (2020). Whale optimization algorithm: A systematic review of contemporary applications, modifications and developments. *Neural Computing and Applications,* 32: 16245–16277.

Romeh, Ali El, and Seyedali Mirjalili. (2023). Multi-robot exploration of unknown space using combined meta-heuristic salp swarm algorithm and deterministic coordinated multi-robot exploration. *Sensors,* 23(4): 2156.

Sadayappan, Balamurugan, Preethiya Balakrishnan, Vishal C.R., Neethu T. Vijayan, Mahendran Subramanian, and Mangesh U. Gauns. (2023). Applications of machine learning in chemical and biological oceanography. *ACS Omega,* 8(18): 15831–15853.

Sahoo, Saroj Kumar, Apu Kumar Saha, Sukanta Nama, and Mohammad Masdari. (2023). An improved moth flame optimization algorithm based on a modified dynamic opposite learning strategy. *Artificial Intelligence Review,* 56(4): 2811–2869.

Sangwan, Saurabh Raj, and M.P.S. Bhatia. (2020). Denigration bullying resolution using wolf search optimized online reputation rumour detection. *Procedia Computer Science,* 173: 305–314.

Storms, Mona, Aryan Jakhar, Oliver Mitesser, Andreas Jechow, Franz Hölker, Tobias Degen, Thomas Hovestadt, and Jacqueline Degen. (2022). The rising moon promotes mate finding in moths. *Communications Biology,* 5(1): 393.

Su, Yuchao, Qiuzhen Lin, Zhong Ming, and Kay Chen Tan. (2022). Adapting decomposed directions for evolutionary multiobjective optimization. *IEEE Transactions on Cybernetics,* 53(10): 6289–6302.

Tubishat, Mohammad, Salinah Ja'afar, Mohammed Alswaitti, Seyedali Mirjalili, Norisma Idris, Maizatul Akmar Ismail, and Mardian Shah Omar. (2021). Dynamic salp swarm algorithm for feature selection. *Expert Systems with Applications,* 164: 113873.

Wolf, Thomas. (2022). *Managing a Nonprofit Organization: 40th Anniversary Revised and Updated Edition.* Simon and Schuster.

Wu, Huiying, Meihua Zou, Ye Ke, Wenqi Ou, Yonghong Li, and Minquan Ye. (2022). Effect evaluation and intelligent prediction of power substation project considering new energy. *CMES – Computer Modeling in Engineering & Sciences,* 132(3).

Yang, Xin-She. (2020). *Nature-inspired Optimization Algorithms*. Academic Press.

3 | AI Advancements in Understanding Animal Behavioral Paradigms

Antara Ghosh[1] and Dipti Debnath[2*]

[1] Department of Computer Application, RCC Institute of Information Technology, Kolkata, West Bengal, India

[2] Biochemistry and Cancer Biology Program, College of Biomedical and Translational Sciences, Texas, USA

The intersection of artificial intelligence (AI) and animal behavior studies is transforming our understanding and management of animal welfare. This chapter explores the advancements in AI, particularly in machine learning, deep learning, and neural networks, and their applications in identifying and analyzing animal behavior. Traditional visual recognition methods have limitations that AI-powered solutions can overcome, enhancing accuracy and capability in individual animal identification and behavior monitoring. The use of Convolutional Neural Networks (CNNs) has shown significant promise, particularly in tasks like identifying individual giant pandas, which aids in wildlife monitoring and conservation efforts. Sensor technologies such as accelerometers, gyroscopes, GPS trackers, and RFID tags, combined with AI, provide detailed insights into various animal behaviors, improving farm management practices, breeding programs, and overall animal welfare. Furthermore, AI's role in recognizing animal emotions is highlighted, showing its potential in enhancing animal quality of life through better understanding of their emotional states. Case studies on automated pain recognition in domestic cats and other species demonstrate the effectiveness of AI in veterinary care, paving the way for improved animal welfare through timely and accurate emotion detection. This integration of AI into animal behavior studies supports sustainable management of animal resources and promotes biodiversity conservation for future generations.

*Corresponding author: diptidebnath@rocketmail.com

Introduction

Animals are essential to our ecosystem, deserving care and attention to their welfare. Concurrently, artificial intelligence (AI) has advanced significantly, particularly in machine learning and deep learning, showcasing potential across various domains (Obaid, 2023; Baduge et al., 2022; Arrieta et al., 2020; Chan et al., 2023; Ajagbe et al., 2023). AI enhances our ability to identify and understand animal behavior, benefiting animal welfare, breeding, research, and farm management. Traditional visual recognition methods (Zhang et al., 2024) have limitations, but AI-powered solutions improve accuracy in monitoring animal behavior and individual identification (Neethirajan, 2022) .

Visual recognition has typically relied on human assistance, lacking accuracy and long-term effectiveness (Ketkar et al., 2021; Han et al., 2020; Cheeseman et al., 2022). Efficient individual recognition can swiftly provide critical information for animal welfare. For instance, deep learning techniques like Convolutional Neural Networks (CNNs) (Ketkar et al., 2021) have proven effective in identifying individual giant pandas, a species previously on the brink of extinction due to limited breeding and habitat (Duan et al., 2020; James et al., 2023). CNNs excel at extracting features from images, enabling the differentiation of subtle intra-species differences (Liu et al., 2021). A model recently achieved 95% accuracy in identifying individual giant pandas (Swarup et al., 2021).

Deep learning accelerates animal identification and monitoring, yielding vital data for conservation planning (Nathan et al., 2022). This data aids in understanding migratory patterns, significant habitats, and conservation effectiveness. Accurate identification of individual research animals allows for in-depth analyses of behavior and ecology (Raihan, 2023) . By automating identification, deep learning models enable researchers to focus on data interpretation, enhancing insights into social structures and species dynamics. Integrating deep learning into wildlife management promises advancements in animal population understanding and welfare (Zhang et al., 2024).

Sensor technologies, such as accelerometers, GPS trackers, and RFID (Radio Frequency IDentification) tags, provide detailed behavioral information, including feeding habits and social interactions (Carlslake et al., 2020). For example, accelerometers can detect subtle motions, while GPS monitors animal movements over distances. AI systems can optimize farm management using this data to improve feeding schedules and predict disease outbreaks (Biase et al., 2022).

Combining sensor technology with AI enhances productivity, animal welfare, and sustainability in agriculture and wildlife management (Lockie et al., 2020). This technology supports ethical resource stewardship and fosters resilience in agricultural systems. AI has transformed animal behavior assessment, making processes faster and more accurate, though challenges remain in visually identifying individual animals on farms.

Recent research has leveraged deep learning to develop a face recognition model using CNNs for giant pandas (Chen et al., 2023). These models enhance

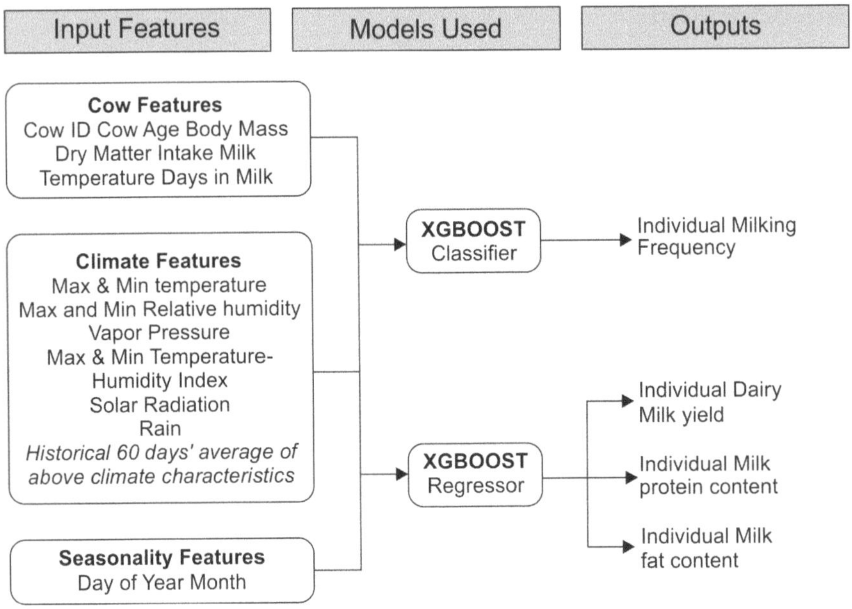

Figure 1: AI-integrated robotic milking systems to predict milk production with quality checking, optimize milking strategies and enhance cow welfare.

animal detection accuracy. AI also improves animal management efficiency by analyzing behavior data from sensors. For instance, a machine learning system (Figure 1) analyzed data on 80 cows over five years, predicting milk yield and composition with over 80% accuracy, enhancing dairy operations (Ji et al., 2022). Similarly, a poultry welfare approach using robots and sensors ensures the well-being of chickens (Park et al., 2022).

K. Jiang et al. (2022) improved a deep learning algorithm (CBAM-YOLOv7) for intelligent duck counting, demonstrating its potential for real-time monitoring in agriculture (Jiang et al., 2022). These advancements present opportunities to enhance animal husbandry, contributing to sustainable and ethical resource management. The upcoming sections will explore AI applications in ecological issues like animal emotion, health, nutrition, and disease prevention.

AI in Animal Emotion Recognition

AI applications in animal emotion recognition offer significant benefits for understanding animal behavior, human-animal ecological interactions, and animal welfare. By utilizing image processing and machine learning, including CNN models like ResNet50, AI can identify animal emotions, thereby aiding in addressing their emotional states and improving quality of life (Neethirajan et al., 2021). In a study on domestic shorthair cats, ResNet50 combined with catFACS-based geometric analysis achieved over 72% accuracy in recognizing pain levels post-Ovariohysterectomy (Feighelstein et al., 2022; Neethirajan et al., 2021). Similar methods, such as the Horse Grimace Scale (HGS), achieved 75.8% accuracy in categorizing pain in horses and 88.3% accuracy in distinguishing pain presence (Neethirajan et al., 2021). These AI-driven advancements facilitate early disease identification and improve animal management, opening new opportunities for enhancing animal welfare.

Overview of Traditional Methods vs. AI-based Approaches

Surveys and questionnaires have traditionally assessed animal emotions through potentially biased human observations (Meagler, 2009). While some studies use blood tests for hormonal changes related to stress (Man et al., 2023; James et al., 2023), this method is invasive and impractical for large-scale assessments. Researchers often analyze several minutes of animal activity to infer emotional states, a time-consuming process that may miss subtle cues. Regular caregiver interactions can mask true emotional signs, while individual interviews disrupt farming practices.

AI-based methods offer an alternative by using machine learning to identify and analyze animal emotions (Zhang et al., 2024). These approaches collect substantial labeled data, including images, videos, sound recordings, and physiological measurements. Features such as facial expressions and body gestures are extracted using video imaging technology, with deep learning techniques like CNNs and RNNs (Recurrent Neural Networks) enhancing model accuracy. Evaluation metrics such as accuracy, precision, and recall are used to assess effectiveness.

AI methods provide advantages over traditional approaches by capturing data objectively, handling large volumes efficiently, and reducing costs through automation. Research shows that trained AI models can achieve high accuracy in emotion recognition and are applicable across various species and environments. However, challenges like data availability, annotation quality, and concerns over data privacy and animal protection remain (Marzi et al., 2023). Therefore, further research and collaboration among experts in animal behavior, computer technology, and ethics are essential.

Case Study 1: Automated Pain Recognition in Domestic Cats

Analyzing Domestic Shorthair Cat Facial Image Data Using Machine Learning Models

Identifying pain in animals and evaluating stimuli are crucial for effective pain management and assessing well-being. Non-verbal cues, particularly facial expressions, are key indicators of pain in both animals and humans. Building on Langford et al.'s pioneering work, various methods have been developed to assess painful facial expressions in species such as mice, rats, rabbits, horses, pigs, sheep, ferrets, and cats (Domínguez-Oliva et al., 2022; Fischer-T. et al., 2022; Onuma et al., 2024; Whittaker et al., 2023).

Pain management in cats poses unique challenges due to limited research on specific painful conditions, potential analgesic side effects, difficulty in assessing ambiguous pain behaviors, and humans' struggle to interpret cats' body language accurately. Consequently, cats often receive fewer analgesic drugs than dogs. To address this, Finka and colleagues developed a method using geometric facial landmarks to detect changes in facial expressions due to pain in short-haired domestic cats (Bonesh-S. et al., 2022)]. These landmarks were derived from facial musculature and the Cat Facial Action Coding System (catFACS), considering variations in facial structures across species and differences in stance.

Facial images from 29 cats were captured at four time points: before surgery, immediately post-surgery, before post-operative pain medication, and during peak post-surgery pain (Feighelstein et al., 2023). Principal Components analysis assessed facial shape variation related to pain intensity, revealing a significant correlation between PC scores and the UNESP-Botucatu MCPS tool, a standard post-surgery pain measure in cats (Finka et al., 2019). This correlation supports the validity of the geometric face model for detecting pain, indicating that these facial landmarks provide essential signals for machine classification. The method's successful validation suggests potential for new, objective pain assessment approaches, enhancing pain management and animal welfare.

Specific Techniques and Algorithms used for Pain Recognition

The automated recognition of emotions and pain in animals remains underexplored, focusing mainly on a few species. A key component in this area is the single-frame Inception V3 CNN, trained to detect Action Units (AUs) relevant to the grimace scale for pain detection, achieving 94% sensitivity in identifying pain in mice (Broome et al., 2022). Mahmoud et al. presented a pipeline for automating pain detection in sheep by identifying nine specific AUs. This method involved facial detection, landmark detection, feature extraction, and face recognition using a Support Vector Machine (SVM) based on Histogram of Oriented Gradients (HOG) features (Chandrakala & Durga Devi, 2021). Broome et al. applied a CNN to automate pain evaluation in horses during castration, achieving an overall accuracy of 75.8% in classifying pain into three levels: absent, moderately present, and obviously present, with an 88.3% accuracy in distinguishing between present and absent pain.

Various histogram-related features commonly used in image processing, such as HOG, Local Binary Patterns (LBP), and Scale-Invariant Feature Transform (SIFT), were extracted along with key features from a VGG16 deep CNN model (Ahadit et al., 2022). Combining these classifiers enhanced outcomes, achieving a response time of 0.51–0.88 seconds for pain estimation in tilted poses and an F1 score of 0.53–0.87 for correctly classified images after decision fusion.

However, automatic landmarking and pose estimation methods were not entirely feasible for donkeys.

Correia-Caeiro et al. developed a prototype Automatic MaqFACS coding system using 53 videos of five Rhesus macaques, manually coded with AUs for each frame. The system, trained with an average of six MacFACS AUs, achieved a categorization accuracy of 89% (Correia-C. et al., 2021). These studies highlight the potential of AI and machine learning in automating pain and emotion recognition in animals, paving the way for improved welfare through more accurate pain assessment methods.

Results and Implications for Veterinary Care

In Feighelstein et al. (2022), cats' emotions were recorded at different time points corresponding to varying intensities of pain:

i. Pre-surgery (18–24 hours prior to surgery)
ii. 1-hour post-surgery (30 minutes to 1 hour after the end of the surgical procedure, before the administration of additional analgesics)
iii. After rescue analgesia (about 4 hours after post-operative analgesia)

These landmarks were selected based on their connection with the musculature and the Cat Facial Action Units (catFACS), which were included as additional labels. Figure 2 illustrates the placement of 48 such landmarks on a cat's face.

Figure 2: Placement of facial landmarks on different animals' face appearing contralateral to their origin.

We employed two methods to create models: a landmark-based approach (LDM) and a deep learning approach (DL). The LDM uses predefined landmarks on the cat's face to analyze expressions and detect pain, chosen based on their relevance to facial musculature and specific facial action units (catFACS). For the DL approach, we utilized the ResNet50 architecture after appropriate training. In the LDM, we developed a Multilayer Perceptron (MLP) neural network with three hidden layers. Both methods were trained with and without data augmentation and feline face alignment to assess the impact of these actions. To validate our models, we applied standard performance metrics:

Accuracy: The percentage of correctly classified instances.
Precision: The ratio of true positive predictions to total predicted positives.
Recall: The ratio of true positive predictions to all actual positives.
F1 Score: The weighted average of precision and recall, useful for imbalanced class distributions.

These metrics evaluate our models' effectiveness in detecting and classifying pain levels in cats, as summarized in Table 1. It is evident that the DL approach generally outperforms the LDM. In Table 1, "Alignment YES" denotes the spatial alignment of facial landmarks through rotation, scaling, and translation to reduce geometric variations, enhancing face processing tasks like recognition. Conversely, "Alignment NO" indicates unaltered images without these adjustments, potentially retaining natural geometric variations. "Data augmentation YES" refers to modifying existing data to create variations that still represent the original data, enhancing the training set's diversity. "Data augmentation NO" means no such techniques are applied, leaving the training set unchanged.

Table 1: Comparison between DL and LDM approach to detect different pain levels of cats

Approach	Alignment	Data Augmentation	Accuracy	Precision	Recall
DL	YES	NO	**0.7239 (± 0.1837)**	**0.7526 (± 0.2139)**	**0.7353 (± 0.3215)**
DL	YES	YES	**0.7051 (± 0.1855)**	**0.7725 (± 0.2385)**	**0.6853 (± 0.3195)**
DL	NO	NO	**0.7360 (± 0.1782)**	**0.8186 (± 0.2045)**	**0.7010 (± 0.2889)**
DL	NO	YES	**0.7344 (± 0.1780)**	**0.84512 (± 0.1948)**	**0.6636 (± 0.3614)**
LDM	YES	NO	0.7196 (± 0.1464)	0.7441 (± 0.1600)	0.7457 (± 0.1943)
LDM	YES	YES	0.7239 (± 0.1290)	0.7315 (± 0.1451)	0.7512 (± 0.1955)
LDM	NO	NO	0.6747 (± 0.1151)	0.7056 (± 0.1442	0.6892 (± 0.2639)
LDM	NO	YES	0.6805 (± 0.1087)	0.6807 (± 0.1103)	0.6933 (± 0.2278)

Case Study 2: Intelligent Detection of Pain Signals

Training Vision Algorithms with Automatic Computational Classifiers to Detect Pain in Horses

Hummel et al. (Ewence & Whitcock, 2024|) applied a landmark-based approach for pain recognition in horses. Unattended pain can significantly impact horses' health, leading to issues like central sensitization, altered pain thresholds, and hyperalgesia, which may result in life-threatening complications. Assessing pain in animals is as crucial as monitoring vital signs like blood pressure and body temperature, as continuous evaluation can enhance pain management and recovery. However, frequent pain assessments can increase stress and potentially elevate pain levels due to constant physical interventions, adversely affecting overall health. Assessing pain presents challenges, as it requires skilled observers to identify behavioral changes or physiological indicators. The practical implementation of these assessments in settings like hospitals or equestrian centers is complicated by the need for trained personnel and the time necessary for thorough observations.

The Facial Action Coding System (FACS) (Ask et al., 2024) is commonly used to measure pain and emotions in non-verbal individuals, such as infants and disabled patients. A recent study using the Equine Facial Action Coding System (EquiFACS) supported the efficacy of horse facial expressions in indicating pain, showing that the facial regions identified by EquiFACS align with those recognized by the Horse Grimace Scale (HGS) and the Equine Pain Face. However, effective pain recognition protocols require training observers to accurately interpret these facial cues to prevent bias, and trained observers must be available for evaluations throughout the day, which can be time-consuming. Furthermore, as prey animals, horses may suppress pain behaviors in the presence of potential threats, revealing pain only when humans are not around.

Description of the Dataset, Model Training, and Performance

Seven horses, including six young Brazilian sport horses and one Mangalarga Marchador, were filmed during benign surgical castration at approximately one year of age. The University of São Paulo required castration for the study, adhering to its anesthesia, surgery, and post-operative pain control protocol. Analgesia and sedation were achieved with intramuscular morphine and intravenous xylazine, while anesthetic induction involved ketamine, diazepam, and glyceryl guaiacol ether. Continuous monitoring ensured stable anesthesia depth and vital signs during the procedure, with heart and breathing rates checked every five minutes and fundamental tests like eye placement, nystagmus movements, and palpebral reflex evaluations confirming optimal anesthesia levels (Eaton et al., 2022).

Local anesthesia involved 100 ml of 2% lidocaine, with 10 ml administered along the scrotal median raphe and 5 cc intra-testicular for each testicle. Orchiectomy was performed using a closed approach with an 8 mm incision perpendicular to the scrotal median raphe. The skin and dartos tunic were opened, the testicle was delivered with intact vaginal tunics, and a ramming pad was applied to the spermatic cord for 5 minutes (King, 2021). This process was repeated for the contralateral testicle. The entire procedure lasted about 40 minutes, with 20 minutes for anesthetic administration and 20 minutes for surgery. The horses were monitored via a camera system positioned in front of the feeder station two days before and four days after the procedure, at four different times of day: 7 am, 10 am, 12 pm, and 4 pm. This setup resulted in the collection of 320 video sequences, each lasting 30 minutes, recorded with Intelbras VHD 1220 B–G4 Multi HD cameras.

Potential Benefits for Equine Health and Welfare

Equine husbandry and nutrition are vital for horses' well-being. Inadequate feeding and care can impair essential bodily systems, jeopardizing their health and welfare, and diminishing their ability to meet human expectations. Horses serve various roles globally, often as working animals crucial for their owners' livelihoods. Poor welfare affects both horses and their owners, particularly when horses cannot perform optimally.

As horses age, they increasingly face conditions like arthritis, underscoring the necessity for tailored care to ensure good welfare. Welfare is defined as an animal's ability to cope with its environment, requiring a thorough understanding of the specific group of horses to thrive in their context (Scialabba, 2021). However, quantifying a horse's thriving under different management practices is challenging due to limited comparative information on their health impacts. Managers often rely on experience and intuition, contrasting with historical practices where specialized personnel cared for horses.

Klecel and Martynuik's article, "Horse Husbandry and Management in the Ancient World", highlights that many effective prehistoric husbandry and breeding practices are still relevant today (Klecel & Martiniuk, 2021). Initially tamed for food, horses became vital for transportation and warfare, paving the way for their role in racing. Iron Age horse racing, akin in scale and significance to modern racing in Great Britain and France, emphasized the importance of factors like rider weight, a topic still under investigation today.

AI in Animal Nutrition and Health

AI has become a tool that refines or even revolutionizes the assessment of how to provide animals with the best conditions for a healthier life. The fields where AI technology revolutionizes the animal nutrition and health as stated in (Zhang et al., 2024), are as follows:

1. **Overcoming Challenges in Tracking Feeding Behaviors Via AI Technology:** Advances in AI enable precise, continuous tracking of individual animal feeding behaviors, including time spent near feeders, feeding frequency, and feed consumption. This monitoring helps detect

changes in feeding patterns, often indicative of health complications, and provides solutions to address these challenges (Ezanno et al., 2021).

2. **AI for Disease Detection and Prevention:** AI methods solve complex problems in disease diagnosis and prevention by using predictive-analytical models that incorporate patient data, hereditary status, and geographical location (Bao & Xie, 2023). This early detection system saves time and costs, allowing for prompt preventive measures.
3. **Pervasive Computing for Health Assessment and Detection:** AI-powered technology allows accurate remote monitoring and diagnosis of animal health conditions, helping farm managers and health professionals identify and treat health complications early, especially in large-scale farming operations (Aharwal et al., 2023).
4. **Benefits of AI in Optimizing Diet and Health:** AI improves animal health and farming productivity through precision nutrition and automated diets (Saha & Pathak, 2023). Technologies like Real Appetite AI and drones optimize feed intake, minimize waste, and maintain appropriate feeding conditions.
5. **Nutrition: The Pareto Principle and Data-Driven Diet Optimization:** AI utilizes data from health, climate, and disease history databases to create targeted animal diets that meet nutritional needs for optimal growth. It analyzes genetic, historical, and environmental data to understand factors impacting health and develop specific feeds. AI also adjusts feeding regimes based on physiological indices like metabolism and nutrient absorption.

Case Study 3: Intelligent Learning Model for Cow Feed Intake

Development of Intelligent Models for Collecting Cow Feed Intake Data

Regarding the structure of the datasets provided, they consisted of tensors containing information about individual meals. Each tensor was created by subtracting a lower weight image from a higher weight image for four attributes (RGB, depth tensors). To achieve this, the following transformations were applied: (a) assembling meals ranging from 0–6 kg per meal or 'as fed' without adjusting for dry matter basis, (b) data augmentation by horizontally and vertically flipping original images, (c) stacking the resulting RGB and depth images to form 4-channel images, and (d) resizing dimensions to preferred sizes of 160, 120, and 4, respectively. Additionally, two categorical variables were derived per tensor: feed type (dichotomized as feed A = 1, feed B = 0) and time period (dummy coded as morning/afternoon/night). Approximately 30% of the images underwent augmentation to enrich the dataset while avoiding redundancy. From manually collected data, around 30,000 RGB D (red, green, blue, depth) tensors were generated per feed type (Bennadji, 2020). These tensors were divided into three datasets for training: 46 tensors each for feed type A and feed type B, and 46 mixed tensors equally distributed between feed types A and B. An additional 7,000 RGBD tensors were generated for model fine-tuning using 300 images acquired in March 2021. The models developed and compared included: (1) a combined model trained on 40,000 tensors from both feed types without specifying the feed type in each tensor, (2) Transfer Learning (TL) models fine-tuned on tensors from each feed type sequentially, and (3) a Multilayer Perceptron and Convolutional Neural Network (MLP-CNN) model incorporating additional categorical variables (Ehteram et al., 2023).

These models utilized an architecture inspired by EfficientNetB0 (Figures 3 a, b), comprising six inverted residual blocks with normalization, convolutional, and depthwise convolutional layers. Early stopping was employed to prevent overfitting during training, using mean squared error as the loss function and root mean square propagation as the optimizer.

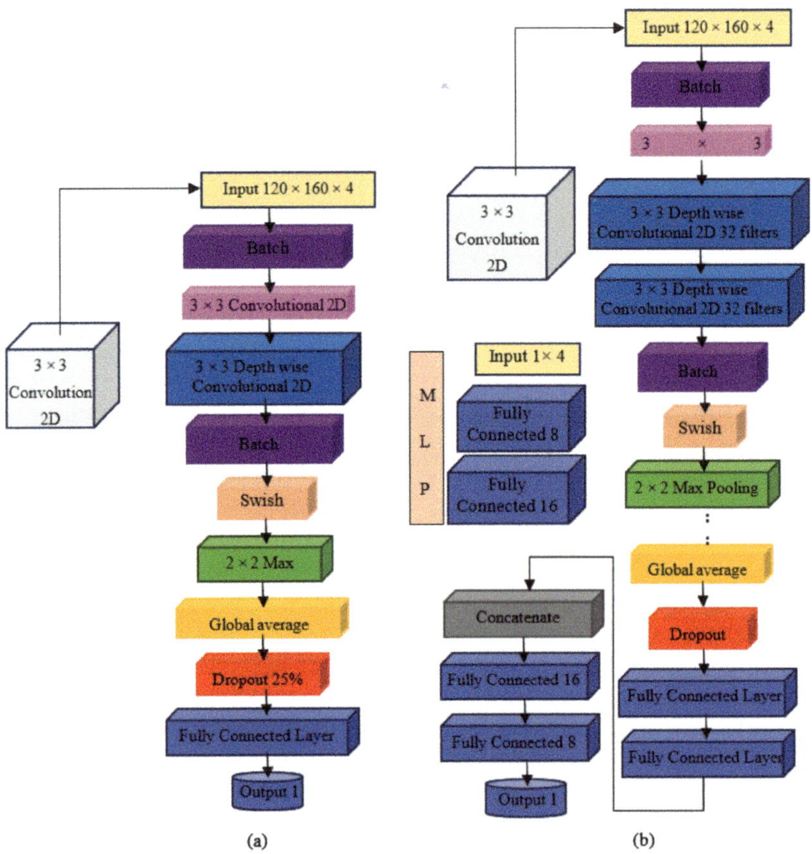

Figure 3: (a) Modified CNN and (b) Modified MLP-CNN model developed following EfficientNet

Mixed data consisting of categorical variables (type of feed and time period) and image data were processed using the MLP-CNN architecture developed specifically for this study, illustrated in Figure 3b. The MLP network handled categorical data transmission, while CNN extracted features from tensors. The MLP network included more Fully Connected layers compared to the CNN, which shared the same architecture as models (a) and (b) except for the final output layer. Finally, outputs from both convolutional networks were merged and passed through several FC layers to predict the weight of each meal. Table 2 depicts the values of the hyperparameters of the modified CNN.

Table 2: Convolutional Neural Network (CNN) models' hyperparameter values

Hyperparameter	Value
Learning Rate (maximum)	0.001
Learning Rate (minimum)	6.25×10^{-5}
Batch Size	16
Dropout Rate	0.25

Application of Automatic Feeding Systems and Their Impact on Feeding Schedules

Over the years, automated feeding systems have seen remarkable advancements, with the latest innovation being on-demand feeding capabilities that provide cows with precise Total Mixed Rations (TMR) instantly as needed, eliminating potential losses compared to conventional once or twice daily feeding (Valoppi et al., 2021). Using advanced laser technology, these systems monitor feed levels continuously and deliver feed to low bunks based on real-time consumption data, ensuring optimal feed management throughout the 24-hour cycle. This approach not only reduces feed shrinkage and enhances accuracy and consistency but also lowers operational costs. Automated feeding significantly decreases feed refusal rates, from 3–5% in TMR-fed groups to 1% or less in automated systems, offering precise timing and eliminating human errors, thereby ensuring high reliability and consistency over time.

Analysis of the Model's Accuracy and Benefits for Dairy Farming

The application of communication technologies in dairy farming, especially through software and hardware aids, has greatly improved decision-making for farmers (Baldin et al., 2021). These technologies enable efficient management and scaling of animal groups without increasing human resources. Their effectiveness is largely due to the high accuracy of machine learning algorithms, which continually adapt to new data inputs. Machine learning enhances decision-making by addressing challenges such as data multicollinearity, variable distributions, and missing values, while also capturing interactions and nonlinear effects in regression and classification problems. Supervised learning algorithms, like random forests, extract patterns from training data, whereas unsupervised methods, such as k-means clustering, identify patterns without specific training sets. The accuracy of on-farm predictions depends on the quality of input data and proper validation methods to prevent overfitting.

The integration of machine learning in dairy farm management is expanding, providing opportunities for future research. Systematic mapping studies, including those by Cockburn and Slob et al. (Cockburn 2020; Slob et al., 2021), have outlined the evolution and application of machine learning in dairy farming. These reviews assess literature across various subdomains, such as animal physiology, reproduction, behavior, and feeding, highlighting methodologies, algorithms, and evaluation metrics used, while identifying challenges and discussing research design implications. Slob et al.'s review (2021) specifically focused on disease detection in milk, milk production forecasting, and milk quality estimation, ensuring robustness through defined search protocols and selection criteria. They compared regression and classification problems, evaluation criteria, validation methods, and algorithm accuracies across studies (Huang et al., 2020), though their scope was limited to key articles within the dairy research domain.

In contrast, this synthesized mapping review covers a broader scope, from January 1999 to December 2021, with an extensive search strategy across scientific databases. It includes all relevant dairy farming subdomains, offering a comprehensive mapping of studies based on geographical distribution and research areas. This study tracks publication behavior over time and ranks evaluation metrics separately for classification and regression problems, providing a detailed analysis of their usage frequencies. Overall, the adoption of machine learning in dairy farming promises advancements in efficiency, accuracy, and sustainability, driven by ongoing research and technological innovation (Neethirajan, 2024).

Case Study 4: Intelligent Feeding System for Pet Obesity Control

Implementation of AI-based Systems for Managing Pet Obesity

The automation of pet obesity management through AI systems offers personalized advice and interventions tailored to each individual pet (Tauseef et al., 2024). This process involves several key steps:

1. **Data Collection:** Collect comprehensive data from various sources including (a) veterinary records such as detailed history from the veterinarian, including the animal's species, weight history, diet, and any existing medical conditions, (b) fitted devices such as Data from activity monitors and other wearable devices that track the pet's daily activities and behaviors, and

(c) owner input like information provided by the pet owners about daily routines, diet specifics, and observed behaviors.

2. **Machine Learning Algorithms:** Utilize machine learning algorithms to analyze the collected data and classify it in relation to pet obesity. These algorithms can (a) analyze feeding patterns by identifying and evaluating feeding habits and diet composition, (b) assess activity levels by monitoring exercise regimens and physical activity, and (c) predict obesity risk by estimating the likelihood of a pet becoming obese based on its current lifestyle and health metrics.
3. **Personalized Recommendations:** Develop an AI solution that provides tailored care and intervention strategies for each pet. This includes: (a) feeding recommendations, i.e., suggesting appropriate diet plans, portion sizes, and feeding schedules, (b) exercise programs, i.e., design exercise routines suitable for the pet's breed, age, and health status, and (c) behavioral management of offering tips for managing behaviors that may contribute to obesity, such as overeating due to stress or boredom.
4. **Monitoring and Feedback**: Implement a system to continuously monitor and provide feedback on the pet's progress. This could involve: (a) platforms like mobile applications/websites, where owners can log daily or weekly updates on their pet's activity, diet, and weight, and (b) automated AI feedback and adjustments to recommendations based on the logged data
5. **Integration with Veterinary Care:** Ensure the AI-based system complements traditional veterinary care by (a) database connectivity by integrating a companion animal database with the veterinary practice management software, and (b) collaborative care by enabling pet owners to share AI-generated insights and recommendations with their veterinarians for further advice and intervention.
6. **Behavioral Insights:** Use AI to analyze behavioral and activity diary data to identify triggers for obesity, such as (a) stress or boredom to determine if emotional factors are influencing overeating or lack of activity, and (b) adjustments to suggest modifications in the pet's environment or routine to mitigate these triggers and promote healthier behaviors.

By automating these processes, managing pet obesity becomes more efficient and effective, leading to better health outcomes for pets and more informed care for pet owners.

Mechanisms of Intelligent Feeding Systems and Their Effectiveness

Smart feeding systems feature mechanisms that dispense pre-programmed amounts of food, allowing pet owners to provide correct portion sizes based on age, weight, activity level, and dietary needs (Valencia et al., 2022). These systems effectively minimize overfeeding, a leading cause of obesity. Some include weight control algorithms that calculate daily caloric needs, which can be adjusted per portion to help pets maintain a healthy weight. Many feeders can also dispense food at specific intervals, aiding pet owners in managing their pets' feeding schedules (Kulaikaret et al., 2023). Scheduled feeding establishes better eating habits and controls calorie intake, contributing to weight management.

In certain regions, intelligent feeding systems calculate daily caloric requirements using measurements like age, weight, activity levels, and metabolic rates to recommend appropriate portion sizes, ensuring precise daily calorie supply and preventing excessive intake that leads to weight gain. Pets have unique energy demands and nutrient needs based on species, breeds, and individual characteristics, so intelligent feeding systems should be flexible to cater to these requirements (Hobbs Jr., 2023).

Successful weight loss for pets relies heavily on owners entering accurate information about their pet's weight, activity level, and diet, and adhering to recommended feeding times and portions. Systems with enhanced functionalities—such as weight management algorithms, preset feeding times, portion control, and internet-based monitoring—are valuable tools for preventing pet obesity. Kulaikar et al. highlighted the need to address behavioral aspects, like food-seeking behavior and emotional eating, which are crucial for regulating ideal body weight in pets. Other feeding methods, such as slow feeders or puzzle feeders, can also promote proper eating habits and prevent overfeeding.

A Comparative Analysis of Traditional and AI-based Models in Understanding Animal Behavior

Conventional techniques in understanding animal behavior rely heavily on manual data collection, which, while offering high ecological validity, is often subject to observer bias and limited in temporal and spatial resolution. These methods are labor-intensive and can be costly, especially for extensive studies. In contrast, AI-based approaches leverage automated data collection, providing higher data quality and the ability to analyze complex behaviors with advanced algorithms. These methods offer continuous monitoring, improving both temporal and spatial resolution. However, they require significant initial investment and expertise in AI and data analysis. While AI-based methods can sometimes struggle with ecological validity, they hold the potential for more objective and comprehensive insights into animal behavior. A comparative analysis of traditional vs AI-based approaches has been depicted in Table 3.

Conclusion

This chapter explored some of the significant ways AI benefits animal welfare. AI enables precise identification of animal behaviors and emotional states, and accurate diagnostics of diseases and their potential progression. This level of accuracy facilitates rapid responses, enhancing overall animal healthcare. Additionally, AI systems efficiently manage operational tasks like environment

Table 3: Comparative analysis of traditional and AI-Based approaches

Aspects	Traditional Methods	AI-Based Approaches
Data Collection	Manual data collection, surveys, experiments, and field studies play essential roles in animal behavior research, providing valuable insights into the behavior, ecology, and conservation of animals in their natural habitats.	May involve manual data collection initially but can also leverage automated data collection methods such as sensor networks or camera traps.
Data Quality	Due to human intervention, it can be subject to observer bias. It is complementary where the access is limited to human capacity (Blanco, 2022).	In remote or challenging environments where automated data collection methods may be impractical or unfeasible, manual observation remains a valuable tool for gathering data on animal behavior (Nazir & Md. Kaleem, 2021).
Interpretation	Relies on human expertise and interpretation.	Requires expertise in AI algorithms and data analysis (Bao & Xie, 2022).
Ecological Validity	High, reflects the actual behavior experienced in daily life (Hertel et al., 2020).	Variable, unpredictable depending on the data used for the analysis.
Temporal Resolution	Depends on the observation period or studies.	It can monitor continuously given the use of sensors and algorithms (Bownik & Wlodkowic, 2021).
Spatial Resolution	Depends on the observation period or studies.	It can monitor continuously given the use of sensors and algorithms (Carlslake et al., 2020).
Complexity of Behavior	Pertaining to complexity restricted only to the cabaret of behavior.	Possibility to analyze even complex behaviors using intelligent algorithms (Bao & Xie, 2022) .
Cost	Cost variable, contingent on equipment and human resources.	Immediate cost for acquiring equipment, possible savings over recurrent tests.

regulation, facility cleaning, and standardizing animal care services, minimizing errors. AI also advances animal health and nutrition by developing tailored feeding plans and predicting diseases before they become problematic. However, there are inherent limitations in applying AI techniques. For example, CNNs are effective for animal detection and action recognition but may struggle with images captured at nighttime or complex scenes. Sensor-based approaches are more effective for collecting behavioral data but may fail to identify complex emotions, due to lack of documented varieties of behavioral patterns. Machine learning algorithms can identify nutrition and health variables but may lack an understanding of animal needs, since the animal farming environment overlaps only partially to the natural habitats. Artificial Intelligence Models (AIMs) are effective in disease diagnosis but rely on rule-based systems and have limited learning capabilities. Thus, no single AI method is without weaknesses or risks.

To advance AI for animal care, several areas need consideration. Integration of AI with the Internet of Things (IoT) can create more advanced animal care facilities, enabling continuous monitoring and improvement of care aspects like feeding and environment. AI algorithms can analyze real-time data from connected IoT devices to ensure animals receive optimal care. Future AI models could enhance care plans based on individual animal behaviors, emotions, and health states, personalizing care for better outcomes. Interdisciplinary cooperation between technologists, veterinarians, and animal behavior researchers is crucial for these advancements.

REFERENCES

Ahadit, Alagesan Bhuvaneswari, and Ravi Kumar Jatoth. (2022). A novel multi-feature fusion deep neural network using HOG and VGG-Face for facial expression classification. *Machine Vision and Applications,* 33(4): 55.

Aharwal, Bhavna, Biswajit Roy, Somesh Meshram, and Aayush Yadav. (2023). Worth of artificial intelligence in the epoch of modern livestock farming: A review. *Agricultural Science Digest – A Research Journal,* 43(1): 1–9.

Ajagbe, Sunday Adeola, Matthew O. Adigun, Joseph B. Awotunde, John B. Oladosu, and Yetunde J. Oguns. (2023). Internet of things enabled convolutional neural networks: Applications, techniques, challenges, and prospects. *IoT-enabled Convolutional Neural Networks: Techniques and Applications,* 2023: 27–63.

Arrieta, Alejandro Barredo, Natalia Díaz-Rodríguez, Javier Del Ser, Adrien Bennetot, Siham Tabik, Alberto Barbado, Salvador García et al. (2020). Explainable Artificial Intelligence (XAI): Concepts, taxonomies, opportunities, and challenges toward responsible AI. *Information Fusion,* 58: 82–115.

Ask, Katrina, Marie Rhodin, Maheen Rashid-Engström, Elin Hernlund, and Pia Haubro Andersen. (2024). Changes in the equine facial repertoire during different orthopedic pain intensities. *Scientific Reports,* 14(1): 129.

Azad, K.R., and M.C. Valson. (2020). *Trade on Wild Animals and Animal Products: Effectiveness of Law in India a Critical Evaluation.* The National University of Advanced Legal Studies, India.

Baduge, Shanaka Kristombu, Sadeep Thilakarathna, Jude Shalitha Perera, Mehrdad Arashpour, Pejman Sharafi, Bertrand Teodosio, Ankit Shringi, and Priyan Mendis. (2022). Artificial intelligence and smart vision for building and construction 4.0:

Machine and deep learning methods and applications. *Automation in Construction,* 141: 104440.

Baldin, Michel, Tom Breunig, Roger Cue, Albert De Vries, Mark Doornink, Jan Drevenak, Robert Fourdraine et al. (2021). Integrated decision support systems (IDSS) for dairy farming: A discussion on how to improve their sustained adoption. *Animals,* 11(7): 2025.

Bao, Jun, and Qiuju Xie. (2022). Artificial intelligence in animal farming: A systematic literature review. *Journal of Cleaner Production,* 331: 129956.

Bennadji, Bilel. (2020). *Using Encoder-decoder Architecture for Material Segmentation Based on Beam Profile Analysis.* Master's thesis. University of Oulu, Finland.

Biase, Adriele Giaretta, Tiago Zanett Albertini, and Rodrigo Fernandes de Mello. (2022). On supervised learning to model and predict cattle weight in precision livestock breeding. *Computers and Electronics in Agriculture,* 195: 106706.

Blanco, Fernando. (2022). Cognitive bias. In: *Encyclopedia of Animal Cognition and Behavior*, pp. 1487–1493. Cham: Springer International Publishing.

Boneh-Shitrit, Tali, Marcelo Feighelstein, Annika Bremhorst, Shir Amir, Tomer Distelfeld, Yaniv Dassa, Sharon Yaroshetsky et al. (2022). Explainable automated recognition of emotional states from canine facial expressions: The case of positive anticipation and frustration. *Scientific Reports,* 12(1): 22611.

Bownik, Adam, and Donald Wlodkowic. (2021). Advances in real-time monitoring of water quality using automated analysis of animal behavior. *Science of the Total Environment,* 789: 147796.

Broome, Sofia, Marcelo Feighelstein, Anna Zamansky, Gabriel Carreira Lencioni, Pia Haubro Andersen, Francisca Pessanha, Marwa Mahmoud et al. (2022). Going deeper than tracking: A survey of computer-vision based recognition of animal pain and affective states. arXiv preprint arXiv:2206.08405.

Carslake, Charles, Jorge A. Vázquez-Diosdado, and Jasmeet Kaler. (2020). Machine learning algorithms to classify and quantify multiple behaviors in dairy calves using a sensor: Moving beyond classification in precision livestock. *Sensors,* 21(1): 88.

Chan, Kit Yan, Bilal Abu-Salih, Raneem Qaddoura, Al-Zoubi Ala'M, Vasile Palade, Duc-Son Pham, Javier Del Ser, and Khan Muhammad. (2023). Deep neural networks in the cloud: Review, applications, challenges, and research directions. *Neurocomputing,* 2023: 126327.

Chandrakala, M., and P. Durga Devi. (2021). Two-stage classifier for face recognition using HOG features. *Materials Today: Proceedings,* 47: 5771–5775.

Cheeseman, Ted, Ken Southerland, Jinmo Park, Marilia Olio, Kiirsten Flynn, John Calambokidis, Lindsey Jones et al. (2022). Advanced image recognition: A fully automated, high-accuracy photo-identification matching system for humpback whales. *Mammalian Biology,* 102(3): 915–929.

Chen, Wen, Bochuan Zheng, Jindong Zhang, Liming Xu, Jin Hou, and Vanessa Hull. (2023). Giant panda face recognition based on PandaFaceNet. *Ecological Informatics,* 77: 102225.

Cockburn, Marianne. (2020). Application and prospective discussion of machine learning for the management of dairy farms. *Animals,* 10(9): 1690.

Correia-Caeiro, Catia, Kathryn Holmes, and Takako Miyabe-Nishiwaki. (2021). Extending the MaqFACS to measure facial movement in Japanese macaques (*Macaca fuscata*) reveals a wide repertoire potential. *PLoS One,* 16(1): e0245117.

Domínguez-Oliva, Adriana, Daniel Mota-Rojas, Ismael Hernández-Avalos, Patricia Mora-Medina, Adriana Olmos-Hernández, Antonio Verduzco-Mendoza, Alejandro

Casas-Alvarado, and Alexandra L. Whittaker. (2022). The neurobiology of pain and facial movements in rodents: Clinical applications and current research. *Frontiers in Veterinary Science,* 9: 1016720.

Duan, Yingying, Haina Rong, Dunwu Qi, Luis Valencia-Cabrera, Gexiang Zhang, and Mario J. Pérez-Jiménez. (2020). A review of membrane computing models for complex ecosystems and a case study on a complex giant panda system. *Complexity,* 2020(1): 1312824.

Eaton, Samantha L., Fraser Murdoch, Nina M. Rzechorzek, Gerard Thompson, Claudia Hartley, Benjamin Thomas Blacklock, Chris Proudfoot et al. (2022). Modeling neurological diseases in large animals: Criteria for model selection and clinical assessment. *Cells,* 11(17): 2641.

Ehteram, Mohammad, Ali Najah Ahmed, Zohreh Sheikh Khozani, and Ahmed El-Shafie. (2023). Graph convolutional network – Long short term memory neural network-multi layer perceptron-Gaussian progress regression model: A new deep learning model for predicting ozone concentration. *Atmospheric Pollution Research,* 14(6): 101766.

Ewence, Heather, and Fleur Whitlock. (2024). The Horse. In: *The UFAW Handbook on the Care and Management of Laboratory and Other Research Animals*, pp. 628–661. Research Gate.

Ezanno, Pauline, Sébastien Picault, Gaël Beaunée, Xavier Bailly, Facundo Muñoz, Raphaël Duboz, Hervé Monod, and Jean-François Guégan. (2021). Research perspectives on animal health in the era of artificial intelligence. *Veterinary Research,* 52: 1–15.

Feighelstein, Marcelo, Ilan Shimshoni, Lauren R. Finka, Stelio P.L. Luna, Daniel S. Mills, and Anna Zamansky. (2022). Automated recognition of pain in cats. *Scientific Reports,* 12(1): 9575.

Feighelstein, Marcelo, Lea Henze, Sebastian Meller, Ilan Shimshoni, Ben Hermoni, Michael Berko, Friederike Twele et al. (2023). Explainable automated pain recognition in cats. *Scientific Reports,* 13(1): 8973.

Finka, Lauren R., Stelio P. Luna, Juliana T. Brondani, Yorgos Tzimiropoulos, John McDonagh, Mark J. Farnworth, Marcello Ruta, and Daniel S. Mills. (2019). Geometric morphometrics for the study of facial expressions in non-human animals, using the domestic cat as an exemplar. *Scientific Reports,* 9(1): 9883.

Fischer-Tenhagen, Carola, Jennifer Meier, and Alina Pohl. (2022). Do not look at me like that: Is the facial expression score reliable and accurate to evaluate pain in large domestic animals?: A systematic review. *Frontiers in Veterinary Science,* 9: 1002681.

Hertel, Anne G., Petri T. Niemelä, Niels J. Dingemanse, and Thomas Mueller. (2020). A guide for studying among-individual behavioral variation from movement data in the wild. *Movement Ecology,* 8: 1–18.

Hobbs Jr, Lonnie. (2023). *Essays on the Economics and Management of Pet Food.* PhD diss. Kansas State University.

Hou, Jin, Yuxin He, Hongbo Yang, Thomas Connor, Jie Gao, Yujun Wang, Yichao Zeng et al. (2020). Identification of animal individuals using deep learning: A case study of giant pandas. *Biological Conservation,* 242: 108414.

Huang, Jui-Chan, Kuo-Min Ko, Ming-Hung Shu, and Bi-Min Hsu. (2020). Application and comparison of several machine learning algorithms and their integration models in regression problems. *Neural Computing and Applications,* 32(10): 5461–5469.

James, Katharine Ann, Juliet Ilena Stromin, Nina Steenkamp, and Marc Irwin Combrinck. (2023).Understanding the relationships between physiological and psychosocial stress, cortisol, and cognition. *Frontiers in Endocrinology,* 14: 1085950.

Ji, B., Banhazi, T., Phillips, C.J., Wang, C., and Li, B. (2022). A machine learning framework to predict the next month's daily milk yield, milk composition, and milking frequency for cows in a robotic dairy farm. *Biosystems Engineering,* 216: 186–197. https://doi.org/10.1016/j.biosystemseng.2022.02.

Jiang, K., Xie, T., Yan, R., Wen, X., Li, D., Jiang, H., Jiang, N. et al. (2022). An attention mechanism-improved YOLOv7 object detection algorithm for hemp duck count estimation. *Agriculture,* 1210: 1659. https://doi.org/10.3390/agriculture12101659.

Ketkar, Nikhil, Jojo Moolayil, Nikhil Ketkar, and Jojo Moolayil. (2021). Convolutional neural networks. In: *Deep Learning with Python: Learn Best Practices of Deep Learning Models with PyTorch,* pp. 197–242. Springer.

King, E. Heath. (2021). Surgery of the scrotum and its contents. *Bovine Reproduction,* 2021: 200–209.

Klecel, Weronika, and Elżbieta Martyniuk. (2021). From the Eurasian steppes to the Roman circuses: A review of early development of horse breeding and management. *Animals,* 11(7): 1859.

Kulaikar, Jiten, Dhanashree Kurade, Anamika Sawant, Pooja Sthawarmath, and Anupama Chaurasia. (2023). IoT based automatic pet feeding and monitoring system. *International Journal of Modern Developments in Engineering and Science,* 2(4): 24–27.

Liu, Yao, Hongbin Pu, and Da-Wen Sun. (2021). Efficient extraction of deep image features using convolutional neural network (CNN) for applications in detecting and analyzing complex food matrices. *Trends in Food Science & Technology,* 113: 193–204.

Lockie, Stewart, Kate Fairley-Grenot, Rachel Ankeny, Linda Botterill, Barbara Howlett, Alex Mcbratney, Elspeth Probyn, et al. (2020). *The Future of Agricultural Technologies.* Australian Council of Learned Academies (ACOLA).

Man, Idy S.C., Robin Shao, Wai Kai Hou, Shirley Xin Li, Fiona Yan Liu, Maggy Lee, Yun Kwok Wing et al. (2023). Multi-systemic evaluation of biological and emotional responses to the Trier Social Stress Test: A meta-analysis and systematic review. *Frontiers in Neuroendocrinology,* 68: 101050.

Marzi, Sarah J., Brian M. Schilder, Alexi Nott, Carlo Sala Frigerio, Sandrine Willaime-Morawek, Magda Bucholz, Diane P. Hanger et al. (2023). Artificial intelligence for neurodegenerative experimental models. *Alzheimer's & Dementia,* 1912: 5970–5987.

Meagher, Rebecca K. (2009). Observer ratings: Validity and value as a tool for animal welfare research. *Applied Animal Behaviour Science,* 119(1–2): 1–14.

Nathan, Ran, Christopher T. Monk, Robert Arlinghaus, Timo Adam, Josep Alós, Michael Assaf, Henrik Baktoft et al. (2022). Big-data approaches lead to an increased understanding of the ecology of animal movement. *Science,* 3756582: eabg 1780.

Nazir, Sajid, and Muhammad Kaleem. (2021). Advances in image acquisition and processing technologies transforming animal ecological studies. *Ecological Informatics,* 61: 101212.

Neethirajan, Suresh, Inonge Reimert, and Bas Kemp. (2021). Measuring farm animal emotions: Sensor-based approaches. *Sensors,* 21(2): 553. https://doi.org/10.3390/s21020553.

Neethirajan, Suresh. (2022). Affective state recognition in livestock: Artificial intelligence approaches. *Animals,* 12(6): 759.

Neethirajan, Suresh. (2024). Net Zero Dairy Farming: Advancing Climate Goals with Big Data and Artificial Intelligence. *Climate,* 12(2): 15.

Obaid, Omar Ibrahim. (2023). From machine learning to artificial general intelligence: A roadmap and implications. *Mesopotamian Journal of Big Data,* 2023: 81–91.

Onuma, Kenta, Masaki Watanabe, and Nobuya Sasaki. (2024). The grimace scale: A useful tool for assessing pain in laboratory animals. *Experimental Animals,* 2024: 24–0010.

Park, M., Britton, D., Daley, W., McMurray, G., Navaei, M., Samoylov, A., Usher, C., and Xu, J. (2022). Artificial intelligence, sensors, robots, and transportation systems drive an innovative future for poultry broiler and breeder management. *Animal Frontiers,* 12(2): 40–48. https://doi.org/10.1093/af/vfac001.

Raihan, Asif. (2023). Artificial intelligence and machine learning applications in forest management and biodiversity conservation. *Natural Resources Conservation and Research,* 6(2): 3825.

Saha, Subodh Kumar, and Nitya Nand Pathak. (2023). *Fundamentals of Animal Nutrition.* Springer.

Scialabba, Nadia El-Hage (Ed.). (2021). *Managing Healthy Livestock Production and Consumption.* Academic Press.

Sharma, Abhishek, Vaidehi Sharma, Mohita Jaiswal, Hwang-Cheng Wang, Dushantha Nalin K. Jayakody, Chathuranga M. Wijerathna Basnayaka, and Ammar Muthanna. (2022). Recent trends in AI-based intelligent sensing. *Electronics,* 11(10): 1661.

Slob, Naftali, Cagatay Catal, and Ayalew Kassahun. (2021). Application of machine learning to improve dairy farm management: A systematic literature review. *Preventive Veterinary Medicine,* 187: 105237.

Swarup, Pranjal, Peng Chen, Rong Hou, Pinjia Que, Peng Liu, and Adams Wai Kin Kong. (2021). Giant panda behaviour recognition using images. *Global Ecology and Conservation,* 26: e01510.

Tauseef, Md, Esha Rathod, S.M. Nandish, and M.G. Kushal. (2024). Advancements in pet care technology: A comprehensive survey. In: *2024 4th International Conference on Data Engineering and Communication Systems (ICDECS),* pp. 1–6. IEEE.

Valencia, Frechelyn L., Jerlyn N. Brital, Kristian Dave Aranas, and Ronieto Mendoza. (2022). *Solar-powered Fishpond Feeder with Real-time Water Quality Monitoring System.* ResearchGate.

Valoppi, Fabio, Melissa Agustin, Felix Abik, Danila Morais de Carvalho, Jaison Sithole, Mamata Bhattarai, Jutta J. Varis et al. (2021). Insight on current advances in food science and technology for feeding the world population. *Frontiers in Sustainable Food Systems,* 5: 626227.

Whittaker, Alexandra L., and Chalette Brown. (2023). Pain assessment in laboratory species. In: *Anesthesia and Analgesia in Laboratory Animals*, pp. 261–275. Academic Press.

Yu, Jiyeon, Angelica de Antonio, and Elena Villalba-Mora. (2022). Deep learning (CNN, RNN) applications for smart homes: A systematic review. *Computers,* 11(2): 26.

Zhang, Li, Wenqiang Guo, Chenrui Lv, Meng Guo, Mei Yang, Qiuyue Fu, and Xiaomeng Liu. (2024). Advancements in artificial intelligence technology for improving animal welfare: Current applications and research progress. *Animal Research and One Health,* 2(1): 93–109.

4 | Vocal Communications in the Animal Kingdom

Naureen Afrose, Rideb Chakraborty, Pratibha Bhowmick and Mithun Bhowmick*

Bengal College of Pharmaceutical Sciences and Research, Basu Sarani, Bidhannagar, Durgapur, 713212, West Bengal, India

Communication in all living organisms is a basic form of survival criteria, amongst which vocal communication is a significant type. Animal vocalization triggers specific socio-sexual behavior that reveals information about the evolutionary, ecological, and social context of the species. In nature, sound is widely used for communication. It is important for species specific interactions, mating attraction, defense of territories, predator avoidance, warning signals, and social cohesiveness. The vocal communication for different species is not majorly comparable, though they have similarities in syllable structures and some patterns. The patterns in vocal repertoire across various animal kingdoms reveal their evolutionary distributions. The vocalization pattern and an animal's capacity for auditory communication are determined by complex brain connections and physiological adaptations. The vocal repertory has also been greatly impacted by both physiological and ecological influences. Through numerous examples of species-species interaction, the chapter provides a clear understanding of the various vocal repertoire forms and the ecological significance of vocal signals in response to conspecific and heterospecific interactions and effects of environmental noise. The pattern in which various animal species communicate is also covered, emphasizing the value of vocal communication for everything right from mating calls to emotion exchange. Artificial Intelligence (AI) has nowadays revolutionized the recording of animal interaction patterns and is helping vastly in analyzing these vocalizations. Cutting-edge AI is aiding in deciphering the vocal signals and AI algorithms are being utilized to automate the process of data collection. The potential of AI in deciphering vocal communication in the animal kingdom is also elaborated. The chapter thus concludes with the

*Corresponding author: drmithunbhowmick@gmail.com

discussion of the various forms of animal vocal signals showing a contrast between the traditional methods of interpretation and the modern transformative impact of AI in understanding the vocal communication of the animal kingdom.

Introduction

Vocal communication results from macro-evolutionary events that lead to neuronal innovations for social interaction. While humans instinctively associate vocal communication with speech, animal vocalizations are equally complex and vital for various species. Such behaviors arise from specific sounds, gestures, and patterns essential for interactions, mating, or foraging, providing evolutionary advantages for survival and growth (Bass et al., 2010). Vocal learning, voluntary control over vocal apparatus, and diverse vocal repertoires are crucial for the evolution of human vocal communication. In contrast, non-human animals may have undergone significant vocal brain changes during evolution, lacking these capabilities.

Animal vocal patterns are genetically pre-programmed, with selective pressures influencing how genes and environments shape communication and species adaptation to ecological niches (Belyk & Brown, 2017). Through vocalizations, animals express diverse information, including warning signals, territory claims, mating calls, and social relationships. This rich tapestry of expressive behaviors is evident in primates, reptiles, and birds. Non-human primates, particularly apes and monkeys, like *Callimico goeldii, Macaca radiata, Miopithecus talapoin, Macaca fuscata, Macaca silenus*, and *Daubentonia madagascariensis*, utilize intricate vocalizations for social functions like marking territory and group coordination (McComb & Semple, 2005; Ghazanfar, 2013; Fischer & Price, 2017). Their vocal structures are closely linked to emotional and motivational states, although their understanding of communicative intent is limited (Snowdon, 2017). Reptiles like *Chelonoidis carbonarius, Caiman crocodiles, Micrurus lemniscatus, Testudo horsfieldii, Calotes versicolor, Chrysemys picta, Anolis chlorocyanus, Eublepharis macularius, Alligator mississippiensis, Crocodylus acutus, Vipera berus, Acanthodactylus erythrurus, Hemidactylus mabouia, Ptyodactylus guttatus,* and *Crotalus durissus*, despite lacking vocal cords, use acoustic signals for mating and territory defense (Frankenberg, 1975; Vliet, 1989; Macedonia & Stamps, 1994; Gagno, 2013; Russell & Bauer, 2021). Birds like *Sayornis Phoebe, Taeniopygia guttata, Zonotrichia leucophrys, Procnias tricarunculatus, Procnias nudicollis* exhibit a remarkable variety of vocalizations, including complex calls and songs, crucial for communication and mate selection (Read & Weary, 1992; Pepperberg, 2013; Loo & Cain 2021).

Since Darwin's 1871 observations, scientists have sought to understand animal communication mechanisms (Darwin, 1888). Recent research, particularly by Sueur and Farina, has focused on the ecological significance of vocalizations (Sueur & Farina, 2015). Understanding vocal communication requires an integrated approach that combines behavioral observations, acoustic analyses, and advanced computational tools, including bioacoustic monitoring and spectrograms (Takahashi et al., 2021). Decoding the intricate interactions of sounds, gestures, and social contexts across avian, reptilian, and primate populations offers vital insights into ecology, conservation, and animal behavior.

Artificial intelligence (AI) equips researchers with advanced tools for analyzing animal vocal communication, enhancing our understanding of its diversity and complexity. By utilizing AI, we can uncover insights into the behavioral, ecological, and evolutionary dynamics of vocal communication across the animal kingdom (Suzuki et al., 2020; Congdon et al., 2022). Current research explores the intersection of biology and technology to deepen our understanding of the intricate language of the natural world.

The Ecological Significance of 'Eco-Symphonies'

Animal vocalizations, often called 'eco-symphonies', are vital to communication networks across various species. These vocalizations serve multiple functions, including mate attraction, territorial defense, and social cohesion among offspring, exemplified by birds' melodic songs and whales' complex cries (Janik, 2014; Verpoten, 2021).

The diversity of animal vocalizations is remarkable, with different species utilizing various vocal signals characterized by unique frequencies, lengths, volumes, and patterns to fit their social systems and ecological niches. For instance, mice produce ultrasounds for social interactions, mating, and isolation of pups, while mammals like wolves, elephants, and primates use vocalizations to communicate reproductive intentions and maintain social relationships. These signals are crucial for announcing dominance and facilitating mate selection within groups.

Hunting animals, such as the African wild dog (*Lycaon pictus*), use vocal sounds to coordinate efforts with their pack. Similarly, bats employ vocal communication during hunting and in social settings. Birds like the zebra finch exhibit elaborate vocal patterns for courtship, while whales and dolphins utilize vocalizations for communication, social bonding, and navigation. Songbirds, for instance, engage in intricate vocal routines during courting rituals (Carouso-Peck et al., 2021), and primates use distinct calls for group coordination and predator warnings. Advances in bioacoustics allow scientists to decode these vocal repertoires, providing insights into their development and function.

Vocalizations are essential for social interactions and have implications for reproductive success, as shown in birds where song complexity correlates with male quality and territory. Male birds often use various tunes to attract mates, indicating their genetic quality and health. This dynamic of mate attraction through vocal communication is well-documented. In marine environments, vocalizations maintain group cohesion and facilitate activities like hunting and navigation among dolphins and whales (King & Jensen, 2023). Whales produce a wide range of sounds, from eerie groans to beautiful melodies, used for social bonding and navigation. Notably, different humpback whale populations exhibit unique song patterns, suggesting cultural transmission akin to human societies (Schall et al., 2020; Whitehead et al., 2023).

Dolphins use various vocalizations—whistles, clicks, and pulsed calls—for communication, aiding in locating prey and expressing emotions. AI-powered acoustic analysis enhances our understanding of these communication patterns, which is crucial for dolphin conservation (Huijser et al., 2020; Premoli et al., 2023). Similarly, marine mammals use high-frequency clicks to navigate and locate prey (Clink et al., 2020; Brualla et al., 2023; Jordan et al., 2023). In dense jungles, frogs and insects rely heavily on audio signals for mate identification and territory establishment. On land, carnivores and birds of prey employ vocal communication to coordinate hunting strategies. African wild dogs use complex exchanges to crowd and isolate prey, while bats utilize echolocation and vocalizations for hunting. Raptors like eagles and falcons use piercing calls to coordinate aerial attacks, showcasing the relationship between vocal communication and predatory behavior.

Dawn bird choruses influence plant community dynamics and ecosystem structure by controlling seed distribution and insect populations (Farina et al., 2014). Predator vocalizations can impact energy flows and trophic cascades, affecting ecosystem resilience and stability. Understanding the ecological relevance of animal vocalizations has significant conservation implications (Guyette & Post, 2023). Recognizing the importance of eco-symphonies can guide conservation initiatives aimed at preserving acoustic environments and mitigating human impacts on vocalizations.

The study of animal vocalizations offers a glimpse into the intricate ecological interactions and adaptive mechanisms that shape life on Earth. Eco-symphonies resonate throughout diverse habitats, creating a tapestry of sounds that reflects the complexity of the natural world. By deciphering the ecological significance of these vocalizations, scientists deepen our understanding of animal behavior and the intricacies of life.

Evolutionary Perspectives on Vocal Communication

Cortical mechanisms for producing and learning vocalizations vary across species, reflecting selective pressures shaped by different ecological and social niches. Vocal adaptations are species-specific, indicating that animals possess varying singing abilities based on their environments. Recent research suggests that high intelligence does not exclusively correlate with rich vocalization, as some reptiles exhibit vocal signaling. Neuroimaging studies have identified brain regions in reptiles, such as green tree pythons, responsible for processing diverse vocalizations, indicating independent evolution of vocal communication (Armstrong et al., 1991). Comparative genomic studies confirm the conservation of genes related to vocal behavior across reptile groups.

Monkeys exhibit extraordinary vocal repertoires, with technological advances like computational modeling revealing the social meanings behind their cries (Moore & Gockel, 2012; Mello et al., 2015; Garcia & Ravignani, 2020). Neuroimaging techniques, including EEG and fNIRS, have identified networks in signal voices and socio-emotional memory, highlighting the involvement of the prefrontal cortex and anterior cingulate cortex (ACC) in vocal modulation (Klink et al., 2021; Searcy & Nowicki, 2023). In birds, vocal communication research has provided insights into the neurology of song production (Coffey et al., 2019; Elemans et al., 2008). Recent mapping experiments have clarified the neuronal architecture in songbirds that underpins vocalization, revealing the role of neurotrophic factors and epigenetic processes in avian vocal circuits. Comparative genomic analyses have identified common genetic modules associated with vocal learning across bird lineages, enhancing our understanding of the evolution of vocal communication abilities (Hage et al., 2013; Mello & Clayton, 2015).

AI-infused Interface in Deciphering Animal Ecology

AI addresses complex problems across various fields, including statistics, information science, software development, computational modeling, and data analysis. Evolving since the 1980s, machine learning (ML) has become a key technique, while deep learning has gained popularity for managing large datasets since the 2000s. Although AI techniques can solve problems and integrate information, they are often underutilized in animal health research. The capacity to collect and share extensive information has heightened the need for effective data analysis methods. ML has emerged as a valuable strategy in ecology, bridging large datasets with meaningful ecological insights. Recent advancements in ML technology have enhanced the traditional ecological study pipeline. The rising demand for ML in animal ecology and conservation stems from the challenges posed by complex ecological data. Thus, the relationship between ecology and ML should be reciprocal, as accurate model creation requires integrating ecological knowledge into ML techniques (Saareenma et al., 1988; Matsuzawa, 2003; Tabak et al., 2019; Ditria et al., 2020; Eikelboom et al., 2021; Maharani et al., 2021; Han et al., 2023).

The Benefits and Drawbacks of AI vs. Conventional Approaches in Animal Communication Research

Animals communicate in intricate and fascinating ways, using vocalizations, gestures, postures, and chemical signals. Gaining an understanding of these exchanges can provide novel perspectives on animal cognition, social connections, and behavior. However, it can be difficult to understand animal communication; it requires careful monitoring, evaluation, and evaluation of data. This is where the potential of AI comes into play. While traditional methods have established the foundation for studying animal communication, AI provides new resources and opportunities to further explore this intricate field (Christin et al., 2019).

Conventional Approaches

Animal communication research has typically relied on observational studies, ethnographic methods, bioacoustics, and field experiments. Observational studies capture animal postures, vocalizations, and interactions, offering insights into communication systems. Ethnographic methods immerse researchers in animal cultures, uncovering nuances and social contexts often overlooked in controlled settings. Bioacoustics analyzes animal sounds through spectrographs and sound localization techniques, quantifying complex vocalizations. Field experiments test hypotheses by manipulating environmental factors or animal behavior. However, these conventional approaches face limitations, including subjectivity, restricted data, and scalability challenges, which can result in incomplete information, time-consuming manual analyses, and difficulties in handling large, complex multimodal datasets (Dall et al., 2005; Sutherland, 2006; Thessen, 2016; Valletta et al., 2017; Christin et al., 2019; Khalighifar, 2020; Droge et al., 2021).

AI-based Approach

AI offers a powerful toolbox for overcoming these limitations and revolutionizing animal communication research. Machine learning algorithms can analyze vast amounts of data, including audio, video, and sensor data, with unprecedented speed and accuracy (Dall et al., 2005).

Automatic Detection and Classification

AI algorithms can automatically detect and classify animal communication signals, reducing observer bias and allowing for large-scale analysis of communication events. For example, deep learning models can identify specific bird calls within hours-long recordings, a task that would take humans significantly longer (Ditria et al., 2020).

Pattern Recognition

AI excels at uncovering hidden patterns and relationships within complex data. Algorithms can identify subtle variations in vocalizations, postures, or behaviors that might be missed by human observers, potentially revealing new information about communication intent or individual differences (Petso et al., 2021).

Real-time Analysis

AI algorithms can analyze data in real-time, enabling researchers to track and respond to ongoing communication events in the field. This opens up possibilities for dynamic experiments and interactive studies of animal communication (Panigrahi et al., 2023).

Integration of Multiple Data Sources

AI can analyze data from diverse sources such as acoustics, video, and GPS tracking, offering a comprehensive understanding of communication signals (Freenders et al., 2008). This multimodality provides richer insights into animal interactions than a single data type alone. However, AI has several drawbacks in animal communication research:

- **Data dependency:** AI algorithms require extensive training data, which can be costly and time-consuming to collect, especially for non-vocal communication signals (Weber et al., 2023).
- **Interpretability:** Understanding how "black box" AI models reach conclusions can be difficult, potentially hindering scientific insight and researchers' ability to refine their hypotheses (Aamodt & Nygard, 1995).
- **Overfitting:** AI models may perform well on training data but struggle to generalize to new data, leading to inaccurate conclusions and limiting the broader applicability of research findings (Rang et al., 2021).

Primate, Reptiles, and Avian Vocalization Patterns

The capacity for vocalization-based communication is essential for social relationships, reproduction, and survivability in the animal kingdom. This chapter explores the rich and varied vocal repertoires of three different animal groups: birds, reptiles, and primates. Each group has evolved distinct and sophisticated vocalizations that fulfill a range of purposes, despite the fact that their evolutionary trajectories separated millions of years ago (Petkov & Jarvis, 2012).

Primate Vocalization Pattern

Primates, our closest living relatives, exhibit a diverse range of flexible and complex vocalizations. For example, chimpanzees use around 30 different vocalizations, each with distinct meanings and contexts (Zimmermann, 2017). Their notable pant hoots serve to warn of danger or summon group members, while whimpers indicate submission, and barks signal hostility. Chimpanzees are also skilled vocal mimics, imitating not only their calls but also human speech and ambient sounds (Filippi et al., 2017).

Despite their peaceful social relationships, bonobos primarily use vocalizations to maintain harmony. High-pitched trills, resembling laughter, occur during play and reconciliation, promoting social bonding. In contrast, gorillas rely on low-frequency vocalizations, such as grunts and roars, to assert dominance and protect their territory (Seyforth & Cheney, 2010).

Monkeys and other non-human primates also engage in complex vocal communication. Vervet monkeys have a sophisticated warning system where distinct cries signal different predators, enhancing group survival by enabling effective responses to threats (Newman, 2004). Gibbons possess a complex singing system, utilizing elaborate choruses and duets for social cohesion, mate attraction, and territorial defense (Lierbal & Kaminski, 2012). Solitary orangutans use long-distance calls to find mates and alert others to danger, reflecting their intricate social lives (Nurcahyo et al., 2017). Studies suggest that primates like chimpanzees and bonobos can adapt their vocalizations in response to social cues and environmental changes, demonstrating vocal learning (Capshaw et al., 2021).

Reptile Vocalization Pattern

Reptiles have long been thought to possess a limited range of vocalizations, although they do have some. For instance, lizards produce whistles, chirps, and clicks to mark territory, attract mates, and interact with each other. Despite lacking vocal cords, snakes use hisses, thumps, and vibrations for signaling hostility and attracting mates. Rattlesnakes 'vocalize' using their distinctive rattles to warn off predators and competitors (Russell & Bauer, 2021).

Crocodilians boast a diverse vocal repertoire for various life stages. They use softer sounds for courting and social interactions, while bellows and roars serve territorial purposes. Interestingly, crocodile hatchlings 'chirp' in unison during hatching, suggesting early communication (Cooke, 1893). Some turtles, like snapping turtles, vocalize and hiss to communicate with mates and deter predators (Baker, 2022). Additionally, certain snakes and lizards produce infrasonic vocalizations that may aid in social interactions and territory defense (Blythe, 2020).

Avian Vocalization Pattern

Birds are renowned for their complex songs, crucial for maintaining social bonds, attracting mates, and defending territories. Songbirds, like mockingbirds and nightingales, exhibit exceptional vocal control, allowing them to imitate human speech and other birds' melodies (Anastasi, 2017). The purpose of bird songs varies by species. Male songbirds often use elaborate songs to signal health and reproductive fitness to attract females (Liu et al., 2013). Vocalizations also serve to defend territories, alert others to danger, and facilitate flock communication.

Other bird groups use different vocalizations. Parrots are particularly adept at mimicking noises, including human speech, while owls hoot to attract mates and establish territory (Anastasi, 2017). Penguins employ diverse vocalizations, such as growls and trumpets, to assert dominance and coordinate group movements. Hummingbirds produce intricate vocalizations for aggression, territorial defense, and courtship (Anastasi, 2017). Crows and ravens showcase a range of vocalizations, including mimicking human speech, highlighting their cognitive abilities (Liu et al., 2013). Recent research suggests that non-songbirds, like parrots and hummingbirds, may also possess vocal learning skills, adding complexity to avian communication (Anastasi, 2017).

Body Language of Primates, Reptiles, and Avians in Decoding Eco-social Dynamics

The study of vocal communication and body language in animals reveals the complexities of social interactions and ecological adaptations. Primates, reptiles, and birds utilize diverse vocalizations and nonverbal cues for inter- and intra-species communication, providing insights into their eco-social dynamics, including social hierarchies and environmental adaptations. Primate body language is essential for navigating habitats, resource distribution, mate selection, territorial defense, and foraging. Rodents, particularly rats, also rely on body language to regulate social interactions and reproductive strategies, fostering social cohesion.

Vocal communication is crucial for maintaining social order and nurturing offspring. Studies highlight tactile interactions between parent birds and their young, shedding light on social learning and foraging skills passed within avian families. Birds exhibit significant variations in body language across species, with research comparing wild and captive populations revealing differences in vocal dialects, mating rituals, and foraging strategies, underscoring the influence of environmental factors and social learning mechanisms (Arakawa et al., 2008; Liebel & Call, 2012; Fusani et al., 2014; Roberts & Roberts, 2016; Kenny et al., 2017; Liebel & Ona, 2018; Graham et al., 2018; Riters et al., 2019; Demuru et al., 2020; Ebbesen & Froemke, 2021; Jablonsky et al., 2021, Lewis et al., 2021; Knaebe et al., 2022; Kitano et al., 2022; Pereira et al., 2022; Kalan et al., 2023; Petkov & Jarvis, 2023).

AI-driven Tools and Outcomes in Interpreting Primate, Reptiles, and Avian Vocalization Patterns

AI-driven tools analyze acoustic data using advanced algorithms, offering insights into animal vocalizations. These tools efficiently handle large vocalization datasets, employing ML techniques like deep neural networks and support vector machines to classify and categorize calls based on acoustic properties. This capability helps researchers identify unique calls and correlations with behaviors such as social interactions, territorial defense, and mating displays (Picciulin et al., 2013; Zhang et al., 2023; Das et al., 2024).

AI techniques enable automatic vocalization detection and real-time monitoring in natural habitats, allowing extended data collection on animal vocal behavior. This is particularly valuable for studying nocturnal or elusive species. For instance, the software DeepSqueak, developed in 2019, detects and classifies ultrasonic vocalizations in animals like mice and lemurs, facilitating cross-taxa studies (Hoy, 2018; Romero-M. et al., 2021). EAIGLE Inc. has created a tool to analyze Sumatran orangutan vocalizations and gestures in real-time, also applied to various primate species. Passive acoustic monitoring (PAM) enhances data storage for tracking primates in hard-to-reach areas (Lemasson & Hausberger, 2011; Bouchet et al., 2012; Riondato et al., 2013; Valente et al., 2019; Congdon et al., 2022; Clink et al., 2023).

Reptiles typically have low-frequency vocalizations and smaller repertoires. Raven Pro 1.5 was used to study several reptile species, while deep learning tools examined vocal patterns in American alligators and Nile crocodiles (Ligges et al., 2018; Anikin, 2019; Zhou et al., 2023; Jensen et al., 2024). Bird vocalizations are extensively studied using deep learning. BirdNET identifies 984 bird species in North America and Europe, analyzing their social behavior through sound. The program processes audio to create visual sound representations and trains a complex model with about 27 million parameters. Kaleidoscope Pro clusters' sounds using a hidden Markov model, suitable for long-term species monitoring. RavenPro aids in classifying birdsong, interpreting seasonal behavioral changes and daily activity patterns (Ruff et al., 2020; Kahl et al., 2021; Symes et al., 2022).

Comparative Analysis from Empirical Studies

AI technology, including neural networks and ML algorithms, has transformed our understanding of vocal communication across species. These tools have categorized and analyzed monkey vocalizations, revealing social contexts and species-specific patterns. They highlight unique calls in primates, such as the mating and alarm calls of various monkeys and great apes, which are crucial for interpreting social interactions and territorial behaviors. AI algorithms have shown that primates possess diverse vocalizations—each serving distinct communicative purposes within their social groups. Convolutional Neural Networks (CNNs) (Hoy, 2018; Clink et al., 2023) utilize audio analysis to classify these vocalizations and elucidate their roles in social cognition.

Reptile vocalizations indicate mating displays, territorial conflicts, and predator-prey interactions. AI tools enhance the detection of subtle acoustic variations in reptilian vocalizations, producing clearer results than traditional methods. By employing spectrogram analysis, machine learning, and audio recording, AI-based research has improved the understanding of these weak acoustic signals. In avian studies, AI methods allow researchers to analyze large datasets and identify subtle auditory patterns, facilitating insights into the evolutionary origins of bird vocalizations. However, significant variations exist in the vocalizations and behaviors of different species, affecting methodologies. Primates produce complex vocalizations, reptiles offer basic hisses and clicks, and birds display varied harmonic patterns. Deep learning techniques like CNNs and Recurrent Neural Networks are commonly used for primate analysis, while Support Vector Machines (SVMs) and Decision Trees are applied to reptiles. AI systems for bird vocalization analysis often incorporate advanced deep learning models to capture intricate spectral and temporal patterns (Noda et al., 2017; Corneanu, 2019).

Conclusion

Animals rely heavily on vocal communication for various aspects of their lives, including mating, territory defense, danger alerts, and social bonding. They use a wide range of vocalizations, from simple alarm calls to complex songs. Primate vocalizations, including those of monkeys and apes, are particularly intricate, encompassing screams, grunts, calls, and gestures that help organize group activities and express emotions like anger, fear, or affection. While birds utilize diverse vocalizations such as calls and mimicry, reptiles, despite lacking vocal cords, also communicate through various sounds and behaviors. Traditionally, interpreting these acoustic signals has been challenging, but the advent of AI has revolutionized the study of animal vocal communication. AI provides advanced tools for analyzing large volumes of acoustic data, enabling researchers to identify species-specific vocalizations, measure vocal characteristics, and detect subtle variations in communication patterns. Overall, AI's application has enhanced our understanding of the complex communication patterns in primates, reptiles, and birds, revealing the ecological roles these acoustics play in animal behavior and survival.

References

Aamodt, A., and M. Nygård. (1995). Different roles and mutual dependencies of data, information, and knowledge: An AI perspective on their integration., *Data & Knowledge Engineering*, 16: 191–222.

Anastasi, A. (2017). Biology, learning, and evolution of vocality: Biosemiotics of birdsong. *Cognitive Semiotics*, 10.

Arakawa, H., D.C. Blanchard, K. Arakawa, C. Dunlap, and R.J. Blanchard. (2008). Scent marking behavior as an odorant communication in mice. *Neuroscience & Biobehavioral Reviews*, 32(7): 1236–1248.

Anikin, A. (2019). Soundgen: An open-source tool for synthesizing nonverbal vocalizations. *Behavior Research Methods*, 51: 778–792.

Armstrong, D.M., and B.M. Bush (Eds.). (1991). *Locomotor Neural Mechanisms in Arthropods and Vertebrates* (Vol. 16). Manchester University Press.

Baker, A.M. (2022). *Mimicry, Manipulation, and Mutation: A Portfolio of Eclectic Compositions*. PhD Thesis. University of Birmingham.

Bass, A.H., E.D. Gilland, and R. Baker. (2010). Evolutionary origins for social vocalization in a vertebrate hindbrain–spinal compartment. *Science*, 321(5887): 417–421.

Belyk, M., and S. Brown. (2017). The origins of the vocal brain in humans. *Neuroscience & Biobehavioral Reviews*, 77: 177–193.

Blythe, B. (2020). Petronius' talking birds: Mimicry and death in the *Cena Trimalchionis*. *Classical Philology*, 115: 47–69.

Bouchet, H., C. Blois-Heulin, A.S. Pellier, K. Zuberbühler, and A. Lemasson. (2012). Acoustic variability and individual distinctiveness in the vocal repertoire of red-capped mangabeys (*Cercocebus torquatus*). *Journal of Comparative Psychology*, 126(1): 45.

Brualla, N.L., L.A. Wilson, M. Doube, R.T. Carter, A.G. McElligott, and D. Koyabu. (2023). The vocal apparatus: An understudied tool to reconstruct the evolutionary history of echolocation in bats? *Journal of Mammalian Evolution*, 30(1): 79–94.

Buchelt, A., A. Adrowitzer, P. Kieseberg, C. Gollob, A. Nothdurft, S. Eresheim, S. Tschiatschek et al. (2024). Exploring artificial intelligence for applications of drones in forest ecology and management. *Forest Ecology and Management*, 551: 121530.

Capshaw, G., K.L. Willis, D. Han, and H.S. Bierman. (2021). Reptile sound production and perception. In: *Neuroendocrine Regulation of Animal Vocalization*, pp. 101–118. Elsevier.

Carouso-Peck, S., M.H. Goldstein, and W.T. Fitch. (2021). The many functions of vocal learning. *Philosophical Transactions of the Royal Society B*, 376(1836): 20200235.

Christin, S., É. Hervet, and N. Lecomte. (2019). Applications for deep learning in ecology. *Methods Ecol. Evol.*, 10: 1632–1644.

Clink, D.J., Ahmad, A.H., and Klinck, H. (2020). Brevity is not universal in animal communication: Evidence for compression depends on the unit of analysis in small ape vocalizations. *Royal Society Open Science*, 7(4): 200151.

Clink, D.J., I. Kier, A.H. Ahmad, and H. Klinck. (2023). A workflow for the automated detection and classification of female gibbon calls from long-term acoustic recordings. *Frontiers in Ecology and Evolution*, 11: 1071640.

Coffey, K.R., R.E. Marx, and J.F. Neumaier. (2019). DeepSqueak: A deep learning-based system for detection and analysis of ultrasonic vocalizations. *Neuropsychopharmacology*, 44(5): 859–868.

Congdon, J.V., M. Hosseini, E.F. Gading, M. Masousi, M. Franke, and S.E. MacDonald. (2022). Future of artificial intelligence in monitoring animal identification, health, and behaviour. *Animals*, 12(13): 1711.

Cooke, M.C. (1893). *Our Reptiles and Batrachians: A Plain and Easy Account of the Lizards, Snakes, Newts, Toads, Frogs and Tortoises Indigenous to Great Britain*. WH Allen & Company, Limited.

Corneanu, C.A. (2019). *Improving Performance and Interpretability in Recognizing Facial Action Units with Deep Neural Networks*. PhD Thesis. University of Barcelona.

Dall, S.R., L.-A. Giraldeau, O. Olsson, J.M. McNamara, and D.W. Stephens. (2005). Information and its use by animals in evolutionary ecology. *Trends in Ecology & Evolution*, 20: 187–193.

Darwin, C. (1888). *The Descent of Man: And Selection in Relation to Sex*. John Murray, Albemarle Street.

Das, N., N. Padhy, N. Dey, H. Paul, and S. Chowdhury. (2024). Exploring explainable AI methods for bird sound-based species recognition systems. *Multimedia Tools and Applications*, 1–31.

Dell, A.I., J.A. Bender, K. Branson, I.D. Couzin, G.G. de Polavieja, L.P. Noldus, A. Pérez-Escudero et al. (2014). Automated image-based tracking and its application in ecology. *Trends in Ecology & Evolution*, 29: 417–428.

Demuru, E., F. Pellegrino, D. Dediu, and F. Levréro. (2020). Foraging postures are a potential communicative signal in female bonobos. *Scientific Reports*, 10(1): 15431.

Ditria, E.M., S. Lopez-Marcano, M. Sievers, E.L. Jinks, C.J. Brown, and R.M. Connolly. (2020). Automating the analysis of fish abundance using object detection: Optimizing animal ecology with deep learning. *Frontiers in Marine Science*, 429.

Dröge, S., D.A. Martin, R. Andriafanomezantsoa, Z. Burivalova, T.R. Fulgence, K. Osen, E. Rakotomalala et al. (2021). Listening to a changing landscape: Acoustic indices reflect bird species richness and plot-scale vegetation structure across different land-use types in north-eastern Madagascar. *Ecological Indicators*, 120: 106929.

Ebbesen, C.L. and R.C. Froemke. (2021). Body language signals for rodent social communication. *Current Opinion in Neurobiology*, 68: 91–106.

Eikelboom, J.A. (2021). *Sentinel Animals: Enriching Artificial Intelligence with Wildlife Ecology to Guard Rhinos*. PhD Thesis. Wageningen University and Research.

Elemans, C.P., A.F. Mead, L.C. Rome, and F. Goller. (2008). Superfast vocal muscles control song production in songbirds. *PloS One*, 3(7): e2581.

Farina, A., N. Pieretti, and R. Malavasi. (2014). Patterns and dynamics of (bird) soundscapes: A biosemiotic interpretation. *Semiotica*, 198.

Feenders, G., M. Liedvogel, M. Rivas, M. Zapka, H. Horita, E. Hara, K. Wada et al. (2008). Molecular mapping of movement-associated areas in the avian brain: A motor theory for vocal learning origin. *PLoS One*, 3(3): e1768.

Filippi, P., J.V. Congdon, J. Hoang, D.L. Bowling, S.A. Reber, A. Pašukonis, M. Hoeschele et al. (2017). Humans recognize emotional arousal in vocalizations across all classes of terrestrial vertebrates: Evidence for acoustic universals. *Proc. R. Soc. B.*, 284: 20170990.

Fischer, J., and T. Price. (2017). Meaning, intention, and inference in primate vocal communication. *Neuroscience & Biobehavioral Reviews*, 82: 22–31.

Frankenberg, E. (1975). Distress calls of gekkonid lizards from Israel and Sinai. *Israel Journal of Ecology and Evolution*, 24(1–2): 43–53.

Fusani, L., J. Barske, L.D. Day, M.J. Fuxjager, and B.A. Schlinger. (2014). Physiological control of elaborate male courtship: Female choice for neuromuscular systems. *Neuroscience & Biobehavioral Reviews*, 46: 534–546.

Gagno, S. (2013). *The Auditory Sense in Tortoises, using Hermann's Tor-toise, Testudo Hermanni (GMELIN, 1789) as an Example*. ResearchGate.

Garcia, M. and A. Ravignani. (2020). Acoustic allometry and vocal learning in mammals. *Biology Letters*, 16(7): 20200081.

Ghazanfar, A.A. (2013). Multisensory vocal communication in primates and the evolution of rhythmic speech. *Behavioral Ecology and Sociobiology*, 67: 1441–1448.

Graham, K.E., C. Hobaiter, J. Ounsley, T.R.W. Furuichi, and R.W. Byrne. (2018). Bonobo and chimpanzee gestures overlap extensively in meaning. *PLoS Biology*, 16(2): e2004825.

Guyette, M.Q., and J.C. Post. (2023). Ecomusicology, ethnomusicology, and soundscape ecology. In: *Current Directions in Ecomusicology: Music, Culture, Nature*, pp. 40–56. Routledge.

Hage, S.R., N. Gavrilov, and A. Nieder. (2013). Cognitive control of distinct vocalizations in rhesus monkeys. *Journal of Cognitive Neuroscience*, 25(10): 1692–1701.

Han, B.A., K.R. Varshney, S. LaDeau, A. Subramaniam, K.C. Weathers, and J. Zwart. (2023). A synergistic future for AI and ecology. *Proc. Natl. Acad. Sci. U.S.A.*, 120: e2220283120.

Hoy, M.B. (2018). Alexa, Siri, Cortana, and more: An introduction to voice assistants. *Medical Reference Services Quarterly*, 37(1): 81–88.

Huijser, L.A., V. Estrade, I. Webster, L. Mouysset, A. Cadinouche, and V. Dulau-Drouot. (2020). Vocal repertoires and insights into social structure of sperm whales (*Physeter Macrocephalus*) in Mauritius, southwestern Indian Ocean. *Marine Mammal Science*, 36(2): 638–657.

Jablonszky, M., S. Zsebők, M. Laczi, G. Nagy, E. Vaskuti, and L.Z. Garamszegi. (2021). The effect of social environment on bird song: Listener-specific expression of a sexual signal. *Behavioral Ecology*, 32(3): 395–406.

Janik, V.M. (2014). Cetacean vocal learning and communication. *Current Opinion in Neurobiology*, 28: 60–65.

Jensen, T.R., A. Anikin, M. Osvath, and S.A. Reber. (2024). Knowing a fellow by their bellow: Acoustic individuality in the bellows of the American alligator. *Animal Behaviour*, 207: 157–167.

Jordan, N.R., K.A. Golabek, C.J. Marneweck, D.G. Marneweck, M.M. Mbizah, D. Ngatia, D. Rabaiotti et al. (2023). Hunting behavior and social ecology of African wild dogs. In: *Social Strategies of Carnivorous Mammalian Predators: Hunting and Surviving as Families*, pp. 177–227. Springer International Publishing.

Kahl, S., C.M. Wood, M. Eibl, and H. Klinck. (2021). BirdNET: A deep learning solution for avian diversity monitoring. *Ecological Informatics*, 61: 101236.

Kalan, A.K., R. Nakano, and L. Warshawski. (2023). What we know and don't know about great ape cultural communication in the wild. *American Journal of Primatology*, e23560.

Kenny, E., T.R. Birkhead, and J.P. Green. (2017). Allopreening in birds is associated with parental cooperation over offspring care and stable pair bonds across years. *Behavioral Ecology*, 28(4): 1142–1148.

Khalighifar, A. (2020). *Application of Deep Learning to Automated Species Identification Systems*. PhD Thesis. University of Kansas.

King, S.L., and F.H. Jensen. (2023). Rise of the machines: Integrating technology with playback experiments to study cetacean social cognition in the wild. *Methods in Ecology and Evolution*, 14(8): 1873–1886.

Kitano, K., A. Yamagishi, K. Horie, K. Nishimori, and N. Sato. (2022). Helping behavior in prairie voles: A model of empathy and the importance of oxytocin. *Iscience*, 25(4).

Klink, P.C., X. Chen, W. Vanduffel, and P.R. Roelfsema. (2021). Population receptive fields in nonhuman primates from whole-brain fMRI and large-scale neurophysiology in visual cortex. *Elife*, 10: e 67304.

Knaebe, B., C.C. Weiss, J. Zimmermann, and B.Y. Hayden. (2022). The promise of behavioral tracking systems for advancing primate animal welfare. *Animals*, 12(13): 1648.

Lemasson, A., and M. Hausberger. (2011). Acoustic variability and social significance of calls in female Campbell's monkeys (Cercopithecus campbelli campbelli). *The Journal of the Acoustical Society of America*, 129(5): 3341–3352.

Lewis, R.N., L.J. Williams, and R.T. Gilman. (2021). The uses and implications of avian vocalizations for conservation planning. *Conservation Biology*, 35(1): 50–63.

Liebal, K., and J. Call. (2012). The origins of non-human primates' manual gestures. *Philosophical Transactions of the Royal Society B: Biological Sciences*, 367(1585): 118–128.

Liebal, K., and J. Kaminski. (2012). Gibbons (*Hylobates pileatus, H. moloch, H. lar, Symphalangus syndactylus*) follow human gaze, but do not take the visual perspective of others. *Animal Cognition*, 15: 1211–1216.

Liebal, K., and L. Oña. (2018). Different approaches to meaning in primate gestural and vocal communication. *Frontiers in Psychology*, 9: 478.

Ligges, U., S. Krey, O. Mersmann, and S. Schnackenberg. (2018). *tuneR: Analysis of Music and Speech*. https://CRAN.R-project.org/package=tuneR.

Liu, W., K. Wada, E.D. Jarvis, and F. Nottebohm. (2013). Rudimentary substrates for vocal learning in a suboscine. *Nature Communications*, 4: 2082.

Loo, Y.Y., and K.E, Cain. (2021). A call to expand avian vocal development research. *Frontiers in Ecology and Evolution*, 9: 757972.

Macedonia, J.M., and J.A. Stamps. (1994). Species recognition in Anolis grahami (*Sauria, Iguanidae*): Evidence from responses to video playbacks of conspecific and heterospecific displays. *Ethology*, 98(3–4): 246–264.

Maharani, M.D.D., H.F. Amin, and Y.S. Triana. (2021). Application of artificial intelligence in modern ecology for detecting plant pests and animal diseases. *International Journal of Quantitative Research and Modeling*, 2: 83–90.

Matsuzawa, T. (2003). The AI project: Historical and ecological contexts. *Animal Cognition*, 6: 199–211.

McComb, K., and S. Semple. (2005). Coevolution of vocal communication and sociality in primates. *Biology Letters*, 1(4): 381–385.

Mello, C.V., and D.F. Clayton. (2015). The opportunities and challenges of large-scale molecular approaches to songbird neurobiology. *Neuroscience & Biobehavioral Reviews*, 50: 70–76.

Moore, B.C., and H.E. Gockel. (2012). Properties of auditory stream formation. *Philosophical Transactions of the Royal Society B: Biological Sciences*, 367(1591): 919–931.

Newman, J.D. (2004). The primate isolation call: A comparison with precocial birds and non-primate mammals. In: L.J. Rogers, and G. Kaplan (Eds.), *Comparative Vertebrate Cognitio*n, pp. 171–187. Springer US, Boston, MA.

Noda, J.J., C.M. Travieso, and D. Sánchez-Rodríguez. (2017). Fusion of linear and mel frequency cepstral coefficients for automatic classification of reptiles. *Applied Sciences*, 7(2): 178.

Nurcahyo, W., V. Konstanzová, and I. Foitová. (2017). Parasites of orangutans (primates: ponginae): An overview. *American J. Primatol.*, 79: e22650.

Panigrahi, S., P. Maski, and A. Thondiyath. (2023). Real-time biodiversity analysis using deep-learning algorithms on mobile robotic platforms. *PeerJ. Computer Science*, 9: e1502.

Pepperberg, I.M. (2013). Evolution of vocal communication: An avian model. In: *Birdsong, Speech, and Language: Exploring the Evolution of Mind and Brain*. MIT Press.

Pereira, T.D., N. Tabris, A. Matsliah, D.M. Turner, J. Li, S. Ravindranath, E.S. Papadoyannis et al. (2022). SLEAP: A deep learning system for multi-animal pose tracking. *Nature Method*s, 19(4): 486–495.

Petkov, C.I., and E.D. Jarvis. (2012). Birds, primates, and spoken language origins: Behavioral phenotypes and neurobiological substrates. *Frontiers in Evolutionary Neuroscience*, 4: 12.

Petkov, C.I., and E.D. Jarvis. (2023). Birds, primates, and spoken language origins: Behavioral phenotypes and neurobiological substrates. *Frontiers in Evolutionary Neuroscience*, 4: 12

Petso, T., R.S. Jamisola Jr, D. Mpoeleng, E. Bennitt, and W. Mmereki. (2021). Automatic animal identification from drone cameras based on point pattern analysis of herd behavior. *Ecological Informatics,* 66: 101485.

Picciulin, M., G. Calcagno, L. Sebastianutto, C. Bonacito, A. Codarin, M. Costantini, and E.A. Ferrero. (2013). Diagnostics of nocturnal calls of *Sciaena umbra* (L., fam. *Sciaenidae*) in a nearshore Mediterranean marine reserve. *Bioacoustics*, 22(2): 109–120.

Premoli, M., S. Pietropaolo, M. Wöhr, N. Simola, and S.A. Bonini. (2023). Mouse and rat ultrasonic vocalizations in neuroscience and neuropharmacology: State of the art and future applications. *European Journal of Neuroscience*, 57(12): 2062–2096.

Rang, W., D. Yang, and D. Cheng. (2021). Dependency-aware tensor scheduler for industrial AI applications: Dymem—An aggressive data-swapping policy for training nonlinear deep neural networks. *IEEE Industrial Electronics Magazine*, 16: 15–23.

Read, A.F., and D.M. Weary. (1992). The evolution of bird song: Comparative analyses. *Philosophical Transactions of the Royal Society of London. Series B: Biological Sciences*, 338(1284): 165–187.

Riondato, I., G. Marco, M. Gamba, and C. Giacoma. (2013). Vocalization of red- and gray-shanked douc langurs (*Pygathrix nemaeus* and *P. cinerea*). *Journal of Primatology*, 2(2): 75–82.

Riters, L.V., J.A. Spool, D.P. Merullo, and A.H. Hahn. (2019). Song practice as a rewarding form of play in songbirds. *Behavioural Processes*, 16: 91–98.

Roberts, A.I., and S.G.B. Roberts. (2016). Wild chimpanzees modify modality of gestures according to the strength of social bonds and personal network size. *Scientific Reports*, 6(1): 33864.

Romero-Mujalli, D., T. Bergmann, A. Zimmermann, and M. Scheumann. (2021). Utilizing DeepSqueak for automatic detection and classification of mammalian vocalizations: A case study on primate vocalizations. *Scientific Reports*, 11(1): 24463.

Ruff, Z.J., D.B. Lesmeister, L.S. Duchac, B.K. Padmaraju, and C.M. Sullivan. (2020). Automated identification of avian vocalizations with deep convolutional neural networks. *Remote Sensing in Ecology and Conservation*, 6(1): 79–92.

Russell, A.P., and A.M. Bauer. (2021). Vocalization by extant nonavian reptiles: A synthetic overview of phonation and the vocal apparatus. *The Anatomical Record*, 304(7): 1478–1528.

Russell, A.P., and A.M. Bauer. (2021). Vocalization by extant nonavian reptiles: A synthetic overview of phonation and the vocal apparatus. *The Anatomical Record*, 304: 1478–1528.

Saarenmaa, H., N.D. Stone, L.J. Folse, J.M. Packard, W.E. Grant, M.E. Makela, and R.N. Coulson. (1988). An artificial intelligence modeling approach to simulating animal/habitat interactions. *Ecological Modelling*, 44: 125–141.

Schall, E., K. Thomisch, O. Boebel, G. Gerlach, S. Spiesecke, and I. Van Opzeeland. (2020). Large-scale spatial variabilities in the humpback whale acoustic presence in the Atlantic sector of the Southern Ocean. *Royal Society Open Science*, 7(12): 201347.

Searcy, W.A., and S. Nowicki. (2023). Birdsong learning, avian cognition, and the evolution of language. *Animal Behaviour*, 151: 217–227.

Seyfarth, R.M., and D.L. Cheney. (2010). Production, usage, and comprehension in animal vocalizations. *Brain and Language*, 115: 92–100.

Snowdon, C.T. (2017). Vocal communication in family-living and pair-bonded primates. In: *Primate Hearing and Communication*, pp. 141–174, Springer.

Sueur, J., and A. Farina. (2015). Ecoacoustics: The ecological investigation and interpretation of environmental sound. *Biosemiotics*, 8: 493–502.

Sutherland, W.J. (2006). Predicting the ecological consequences of environmental change: A review of the methods. *Journal of Applied Ecology*, 599–616.

Suzuki, T.N., D. Wheatcroft, and M. Griesser. (2020). The syntax–semantics interface in animal vocal communication. *Philosophical Transactions of the Royal Society B*, 375(1789): 20180405.

Symes, L.B., K.D. Kittelberger, S.M. Stone, R.T. Holmes, J.S. Jones, I.P. Castaneda Ruvalcaba, M.S. Webster, and M.P. Ayres. (2022). Analytical approaches for evaluating passive acoustic monitoring data: A case study of avian vocalizations. *Ecology and Evolution*, 12(4): e8797.

Tabak, M.A., M.S. Norouzzadeh, D.W. Wolfson, S.J. Sweeney, K.C. Vercauteren, N.P. Snow, J.M. Halseth et al. (2019). Machine learning to classify animal species in camera trap images: Applications in ecology. *Methods Ecol. Evol.*, 10(2019): 585–590.

Takahashi, D.Y., A. El Hady, Y.S. Zhang, D.A. Liao, G. Montaldo, A. Urban, and A.A. Ghazanfar. (2021). Social-vocal brain networks in a non-human primate. *bioRxiv*.

Thessen, A. (2016). Adoption of machine learning techniques in ecology and earth science. *One Ecosystem*, 1: e8621.

Valente, D., C. De Gregorio, V. Torti, L. Miaretsoa, O. Friard, R.M. Randrianarison, C. Giacoma, and M. Gamba. (2019). Finding meanings in low dimensional structures: Stochastic neighbor embedding applied to the analysis of Indri indri vocal repertoire. *Animals*, 9(5): 243.

Valletta, J.J., C. Torney, M. Kings, A. Thornton, and J. Madden. (2017). Applications of machine learning in animal behavior studies. *Animal Behaviour*, 124: 203–220.

Verpooten, J. (2021). Complex vocal learning and three-dimensional mating environments. *Biology & Philosophy*, 36(2): 12.

Vliet, K.A. (1989). Social displays of the American alligator (Alligator mississippiensis). *American Zoologist*, 29(3): 1019–1031.

Weber, M., M. Engert, N. Schaffer, J. Weking, and H. Krcmar. (2023). Organizational capabilities for AI implementation: Coping with inscrutability and data dependency in AI. *Inf. Syst. Front.*, 25: 1549–1569.

Whitehead, H., J.K. Ford, and A.G. Horn. (2023). Using culturally transmitted behavior to help delineate conservation units for species at risk. *Biological Conservation*, 285: 110239.

Zhang, Y.J., Z. Luo, Y. Sun, J. Liu, and Z. Chen. (2023). From beasts to bytes: Revolutionizing zoological research with artificial intelligence. *Zoological Research*, 44(6): 1115.

Zhou, L., J. Lei, X. Zhai, H. Shi, and J. Wang. (2023). Chinese striped-neck turtles vocalize underwater and show differences in peak frequency among different age and sex groups. *PeerJ*, 11: e14628.

Zimmermann, E. (2017). Evolutionary origins of primate vocal communication: Diversity, flexibility, and complexity of vocalizations in basal primates. In: R.M. Quam, M.A. Ramsier, R.R. Fay, A.N. Popper (Eds.), *Primate Hearing and Communication*, pp. 109–140. Springer International Publishing, Cham.

5 | Predicting Interactions Among Species in Ecological Networks: A Roadmap to Address Eltonian Shortfall

Subhajit Roy[1*], Sayantani Saha[1] and Ayan Mondal[2]

[1] Department of Information Technology, Maulana Abul Kalam Azad University of Technology, West Bengal, Simhat, Haringhata, Nadia - 741249, West Bengal, India

[2] Department of Zoology, Government General Degree College, Government of West Bengal, Siyalsai, Mohanpur, Paschim Medinipur - 721436, West Bengal, India

Understanding inter-species interactions within ecological niches is the key to addressing the Eltonian shortfall, predicting novel interactions of introduced species, and understanding endangered species' preferences. Initial models, limited by rigid assumptions, were unable to fully capture the complexities of real-world interaction networks. However, with advances in computing and access to vast databases, machine learning (ML) models like Boosted Regression Trees, Random Forests, k-Nearest Neighbor, SVM, and neural networks now offer the flexibility needed for dynamic species interaction models (SIMs). SIMs leverage morphological, ecological, ethological, geographical, and genetic traits to predict interactions, with evolutionary relationships providing additional predictive power. Integrating functional traits with phylogenetic data and ML could significantly enhance SIM accuracy and ecosystem management strategies.

Despite this progress, SIMs' comprehensiveness across various species groups and regions remains untested. This review provides an analysis of existing ML-based SIMs, organized by application, model parameters, and data types used. Improved understanding of species interactions would link community and network ecology with spatial ecology, bridging gaps in knowledge about inter-species relationships and informing conservation strategies across species, communities, and ecosystems at multiple scales.

*Corresponding author: subhajit.roy@makautwb.ac.in

Introduction

The unbridled ecological destruction and systematic exploitation of nature have jeopardized the natural cycles which are catalysts to regeneration of ecosystems (Clark and Foster, 2010). Change of land/sea uses, pollution, overuse (Hall, 1972), climate change (Weiskopf et al., 2020; Muleneh, 2021) are amongst the drivers of biodiversity loss. Despite global conservation efforts, biodiversity is declining at an unprecedented rate (Ceballos and Ehrlich, 2023; Schickhoff et al., 2023). This is evident by the inability to reach any of the 20 Aichi Biodiversity Targets outlined in 2010. It is imperative to fill the information gaps in biodiversity because appropriate conservation and restoration strategies can never be strategically planned without a thorough knowledge base of the ecosystem (Hortal et al., 2015). Although there has been significant progress in quantifying the different types of organisms (Linnean shortfall), understanding their geographic distributions (Wallacean shortfall) (Diniz-Filho et al., 2023), and understanding their evolutionary relationships (Darwinian shortfall) (Diniz-Filho et al., 2013), the knowledge of the interactions between them (Eltonian shortfall) is trifling primarily due to dearth of the volume of empirical data needed for the purpose. While direct pairwise biotic interactions have failed to address the dynamics of community structures, the ecological networks have provided an alternative. The indeterminately immense number of potential indirect interaction chains in an ecological community (Dodds and Nelson, 2006) makes it almost impossible to gather empirical data.

Advances in Artificial Intelligence (AI) and Machine Learning (ML) are transforming perceptions of ecology (Perry et al., 2022). Species distribution, inter-species interactions, automatic identification from images, camera traps or from call recordings, effects of various climatic parameters on populations of a certain species and modeling the meta-foodweb of an entire ecosystem are being simulated using various ML algorithms, involving both tabular as well as from image, video, audio data.

Modeling Interactions

Biotic interactions vary greatly and can be classified according to their types, strength and symmetry. It was suggested that the population of any organism has to "struggle for existence" and the competitors, antagonists, and pathogens act as limiting checks (Darwin, 1859) to the unrestrained population growth. The initial classification of non-human species interactions refuted the 'struggle for survival and classified relations between species as parasitism, commensalism, and mutualism based on facilitative and antagonistic approaches to resource utilization (van Beneden, 1878). Later on, competition (Gause, 1934; Connell, 1961) and amensalism (Haskell, 1949) are also being accepted as categories of biotic interactions, based on the net balance of trade (benefit-cost) matrix. The co-actions are thus classified on the basis of positive (+), negative (–), or neutral (0) effect on the participants (Lidicker, 1979). The possible co-actions are thus antagonism (prey-predator/parasitism/herbivory) (+, –), mutualism (+, +), commensalism (+, 0), amensalism (–, 0), competition (–, –), and neutralism (0, 0), the last of the lot can be considered to be a lack of effect on either of the interacting species. True neutralism is virtually not possible to prove and hence is ignored for the purpose of this study.

Early studies in animal ecology were either descriptive, classifying animals according to their habitats (Pearse, 1926), or quantitatively analyzed the abiotic environmental factors that restricted their distribution and population (Elton, 1927). The limiting effects of competition, predation, and parasitism were mathematically incorporated into population growth models (Lotka, 1925; Volterra, 1926), which can be considered to be the first attempts at modeling an ecosystem based on the interactions between its constituent components. Most studies at modeling interactions are limited to bipartite networks between two trophic levels, also represented by a biadjacency matrix, where organisms represent two kinds of nodes, and the interactions are restricted only between different kinds of nodes (such as a plant-pollinator or host-parasite). However, given the complexity of multi-trophic relationships in a food web, modeling food chains, food webs, and further complex ecological metawebs consisting of collection of food webs in a given ecosystem need to be studied (Adhurya et al., 2024), and simulations to model them are ongoing.

ML-based Species Interaction Model (SIM) Simulations

Antagonistic interactions, a major focus in ecology, are classified as parasitism, carnivory, and herbivory. Simulations of these interactions are typically trait-based, phylogenetic, or hybrid (combining traits and phylogeny). Additionally, landscape variables like habitat, niche, and altitude contribute to species co-existence and can be modeled using satellite images and remote sensing (Figure 1).

The traditional trait-based approach uses a variety of traits—morphological (size, color), behavioral (feeding habits), ecological (habitat, niche), physiological (immune response), life history (litter size), and evolutionary (genetic, genomic). Advances in molecular biology have enabled the use of molecular traits (e.g., k-mer protein sequences) for predicting interactions, particularly in microscopic organisms where data on other traits is limited. Recent studies (Adhurya and Park, 2024) have also applied unsupervised ML to predict interactions solely from interaction data, independent of phylogeny or traits.

Depending on the dataset scale, simulations may target local, regional, or global interactions. To facilitate understanding, models are categorized by interaction types, with antagonistic interactions further divided into prey-predator, host-parasite, and herbivory for clarity.

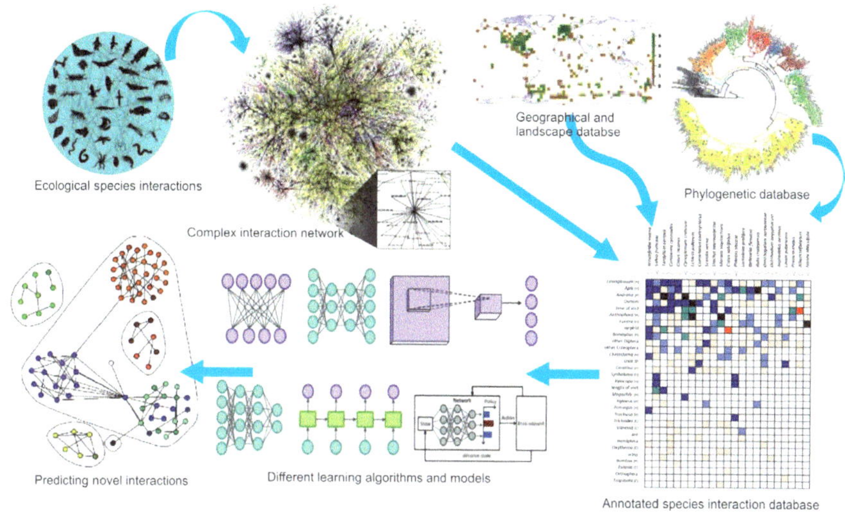

Figure 1: Workflow of ML-based interaction prediction.

Prey-Predator

Prey-predator relationship is one the most documented and most studied of the various interactions (Quiles and Barrientos, 2024). The availability and abundance of prey species is important for any predator to survive in an environment and hence the study of these interactions is a key for understanding the behavior of the predator species. Predator species, being at the top of the trophic level in the food chain, regulate the species in lower trophic levels of the food chain through trophic cascades (Beschta and Ripple, 2009). The intentional or accidental introduction of exotic predators, primarily through anthropogenic means (Olson and James, 1984), has often caused significant disruption in ecosystems. These exotic predators find the native prey species evolutionarily and adaptationally unprepared, frequently leading to the extinction of many native species (Anton et al., 2020) and creating havoc in the ecosystems. These effects are more pronounced in the terrestrial island ecosystems as the suppressed species are left with nowhere to escape, often getting extirpated in the process (Olson and James, 1984; Henderson and Powell, 2001; Platenberg, 2007; Doherty et al., 2016; Woinarski et al., 2024). Thus, apart from the morphological and behavioral traits of the interacting animals, the study of prey-predator interactions is also dependent on geographic and environmental parameters in the location of their interactions.

Trait-based

The predator-prey body size relationship is a key trait for estimating interactions in ecological metaweb models (Table 1) (Gravel et al., 2013). A regional database of the Mediterranean Sea's pelagic macrofauna, excluding cartilaginous

Table 1: A snapshot of ML-based prey-predator interaction prediction models

Ref.	Type of the Model	Data Used	Geographic Scale of Data	ML Model Used	Output Data
Gravel et al., 2013	Trait-based	Morphological trait (body size of both the prey and predator)	Regional (Mediterranean Sea)	Linear quantile regression	Fish food web
Millar, 2019	Trait-based	Morphological trait (length, height, weight, prey size, teeth length, etc.) and physiological trait (bite force, speed, eyesight, prey speed)	Paleontological data	SVM, LDA, Logistic regression, Decision Tree (KNN & naïve Bayes showed less accuracy)	Behavior of the animal (*Tyrannosaurus rex*) as a feeder
Desjardins-Proulx et al., 2017	Hybrid	Morphological traits (body mass), Phylogenetic traits (taxonomic distance) and Binary Ethological and Ecological traits (detritus, above ground, detritivore, carnivore, immobile etc.)	Regional (Germany) - 48 forest soil food webs	KNN with Tanimoto distance (to recommend new preys) and random forest (to predict interaction or not)	Potential novel preys for a predator in a soil food web and logistic prediction of interaction
Llewelyn et al., 2023	Hybrid	Morphological traits (body mass of prey and predator), behavioral and ecological traits (time of activity, resource groups consumed etc.), phylogenetic traits (eigenvector maps)	Global and Local (Simpson Desert in Australia)	Random forest	Predict prey-predator interactions among birds and mammals in Simpson Desert of Australia

Case Study: Prediction of benthic biomass as prey in Bering Sea using environmental traits (Oppel and Huettmann, 2010)

fishes, marine mammals, and turtles due to limited data, included 557 species. Findings showed that predator feeding range increases with body size, although large predators may appear specialized due to fewer prey options within their niche. Ontogenetic dietary shifts, common in fish (Bodner et al., 2021), were modeled using species links or size-based subspecies categories. While the unweighted model could be enhanced with Bayesian weighting for more accurate predictions, it only predicts predator-prey interactions due to its reliance on body size relationships.

This trait-based model also supports predicting palaeoecological traits, such as classifying *Tyrannosaurus rex* as a primary hunter based on traits like teeth length, bite force, and speed. Among various ML methods tested (logistic regression, decision tree, SVM, etc.), naive Bayes and k-nearest neighbors were excluded due to lower accuracy (66%) and misclassifications.

Hybrid

The trait-based model can be enhanced by integrating ecomorphological traits with phylogeny (Llewelyn et al., 2023). Llewelyn et al. used 3,329 prey-predator records from GloBI (Poelen et al., 2014) globally, incorporating 109 Oceania-specific records to model interactions in Australia's Simpson Desert. To mitigate geographic and taxonomic biases, prey-predator records unrelated to the focal predators and interactions of Simpson Desert predators outside this region were appended to the GloBI database. Morphological and behavioral traits (Desjardins-Proulx et al., 2017), along with phylogenetic data via eigenvector mapping (Guénard et al., 2013), served as variables in a random forest model to predict carnivory. Following Gravel et al. (2013), prey size-ranges were identified by loge-transforming body masses, creating a linear predator-prey relationship. Using ecomorphological, phylogenetic, and hybrid methods, quantile regression was applied to seven common Simpson Desert predators, with the hybrid model performing best. Accuracy was higher for introduced predators, likely due to undersampling biases, but random forests effectively predicted interactions even with limited records, showing robustness against random data removal.

Landscape-based

A generalized least square (GLS) model of 12 landscape covariates such as elevation, gradient, habitat type (classified into 10 types), (Euclidean) distances from towns, roads, rivers and ungulate snares, tree cover have been assessed to estimate site-specific prey abundances. Treating livestock depredation count as a proxy for the risk, it is found that livestock depredation risk is more near the ungulate snares, indicating that poaching pressure may be squeezing the tigers towards livestock. Though this study did not employ machine learning instead of statistical modeling, a similar approach has later been executed using machine learning and remote sensing for estimating livestock depredation risk by Bengal Tigers (*Panthera tigris tigris*) in Panna Tiger Reserve of Madhya

Pradesh, central India (Malviya and Krishnamurthy, 2022). Geotagged livestock depredation data for a 6-year time period was collected locally from 156 sites, which has been treated as presence to generate an equal number of random pseudo-absence. Apart from livestock, potential wild prey species such as hare, deer, langur, peacocks were surveyed. Water bodies, vegetation cover, ruggedness were extracted through remote sensing satellite images. Along with these parameters, (Euclidean) distance from roads, vehicular disturbances, land use and land cover data were collected. Univariate logistic regression has been implemented, followed by univariate GAM to select the features and their scales. The finalized geospatial additive model included NDVI, prey encounter rate, human encounter rate, shrub abundance and elevation as the parameters. The results indicate that shrub abundance is most important among the variables, and the depredation risk increases with shrub abundance until a certain threshold point, beyond which more shrub decreases risk of livestock depredation. The study also suggests that livestock depredation risk is more when the prey encounter rate is low, indicating that the tigers look for the domesticated prey when the wild prey base is scarce.

While these landscape-based predictions can predict the probability and quantify depredation by apex predators using traditional statistical or machine learning-based modeling, it is not suitable to identify the possible interacting species, until and unless the method is combined with other methodologies such as morphological or/and phylogenetic traits.

Host-parasite interactions

Host-parasite interactions are crucial for disease surveillance, as many parasites (viruses, bacteria, fungi, worms, protozoa, ticks, and mites) cause diseases in economically important species, including humans, livestock, and pets, as well as in conservation-sensitive species. Like the human-induced introduction of predatory species affecting prey-predator interactions, anthropogenic migration has also led to the introduction of parasitic species (Bataille et al., 2018; Steverding, 2020). The chytrid fungus (*Batrachochytrium dendrobatidis*), infamous for causing amphibian chytridiomycosis, has devastated frog populations globally, except in Asia, where it originated (Scheele et al., 2020). While some parasitic interactions are well-studied, the range of susceptible hosts in wildlife (excluding livestock and pets) remains largely undocumented. Understanding host-parasite interactions is essential for studying macroecological patterns of contagious diseases (Browne et al., 2017) and predicting zoonotic spillover potentials (Tajudeen et al., 2022; Meadows et al., 2023; Escudero-Pérez et al., 2023). Host-parasite associations may stem from inheritance from a common ancestor or host shifts (Brooks and McLennan, 1991; Page, 1993), suggesting that closely related species exhibit similar parasitic associations. Therefore, in addition to physiological and ecological compatibility, understanding the evolutionary relationship is vital for comprehending parasitic interactions.

Trait-based

The life history, physiological, and ecological traits of hosts and parasites have been studied for decades to predict interactions (Steck and Wendeler, 1980). Additional traits derived from demographics and morphology, such as postnatal growth rate and age at first birth, have been included in trait profiles (Han et al., 2015). For bats, factors like diet, activity patterns, migration, and torpor have also been considered (Han et al., 2016).

To identify natural reservoirs of filoviruses, primarily bats, trait variables for 1,116 bat species were compiled from PanTHERIA, assigning a binary status variable based on filovirus positivity. Boosted regression trees generated susceptibility links between bat species and filoviruses, revealing that filovirus-carrier bats are larger at birth, wean at larger sizes, and produce more litters annually compared to other bats. The model achieved 87% accuracy in predicting filovirus positivity, even with 20% of data removed.

Boosted regression trees have also been used for predicting mammal-helminth interactions (Dallas and Becker, 2020), utilizing data from the LNHM database at global, national (USA), and regional (Texas) scales. Host species variables were extracted from PanTHERIA, with 50 models trained on 80% of the data. Both superset and subset models showed that performance depended on host and helminth covariates, with Pearson's correlation indicating consistency across scales. The differences between subset and integrated models were negligible when sufficient data was available, suggesting that with more data, integration across taxa could enhance model accuracy.

Identifying morphological and functional traits of microbes and viruses is challenging. Therefore, molecular, genetic, or genomic data from bioinformatics can help predict microbial interaction networks. Electron microscopy images can predict cytopathic effects in pathogenic microbes (Yakimovich, 2021). While genetic data is often accessible, predictions can vary based on the selected gene, potentially introducing noise into models. Using complete genome sequencing data can improve predictions. A meta-ensemble learning algorithm (Wardeh et al., 2021) predicted novel coronaviruses' emergence through mammalian hosts, highlighting the importance of selecting appropriate traits for modeling bipartite interactions.

A heterogeneous microbial network model (Pan et al., 2024) utilized a knowledge graph-based deep learning technique to predict candidate viruses for target hosts. This model aggregated data on human-virus, bacteriophage-virus-bacteria, and human-bacteria interactions, with a blended deep neural network (DNN) designed and validated. The effectiveness of the model was demonstrated in case studies of pathogenic bacteria, but increasing complexity in the microbiological network may introduce noise that needs addressing. Such models can aid pharmaceutical researchers in developing targeted antibiotics.

In addition to these traits, receptor-ligand binding can predict the effects of pathogenic microorganisms in host cells (Table 2). Both traditional machine

Table 2: A snapshot of ML-based host-parasite interaction prediction models

Ref.	Type of the Model	Data Used	Geographic Scale of Data	ML Model Used	Output Data
Han et al., 2016	Trait-based	Morphological, ecological, life history traits of hosts (bat)	Global – Bat-filovirus interaction data	Boosted regression tree	Prediction of novel filovirus carrier bat species
Dallas and Becker, 2020	Trait-based	Morphological, ethological, life history, and taxonomic traits of host and parasite	Global, national (USA) and local (Texas state of USA) – Mammal host-parasitic helminth (roundworm, flatworm, spiny-headed worm) interaction	Boosted regression tree	Comparison of taxon-specific model with the complete mammal-helminth interaction prediction model (the latter being outperformed)
Wardeh et al., 2021	Trait-based	Genomic traits (genome sequences of viruses and its strains)	Global – Infection of coronaviruses on Terrestrial mammal species	Ensemble learning using Stochastic gradient boosting	Prediction of coronaviruses infecting terrestrial mammals
Pan et al., 2024	Trait-based	Molecular traits (K-mers protein sequences)	Microbe-host dataset separated into Human-virus interactions, Human-bacteria interactions and Phage (virus)-bacteria interactions	Skip-gram model (word2vec), InteractE (Convolution-based Knowledge graph embedding) and blended DNN	Prediction of candidate phages for target bacterial hosts
Dyer et al., 2011	Trait-based	Molecular traits (K-mers protein	Global data of human-HIV interactions	SVM	Protein-protein binding between HIV

		sequences, protein domains and human protein properties)	Local small-scale experiments and manually curated		and human proteins
Cui et al., 2012	Trait-based	Molecular traits (3 consecutive protein sequence – Relative frequency of amino acid triplets)	Global data for Hepatitis C (HCV) and papillomaviruses (HPV) infections in humans	SVM	Protein-protein binding between HCV and human proteins, and between HPV and human proteins
Barman et al., 2014	Trait-based	Molecular traits (domain-domain association, composition of amino acids such as methionine, valine, serine found in viral proteins)	Global virus-protein interaction data from VirusMINT database for Hepatitis B, Hepatitis E, Hepatitis C, simian virus (SV40)	SVM, naïve Bayes, Random forest	Protein-protein binding
Mock et al., 2021	Trait-based	Genomic traits (genome sequences of viruses for each virus-host combination)	Global – Nucleotide sequences of influenza A, rabies lyssavirus and rotavirus A virus and their interactions with hosts	CNN with LSTM	Prediction of viral hosts
Xu et al., 2017	Trait-based	DNA and protein sequences	Global – protein dataset; DNA and protein sequences of influenza viruses and the host proteins they are interacting with	Skip-gram model (word2vec) for NLP; SVM for classification	Prediction of hosts of influenza virus

(*Contd.*)

(*Contd.*)

Ref.	**Type of the Model**	**Data Used**	**Geographic Scale of Data**	**ML Model Used**	**Output Data**
Farrell et al., 2022	Phylogeny-based	Phylogeny of hosts	Global mammal-parasite network (Stepehens et al., 2016)	Bayesian network-based latent score model	Binary interactions between mammal host and parasite
Kitson and Suttle, 2019	Phylogeny-based	Phylogeny of viruses	Global virus taxonomy	Natural language processing	Predict hosts of a virus from its name
Barel et al., 2023	Hybrid	Morphological traits (body length, width, biovolume), ethological traits (lifestyle, nourishment etc.) and taxonomic variables (species, genus, family, order, phylum, kingdom)	Meta foodweb of microbes from 40 sources, including 6 peatlands	A comparison of Random forest, kNN, BRT, GLM, BGLM, NN	Microbial food web in peatlands
Gonzalez-Isunza et al., 2023	Hybrid	Molecular traits (k-mers protein sequences) and phylogenetic tree for viruses	Global	Skip-gram model (word2vec) using a Neural network	Prediction of protein-protein binding (S protein of coronavirus-human host receptor binding)

Case Study: Prediction of protein-protein interactions using genetic and phylogenetic traits (Cuesta-Astroz and Oliveira, 2018) learning (ML) and deep learning algorithms can model protein-protein interactions (Casadio et al., 2022). While most modeling has focused on intraspecies interactions, early attempts also included inter-species interactions, such as HIV with human proteins (Dyer et al., 2011). A model using SVM achieved 70% precision and 40% sensitivity. Other models have improved accuracy with HCV and HPV interactions (Cui et al., 2012) and with HBV and HEV using various supervised learning algorithms (Barman et al., 2014).

Additionally, a deep learning model named VIDHOP predicts virus hosts using genomic sequences (Mock et al., 2021). Two models—one with bidirectional LSTM (long short-term memory) and another combining LSTM with CNN (convolutional neural network) demonstrated similar accuracy, though the CNN-LSTM model slightly outperformed LSTM except for Influenza A virus.

Protein and DNA sequences can be tokenized using natural language processing (NLP) algorithms to extract information (Ofer et al., 2021). An NLP-based method predicted virus hosts using unique DNA and protein sequences (Xu et al., 2017). The skip-gram method applied the word2vec algorithm for protein sequences, followed by SVM classification, achieving good accuracy, particularly for avian influenza virus, the evolutionary ancestor of all influenza viruses.

Phylogeny-based

The necessity for comprehensive data in trait-based networks means they perform better for smaller networks. When sufficient trait data is lacking, performance declines on global-scale networks (Morales-Castilla et al., 2015), and evolutionary relationships shown by phylogenetic trees can serve as proxies. By combining affinity-based modeling with phylogenetic information, a link prediction model for global bipartite mammalian host-parasite networks is simulated using Bayesian networks (Elmasri et al., 2020; Farrell et al., 2022). While evolutionary distances from phylogenetic trees act as proxies for morphological traits, interactions may also be influenced by traits independent of evolutionary relationships. To address this, the phylogeny-only model is being enhanced with node-specific affinity parameters, akin to covariate variable-based network models (Hoff et al., 2002; Hoff, 2005; Bickel and Chen, 2009). The evolutionary distances of mammalian hosts are represented by a phylogenetic tree (Fritz et al., 2009) and converted into a connected weighted network. A host-parasite interaction matrix is created based on the presence or absence of interactions, with higher conditional probabilities assigned to closely related or numerous distantly related hosts. The interaction probability of a parasite is assumed to be influenced by the sum of evolutionary distances to its documented hosts. The full model was compared to two submodels (affinity-only and phylogeny-only), as well as a Jaccard distance-based bilinear latent-distance model and a k-nearest neighbor (kNN) model. Murphy's diagrams indicated that the latent score-based full model outperformed others, with the kNN model as the worst performer; notably, the phylogeny-only model performed similarly to the full model, validating phylogeny as a potential proxy for traits (Elmasri et al., 2020). The results showed that the kNN and bilinear latent-distance models performed equivalently. The weak outputs of the Jaccard distance-based neighborhood models (kNN and bilinear latent-distance) suggested that incorporating phylogeny could enhance their predictive performance for host-parasite associations.

Using the English names of viruses, hosts can be predicted through string matching, and when conclusive predictions are not possible, an NLP-based model identifies a higher taxonomic group of hosts based on the virus type (Kitson and Suttle, 2019). The VHost-Classifier model uniquely extracts interaction information solely from the common name, without incorporating morphological, ecological, or evolutionary variables.

Hybrid

A model combining traits with phylogenetic features (Barel et al., 2023) can be used to infer interactions. An extensive meta food web, illustrating feeding interactions among 164 microbial taxa groups (mainly at the species level) such as nematodes, bacteria, fungi, algae, cyanobacteria, rotifers, flagellates, and ciliates from peatlands, was utilized to train the model. This data was input into several machine learning models, including Random Forest, kNN, GLM (generalized linear model), boosted regression tree, neural network, and Bayesian generalized linear model. Tree-based models and neural networks outperformed others in accuracy, AUC, and TSS; however, the neural network's accuracy was lower, making the Boosted Regression Tree (BRT) the best algorithm for predicting missing interaction links. Although this feeding behavior can be considered carnivory, the inclusion of microbes like bacteria and fungi categorizes them as parasitic for this study.

Combining traits and phylogeny can also predict potential viral hosts (Gonzalez-Isunza et al., 2023). A model was developed to identify non-human animal-hosted coronaviruses likely to infect humans based on a human-binding potential (h-BiP) score. The ML model employs a neural network to convert genomic sequences into vectors that encode the relationship between k-mers. These vectors compute the h-BiP score, serving as the classifier. A phylogenetic analysis reveals the evolutionary proximity of predicted potentially human-infectious coronaviruses to known human pathogens, consistent with the genomic trait-based model.

Herbivory

Herbivores are economically significant as both livestock and pests damaging crops (Deutsch et al., 2018; Drimaj et al., 2023), making the prediction and quantification of herbivorous interactions vital for crop protection. Predation risk negatively impacts the foraging behavior of herbivores across various landscapes, including grasslands and coral reefs (Burkepile and Parker, 2017). Plant and herbivore traits, along with their evolutionary history, influence herbivore plant choices (Pearse et al., 2013). While herbivores tend to use phylogenetically similar plants, this trend varies by scale and species. Traits such as interaction-cost index, risk-index, dietary frequency, and predation pressure are commonly used to model herbivore interactions (Pocock et al., 2021). Seed mass and lipid content significantly influence the feeding preferences of granivorous birds (Díaz, 1996; Gaba et al., 2014) and ground beetles (Gaba et al., 2019). Additionally, food chemical content, floral display traits, and flowering phenology attract herbivores (Fögelstrom et al., 2017; Wu et al., 2021). Muzzle width in megafauna affects dietary choices, except for elephants, which use their trunks for foraging (Lundgren et al., 2024). Leaf traits like water content and surface area influence the diets of various herbivores, especially generalist invertebrates like snails and grasshoppers (Pérez-Harguindeguy et al., 2003). While predictive models for herbivore interactions exist, the application of machine learning in this area is limited. However, pattern recognition has been employed to detect herbivore species (Meineke et al., 2020).

High-resolution, digitized herbarium images of distantly related plant species were obtained from SERNEC and annotated by damage category. Six types of leaf damage caused by insects were categorized: margin feeding, interior feeding, skeletonization, blotch mines, serpentine mines, and stippling, with negative examples of undamaged leaves included for training. Eighty percent of the image data was randomly selected for training, and image segmentation was performed to isolate the leaves. A Single Shot Multibox Detector using a VGG16 base classification network (Liu et al., 2016) detected and classified margin and interior feeding simultaneously. An 18-layer residual net architecture (He et al., 2016) was used for damage classification among eight types. While the model accurately classified ovoid holes, complex-shaped holes and non-insect damages (such as fungal damage) posed challenges, leading to overfitting. The classifier achieved 81.5% accuracy with 1,105 testing images, but inaccuracies arose from limited training data for some categories and significant confusion between margin and interior feeding. A damage mask was proposed to address this confusion, suggesting that tens of thousands of annotated images per category are necessary for improved classification accuracy due to the complexity of box detection compared to species detection. Although this method of pattern recognition cannot precisely predict herbivore species, it can estimate the higher taxonomic group to which the herbivore belongs.

Mutualism

In interaction ecology, mutualism is defined as an inter-species interaction which results in a net benefit for both the interacting species, more precisely, culminating in a reproductive benefit or by boosting survival opportunities in the ecosystem. While the facultative mutualists can thrive even without the presence of the mutualist, the obligate mutualists would essentially go extinct without its mutual partner. While the pollinators such as moths, butterflies, bees, wasps, hoverflies, sunbirds and hummingbirds are obligate mutualists as they need plant nectar to feed, the lichenification of algae and fungi is considered as an example of facultative mutualism, as they either can prepare or gather their own food through autotrophic and saprotrophic nutrition, respectively. While most of the mutualists, such as most of the plants are pollinated by an array of insect species ranging from wasps, moths, butterflies, hoverflies and bees, some specialists such as the plants of *Yucca* genus are obligately pollinated by moths only belonging to the family Prodoxidae, who also happen to feed on these plants in their larval stage.

Trait-based

Plant-pollinator interactions are widely studied bipartite interactions in ecology and culture. Eight plant traits including—such as height, color, floral display size, and nectar volume—determine pollinator visits (Rafferty and Ives, 2013). Pollinator traits like body mass and feeding habits also influence interactions. Trait-based modeling includes complementarity traits (e.g., activity time) and barrier traits (e.g., corolla tube length) (Santamaría and Rodríguez-Gironés, 2007). Using a global dataset, Pichler et al. (2020) simulated plant-bird pollinator networks based on trait-matching. Of the seven ML models, DNN, random forest (RF), and BRT outperformed others, with RF proving the best predictive model. Corolla-bill length matching was the most influential trait. The study was limited to plant-pollinator mutualism but could extend to other networks. Similar results were obtained for bee-plant interactions in regional datasets, identifying floral shape and corolla tube length as key traits (Ornai and Keaser, 2020).

Not all mutualisms depend on these traits; some rely on biochemical or abiotic factors. For example, Rhizobium bacteria associated with legumes via complex NOD proteins (Wang et al., 2012), and lichens depend on soil pH and substrate types (Škvorová et al., 2022). Additionally, plants hosting endophytic microbes can control herbivorous insects (Adeleke et al., 2022). Genomic data of rhizosphere bacteria help predict optimal plant-promoting bacteria, with SVM performing best among KNN, SVM, and LDA (latent dirichlet allocation) models (Indumathi et al., 2021).

Hybrid

Lichens are well-studied symbioses between heterotrophic fungi and autotrophic partners. Research on *Cladonia* fungi shows that photobionts can be specialists, generalists, or intermediates (Yahr et al., 2006). In a study, 1,120 *Cladonia* lichen samples from Europe were analyzed, identifying 181 OTUs of *Cladonia* and 18 OTUs of *Asterochloris*, used to model mycobiont-photobiont interactions based on traits and evolutionary relationships (Škvorová et al., 2022). Soil pH and fertility-affecting radicals were measured, and fungal and algal RNA genes were aligned into two phylogenetic trees. The OTUs were classified using General Mixed Yule Coalescent (GMYC) (Talavera et al., 2013), Bayesian PTP (bPTP)

(Zhang, 2013), and automatic barcode gap discovery (ABGD) (Puillandre et al., 2012). Analysis of lichen morphotypes revealed key mycobiont traits such as thallus type, reproductive structures, podetia, apothecia color, and cortex chemicals. Findings suggested algal variation is driven mainly by climate rather than fungal traits, soil chemistry, or geography. While ML was not used, logistic classifiers could enhance prediction of photobiont-mycobiont interactions using morphological, chemical, and evolutionary data.

Commensalism

Commensalism, like mutualism, is a form of symbiosis between two species where the symbiont benefits, while the host remains unaffected. Often mistaken for mutualism, commensalism is less studied. A common example is African herbivores (e.g., deer, rhino, cattle) and oxpeckers, which feed on ticks from these animals. Similarly, in India, cattle egrets accompany domestic cattle. Though once considered mutualistic, studies (Weeks, 1999; Weeks, 2000; McElligott et al., 2004) indicate that herbivores gain no benefit, as tick loads do not increase without the birds, and the birds often feed on dead skin or wounds, even creating new wounds. This interaction is thus a case of commensalism, not mutualism.

Landscape-based

Geographic overlap is key in symbiotic associations, especially when one partner has a restricted range, enabling habitat-based predictions of presence or absence. The decline of the White Rhinoceros (*Ceratotherium simum*) and the use of chemicals to remove its ectoparasites (mainly hard ticks) have reduced tick populations, impacting tick-feeding birds (Bezuidenhout and Stutterheim, 1980; Mihalca et al., 2011). Cattle dips (acaricides) also threaten oxpecker populations, which rely on African ungulates like rhino, deer, and cattle (Dickman, 1992; Mooring and Mundy, 1996). Key factors for oxpecker populations include tick density, nesting sites, savannah landscapes, water sources, and conservation areas (Kalle et al., 2017). Machine learning models like GLM, GAM, and boosted regression trees were used to predict suitable habitats for reintroducing red-billed oxpeckers in South Africa, with GAM and BRT showing high predictive power in historic ranges.

Amensalism

Amensalism is an inter-species interaction where one species is harmed or inhibited, while the other remains unaffected (Lang and Benbow, 2013; Alhadi and Naji, 2024). The harmed organism, or amensal, may be affected physically or indirectly by competition or by chemicals released by the unaffected species, termed the enemy (Alhadi and Naji, 2024). Although no ML models explicitly target natural amensal interactions, some ML models have been developed to analyze roadkill, which, though not an interspecies interaction, results from human locomotion and is considered amensalism in this study.

Trait-based

Species-specific roadkill rates among vertebrates can be predicted from life-history traits, both morphological and behavioral (González-Suárez et al., 2018; Grilo et al., 2020) (Table 3). Bird and mammal traits from a regional dataset were used, with missing values imputed via random forest. Separate random forest models were developed for mammals and birds using available roadkill data to capture spatial and temporal variations. However, due to limited data on species-specific road-crossing behavior and road traffic, each species was assumed equally likely to cross, adding noise to the predictions. Temporal factors, such as seasonal changes impacting roadkill likelihood, should ideally be integrated to improve model accuracy (Ascensão et al., 2022).

Endophytic microorganisms, known for bioactive metabolites, act as natural biocontrol agents against various pests and pathogens (Gouda et al., 2016). These endophytes, often used as biopesticides, exhibit amensalistic effects by harming pests while remaining unaffected. Studies have shown the impact of Beauveria bassiana on herbivorous insects (Portilla et al., 2017; Kovač et al., 2020) and on disease vectors like Aedes aegypti (Darbro et al., 2012) and Anopheles stephensi (Thomas and Read, 2007). Predictive ML models on amensal interactions are limited, though XGBoost proved effective in modeling the impact of Beauveria bassiana on the rice pest Sesamia calamistis by using fungal strain isolates and rice tissue traits as input variables (Megnidio-Tchoukouegno et al., 2022). An ML model studying Glumon™, a biopesticide, assessed its impact on coffee berry borer (Hypothenemus hampei), integrating ecological traits like shade tree richness and farm proximity to forest, which influences pest control via natural predators (Manson et al., 2022; Karp et al., 2013).

Landscape-based

Landscape-based species distribution modeling (SDM) using only presence-only data is also very well suited to model the roadkill risks and map the potential hotspots (Ha and Shilling, 2018). A state-level (California, USA) identified roadkill data with GPS mapping and timestamps was used for the purpose. A list of environmental parameters and two human population density variables were considered, which were quantified using geospatial data maps and census data of the study area. ML-based MaxEnt (Maximum entropy) model (Phillips and Dudik, 2008), which uses both continuous and categorical variables to understand the environmental and geographical changes, can be used to map the presence of a species and has been found to be one of the best of the SDMs currently in usage (Heumann et al., 2013). By creating separate niche models for four taxonomic groups (ungulates, birds, medium-sized mammals, small mammals), the likelihood scores for their vehicular collisions were predicted and mapped.

Table 3: A snapshot of ML-based non-antagonistic interaction prediction models discussed

Ref.	Type of the Model	Data Used	Geographic Scale of Data	ML Model Used	Output Data	Type of Interaction Predicted
Rafferty and Ives, 2013	Trait-based	Morphological and ecological traits of plants and pollinators along with phenological traits	Local (university arboretum in Wisconsin, USA) and Experimental (in a greenhouse) field data	Linear mixed model	Prediction of plant-pollinator interactions and responses to change in phenology	Mutualism
Pichler et al., 2019	Trait-based	Morphological (length of proboscis, body size, bill curvature, wing length), ethological (feeding behavior, sociality) traits of pollinators Morphological traits of plants (color, shape, inflorescence of flower, shape of corolla, type of plant)	Global (plant-pollinator insect interactions) Regional (plant-hummingbird interactions in Costa Rica)	Random forest (RF) for imputation of missing data A comparison of RF, CNN, DNN, KNN, BRT, GLM, SVM (RF, BRT, DNN outperformed others)	Prediction of plant-pollinator interactions to infer the causally most important trait-matching	Mutualism
Ornai and Keasar, 2020	Trait-based	Morphological traits of flowers (shape, symmetry, depth of flower, etc.)	Local (Mt. Carmel National Park, Israel)	Random forest Logistic regression	Prediction of plant-bee interactions	Mutualism
Škvorová et al., 2022	Hybrid	Environmental traits (pH, chemical composition) of soil, evolutionary traits (OTUs)	Continental (throughout Europe)	GMYC, ABGD and bPTP (for species delineation from DNA sequences)	Prediction of photobiont-mycobiont species interactions in	Mutualism

(*Contd.*)

(*Contd.*)

Ref.	Type of the Model	Data Used	Geographic Scale of Data	ML Model Used	Output Data	Type of Interaction Predicted
				lichen (*Cladonia* spp. and *Asterochloris* spp.)		
Kalle et al., 2017	Landscape-based	Ecological traits (density of mammals which host ticks, presence/absence of starling species), Environmental traits (temperature, precipitation, land cover etc.)	Regional (South Africa)	GLM, GAM, BRT	Prediction of suitable habitats for reintroduction of oxpeckers	Commensalism
Megnidio-Tchoukouegno et al., 2022	Trait-based	Colonization traits (amount of pest colonization in target vegetative tissues of rice cultivars) and treatment traits (treatment types and responses to amounts strains of the fungus used in the treatments)	Experimental	Linear regression, LaSSO, SVM, KNN, XGBoost, Ensemble learning	Prediction of effects of entomopathogenic fungal treatment on the pest colonization on rice tissues	Amensalism
Manson et al., 2022	Trait-based	Ecological traits (shade cover, diversity of shade trees, distance from forest)	Local (Coffee farms in two towns of Indonesia) data collected through field survey	GLMM	Prediction of decline in *Hypothenemus hampei* effect of other parameters in effectiveness of a biopesticide	Amensalism

Kulatunga et al., 2015	IoT-based	Railway accident traits (locomotive type, casualty, location) and locomotive traits (visibility, engine type etc.)	Real-time (video from night vision camera)	DDE filter	Indication of presence of wildlife in the photographs	Amensalism
Ramesh et al., 2017	Computer vision-based	Image traits (hue, saturation, color)	Real-time (still images from live video captured by camera)	HSV image segmentation; SVM	Classification of presence/absence of elephants in the photographs	Competition
Senthamil Selvi et al., 2020	Computer vision-based	Sound traits (high-frequency ultrasound waves to measure distance of the moving object near the train) Image traits (hue, saturation, color)	Real-time (image captured by webcam, triggered by the ultrasonic sensor)	HSV image segmentation	Classification of presence/absence of elephants in the photographs	Amensalism
Kotula et al., 2021	Hybrid	Morphological traits (body size), ecological traits (biogeographic status, generalism, phenology) and phylogenetic traits of host and parasitoid	Local (field data from selected study sites in a locality of New Zealand)	Random forest, KNN	Prediction of effect of biopesticide on apparent competition in the host-parasitoid interaction network	Amensalism
Peters et al., 2016	Trait-based	Ecological traits (area of the forest, number of trees, distance from nearest neighboring tree etc.)	Data generated through simulation of different scenarios	Individual-based model, combined with self-organizing feature maps (SOM)	Types and effects of competition, both underground and above the ground, for resources in plants	Competition

(*Contd.*)

(*Contd.*)

Ref.	Type of the Model	Data Used	Geographic Scale of Data	ML Model Used	Output Data	Type of Interaction Predicted
Barroso-Bergada et al., 2023	Phylogeny-based	Evolutionary traits (OTUs)	Regional (microbial interaction data collected through eDNA from selected grape vineyards of France)	A/ILP	Metaweb of microbial interactions and prediction of change in a microbe abundance in response to variation of abundance of another species	Amensalism, Competition (mainly) Also Predation
Krupa et al., 2020	Trait-based	Morphological traits (diameter of sundew, mean size of plant, web area)	Local field data	Linear mixed effect model	Effect on prey availability, change in web size and decline in abundance of spiders due to sundew plants	Competition
Ahearn et al., 2001	Hybrid	Morphological (age, body mass), ethological (hunting, searching mate, with cubs) and life history (pregnancy, fertility status) traits	Local	GIS	Simulation of human interactions with individual tigers in and around study area	Competition

Case Study: Predictions of zones with high vulnerability of roadkills in an area of Colombia using behavioral and environmental parameters (Perez-Guerra et al., 2024)

Hybrid

Previous studies on roadkill largely ignored driver traits, focusing only on animals crossing roads. Many roadkills, however, result from reckless driving, with train collisions especially fatal for large animals like elephants and rhinos. In India, an estimated 20 elephants are killed yearly by trains. While computer vision-based intrusion detection systems alert loco pilots by processing images near hotspots (Kulatunga et al., 2015; Ramesh et al., 2017; Senthamil Selvi et al., 2020) (Table 3), reaction time remains limited. Integrating satellite GPS collars for tracking animal locations and driver traits, such as drowsiness measured by eye aspect ratio (EAR), could provide a more robust system (Radzali et al., 2023; Albadawi et al., 2023). IoT-based devices could detect drowsy pilots, sound alerts, and modify the train signaling system to help prevent wildlife-vehicle collisions (Samadder et al., 2022). Combining animal and landscape traits with driver drowsiness can identify high-risk hotspots for alerting drivers.

In host-parasitoid dynamics, species and phylogenetic traits predict biocontrol impacts (Kotula et al., 2021). Key host and parasitoid traits, including body size, geographic origin, and phenology, along with phylogenetic data, helped model trophic generalism separately in forests and plantations. Using both RF and KNN algorithms, weighted interaction networks simulated apparent competition and interaction frequencies between hosts and parasitoids. Random forest outperformed KNN due to its ability to model complex, non-linear relationships, whereas KNN's assumption that similar parasitoids interact with similar hosts reduced its accuracy.

Competition

Competition occurs when individuals or species compete for a limited resource, impacting the fitness of one party. Unlike well-studied antagonistic or mutualistic relationships, competition, particularly intra-species, is less explored beyond mathematical models of resource optimization. Symmetric interactions, like competition and mutualism, tend to be less stabilizing than asymmetric ones, such as antagonism or unilateral interactions like amensalism and commensalism (Mougi, 2016). Competition in ecology is further divided into interference competition, where an individual's actions hinder others' resource access, and exploitation competition, where competitors directly limit each other's resource availability (Power, 1992; Holomuzki et al., 2010).

Trait-based

The competition among trees for underground resources is crucial for understanding plant community dynamics, as the most limiting resource constrains individual plant growth (Grace, 1990). By combining an individual-based plant interaction model with self-organizing feature maps (SOM), an unsupervised ANN algorithm (Kohonen, 1982), multidimensional correlations between plant traits and underground resources can be visualized, allowing for predictions of competition outcomes in floral communities (Peters et al., 2016). The Pi Model outlines each plant's zone of influence both above and below ground (May et al., 2009), with both radii increasing with biomass (Enquist, 2002). Resource limitation, four levels of below-ground competition modes (allometric symmetry, size symmetry, complete symmetry, and asymmetry), mortality rate, and CEI (Clark-Evans Index) rate (Clark and Evans, 1954) serve as parameters for the Pi Model. A total of 228 simulations on 6,000-9,000 randomly selected plants generated training datasets for the SOM to predict competition modes. The four parameters are assigned as vectors and comprise the input layer neurons for the SOM. As an unsupervised algorithm, any variable may be predicted after SOM training. The 4-dimensional SOM is depicted by four hexagons representing each variable, with correlations understood through variations in shades of gray within the hexagons. This method estimates hard-to-quantify variables through easily measurable parameters by adjusting input and output variables.

While competition is strongest among closely related plant species or animals at the same trophic level (Darwin, 1859), competition between insectivorous plants and animals, such as spiders, is unique. Insectivorous plants, though capable of autotrophic nutrition through photosynthesis, capture insects and arthropods for nutrients like nitrogen (Darwin, 1875). The effects of sundew (*Drosera brevifolia*) cover on spider prey availability were modeled based on springtail (Collembola) predation, which is a significant prey component for both sundews and certain spider families, using General Linear Mixed Models (Krupa et al., 2020). The study employed ecological traits like sundew cover and web area to model competitive interactions between sundews and spiders.

Phylogeny-based

Operational taxonomic units (OTUs) are utilized in microbial communities to define species-level distinctions based on DNA gene similarity (Escalas et al., 2019; Pauvert et al., 2019). While previous studies focused on bipartite graph interactions, predicting the entire microbiome network is crucial, as interactions like competition, predation, or amensalism imply biological control of microbial pathogens. The microbiome interaction networks in the leaves of cultivated European grapevine, Vitis vinifera, were reconstructed for nine vineyards in France from environmental DNA (eDNA) samples, classified using abductive/inductive logic (A/ILP) based on hypotheses regarding changes in OTU abundance (Barroso-Bergada et al., 2023). In addition to real data, ecological-like data simulated OTU abundance changes for verification. The study predicted several potential antagonists of Plasmopara viticola, many of which are recognized as biocontrol agents in the literature, thus validating the model. The reduction in one OTU's population due to interactions with another is attributed to competition and amensalism, primarily driven by competition.

Hybrid

Unlike most previously discussed ML-based hybrid models that integrate trait data with phylogeny, the following model of exploitation competition between Bengal tigers (*Panthera tigris tigris*) and humans (*Homo sapiens sapiens*) in south-central Nepal combines trait data with landscape data obtained through remote sensing in object-oriented geographic information systems (GIS) (Ahearn et al., 2001). Movement characteristics, including behavior, feeding, mating, fertility, and pregnancy, are assigned default values for direction and rate. The distance tigers travel depends on their state variable values. The simulation model creates objects for male and female tigers, defining home ranges and assigning values for parameters like age, body mass, and fertility status. A male tiger's home range is based on estimates, with two adjacent males circumscribing a female's range. While tigers typically stay near prey for 2–3 days and hunt weekly (Sunquist, 1981), their time near carcasses significantly decreases when the kill is livestock due to human interference, increasing their hunting attempts. The object-oriented TIGMOD model implements relational joins between tigers and prey, as well as between male and female tigers. The 'location' attributes of each individual in the dynamic model change with state changes in the simulation menu and updates to the 'time' attribute. Functional and periodic events are scheduled, triggered by time and life events, respectively. The model was simulated with four wild prey densities and six combinations of wild and domesticated prey, tested against field data. Parameters such as "time villagers remain angry and motivated to poison tigers", "guarding domestic prey", and "delaying the onset of poisoning domestic prey" estimate human reactions. The people's tolerance toward tigers was found to be associated with demographics (Sharma and Neupane, 2023).

Discussion

While morphological, ecological, and ethological traits are easily measurable, many unmeasured traits can be addressed through evolutionary relationships. Phylogenetic trees can serve as proxies for these latent traits, often providing more information than trait sets (Li and Ives, 2017) and enhancing model performance when combined with morphological, ecological, ethological, and biogeographical traits (Llewelyn et al., 2023). Microbial and viral interactions are better predicted through phylogenetic trees or genetic traits, as their morphology has limited ecological impact. Although genomic data is more comprehensive than genetic data, the lack of genome sequences for many species hinders its application. Additionally, many genes have minimal impact on traits (Schnable, 2019), which can introduce bias. Climatic and geographic variables are less significant for microorganisms but are crucial for modeling megafauna and flora interactions.

Statistical models for biotic and abiotic responses often lack the dynamics needed to address real-world complexities. In contrast, ML-based modeling emphasizes dynamic prediction over causal inference, providing better predictive power, though sometimes at the cost of interpretability (Pichler and Hartig, 2023). While statistical models may excel in simulations, their effectiveness diminishes in real-world scenarios (John et al., 2022). ML models have significantly outperformed statistical models like GLM in plant-pollinator networks due to superior identification of causal trait relationships (Pichler et al., 2019).

However, the complexities of ecological interactions increase the risk of overfitting when training data patterns do not generalize to test data (Yang et al., 2020). Among popular ML algorithms for tabular data, tree-based models like random forests (Figure 2) are preferred for their ability to model non-linear relationships (Uddin and Lu, 2024). These models typically favor ensemble approaches like random forests or BRT for regression, while SVM is commonly used for classification tasks due to its efficiency in high dimensions with limited data. However, SVM's popularity has waned in favor of BRT and neural networks since the mid-2010s (Pichler and Hartig, 2023).

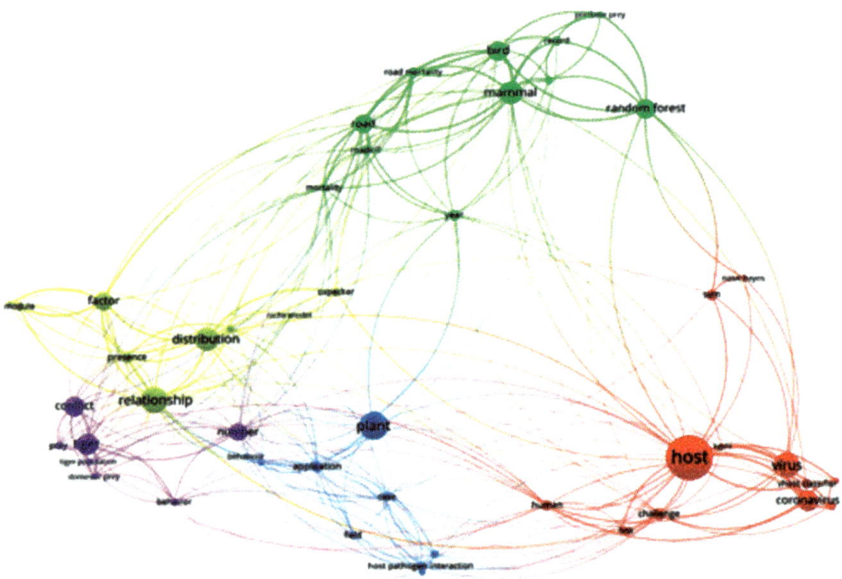

Figure 2: Network visualization of most relevant words from bibliography of ML-based inter-species interaction predictive models.

The evolution of GPUs has enabled DL CNNs to be trained in hours, transforming DL applications (Krizhevsky et al., 2017). DL models leverage multiple processing layers to learn patterns from complex data like images and audio (LeCun et al., 2015), but their effectiveness is limited by the scarcity of ecological data (Strydom et al., 2021), making traditional ML algorithms more suitable. While techniques like transfer learning and data augmentation (Christin et al., 2019) can address data limitations, they may introduce biases (Borowiec et al., 2022).

Current DL applications in ecology focus on multi-dimensional datasets, such as camera trap images and citizen science data (Hansen et al., 2018; Van Horn et al., 2018; Bogucki et al., 2019; Schofield et al., 2019; Clapham et al., 2020; Chen et al., 2020a). DL is also used for sound-based identifications of birds, bats, and mosquitoes (Mac Aodha et al., 2018; Fanioudakis et al., 2018; Chen et al., 2020b). However, DL's performance in structured tabular data is limited (Pichler et al., 2020; Strydom et al., 2021), with sparse applications in predictive ecological modeling, primarily for pest outbreaks (Rammer and Seidl, 2019).

To improve ML model interpretability, explainable artificial intelligence (XAI) is being used (Arrieta et al., 2020) in species distribution models (Ryo et al., 2020) and microbial interaction networks (Barroso-Bergada et al., 2023). Generative AI (GenAI) can generate data to enhance biodiversity monitoring (Rillig et al., 2024) and modeling species interaction matrices (Hirn et al., 2022), using generative adversarial networks (GANs) and variational autoencoders (VAEs).

Natural language processing (NLP) enables machines to analyze human language and extract data from protein sequences in viruses (Yakimovich, 2021). The word2vec method captures contextual information about words (Mikolov et al., 2013; Arora et al., 2016), generating insights about viral proteins. Self-supervised language models have reduced the need for labeled data, making them applicable to large datasets (Chen et al., 2020). However, DL models' sensitivity to hyperparameters and tendency to overfit (Zhang et al., 2021) pose challenges in hyperdiverse ecological applications.

Ecological studies often suffer from sampling biases (Hughes et al., 2021; Carlen et al., 2024) due to socio-economic factors. While ML-based models can address some biases, they also introduce ecological biases. Host-parasite interactions are well-studied, as seen in the 'host' label in Figure 2, but non-antagonistic interactions like amensalism and competition are underrepresented. Most amensalism models are anthropocentric, predicting roadkill incidents, while mutualistic models focus mainly on plant-pollinator interactions, neglecting others like endophytes. Despite a few studies on natural amensalism, ML techniques are underutilized. While ecological models often depict bipartite interaction graphs, the variation in interaction strengths and ecological networks across spatial and temporal dimensions (Strydom et al., 2021) requires further investigation.

Conclusion

Individual-based modeling can simulate community dynamics by incorporating traits, behaviors, interactions, and abiotic factors. Integrating ML algorithms has made ecological network simulation accessible to non-modelers, aiding ecologists and conservationists. ML identifies causal variables of ecological dynamics and helps gather data that is challenging to collect through field observations. However, data scarcity due to sampling biases across spatial, temporal, and geographical scales, along with the importance of certain interactions, skews ML applications in predicting ecological interactions.

Generative AI can generate novel datasets through data-driven learning and probabilistic modeling, compensating for data deficiencies and predicting new inter-species interactions. Most SIMs focus on bipartite graphs, but combining various parameters—morphological, geographical, biochemical, and genetic traits—with phylogenetically computed evolutionary distances can help predict a complete interaction network, albeit with increased computational complexity. To reduce computation costs, evolutionary distances represented by phylogenetic trees can be utilized to predict inter-species ecological interaction networks, addressing the Eltonian shortfall by imputing missing ecological information.

References

Adeleke, B.S., M.S. Ayilara, S.A. Akinola, and O.O. Babalola. (2022). Biocontrol mechanisms of endophytic fungi. *Egyptian Journal of Biological Pest Control*, 32: 46. https://doi.org/10.1186/s41938-022-00547-1.

Adhurya, S., D.-Y. Lee, and Y.-S. Park. (2024). KF-metaweb: A trophic metaweb of freshwater ecosystems of South Korea. *Global Ecology and Biogeography*, 33(7): e13845. https://doi.org/10.1111/geb.13845.

Adhurya, S., and Y.-S. Park (2024). A novel method for predicting ecological interactions with an unsupervised machine learning algorithm. *Methods in Ecology and Evolution*, 15(7): 1247–1260. https://doi.org/10.1111/2041-210X.14358

Ahearn, S.C., J.L.D. Smith, A.R. Doshi, and J. Ding. (2001). TIGMOD: An individual-based spatially explicit model for simulating tiger/human interaction in multiple use forests. *Ecological Modelling*, 140(1–2): 81–97. https://doi.org/10.1016/S0304-3800(01)00258-7.

Albadawi, Y., A. AlRedhaei, and M. Takruri. (2023). Real-time machine learning-based driver drowsiness detection using visual features. *Journal of Imaging*, 9(5): 91. https://doi.org/10.3390/jimaging9050091.

Alhadi, A.R.S.A., and R.K. Naji. (2024). The contribution of amensalism and parasitism in the three-species ecological system's dynamic. *Communications in Mathematical Biology and Neuroscience*, 2024: 33. https://doi.org/10.28919/cmbn/8456.

Anton, A., N.R. Geraldi, A. Ricciardi, and J.T.A. Dick. (2020). Global determinants of prey naiveté to exotic predators. In: *Proceedings of the Royal Society B*, 287(1928): 20192978. https://doi.org/10.1098/rspb.2019.2978.

Arif, S., and M.A. MacNeil. (2022). Predictive models aren't for causal inference. *Ecology Letters*, 25(8): 1741–1745. https://doi.org/10.1111/ele.14033.

Arora, S., Y. Li, Y. Liang, T. Ma, and A. Risteski. (2016). A latent variable model approach to PMI-based word embeddings. *Transactions of the Association for Computational Linguistics*, 4: 385–399. https://doi.org/10.1162/tacl_a_00106.

Arrieta, A.B., N. Díaz-Rodríguez, J. Del Ser, A. Bennetot, S. Tabik, A. Barbado, S. Garcia et al. (2020). Explainable Artificial Intelligence (XAI): Concepts, taxonomies, opportunities, and challenges towards responsible AI. *Information Fusion*, 58: 82–115. https://doi.org/10.1016/j.inffus.2019.12.012.

Ascensão, F., Y.G. Gomes Ribeiro, Z. Campos, D.R. Yogui, and A.L.J. Desbiez. (2022). Forecasting seasonal peaks in roadkill patterns for improving road management. *Journal of Environmental Management*, 321: 115903. https://doi.org/10.1016/j.jenvman.2022.115903.

Bagus, P., J.A. Peña-Ramos, and A. Sánchez-Bayón. (2021). COVID-19 and the political economy of mass hysteria. *International Journal of Environmental Research and Public Health*, 18(4): 1376. https://doi.org/10.3390/ijerph18041376.

Barel, J.M., O.L. Petchey, A. Ghaffouli, and V.E.J. Jassey. (2023). Uncovering microbial food webs using machine learning. *Soil Biology and Biochemistry*, 186: 109174. https://doi.org/10.1016/j.soilbio.2023.109174.

Barman, R.K., S. Saha, and S. Das. (2014). Prediction of interactions between viral and host proteins using supervised machine learning methods. *PLoS One*, 9(11): e112034. https://doi.org/10.1371/journal.pone.0112034.

Barroso-Bergada, D., A. Tamaddoni-Nezhad, D. Varghese, C. Vacher, N. Galic, V. Laval, F. Suffert, and D.A. Bohan. (2023). Unraveling the web of dark interactions: Explainable inference of the diversity of microbial interactions. *Advances in Ecological Research*, 68: 155–183. https://doi.org/10.1016/bs.aecr.2023.09.005.

Bataille, A., I.I. Levin, and E.H.R. Sari. (2017). Colonization of parasites and vectors. In: P. Parker (Ed.), *Disease Ecology: Social and Ecological Interactions in the Galapagos Islands*, pp. 45–79. Springer, Cham, Switzerland. https://doi.org/10.1007/978-3-319-65909-1_3.

Berkson, J. (1944). Application of the logistic function to bio-assay. *Journal of the American Statistical Association*, 39(227): 357–365. https://doi.org/10.2307/2280041.

Beschta, R.L., and W.J. Ripple. (2009). Large predators and trophic cascades in terrestrial ecosystems of the western United States. *Biological Conservation*, 142(11): 2401–2414. https://doi.org/10.1016/j.biocon.2009.06.015.

Bezuidenhout, J.D., and C.J. Stutterheim. (1980). A critical evaluation of the role played by the red-billed oxpecker Buphagus erythrorhynchus in the biological control of ticks. *Onderstepoort Journal of Veterinary Research*, 47(2): 51–75.

Bickel, P.J., and A. Chen. (2009). A nonparametric view of network models and Newman-Girvan and other modularities. In: *Proceedings of the National Academy of Sciences*, 106(50): 21068–21073. https://doi.org/10.1073/pnas.0907096106.

Bodner, K., C. Brimacombe, M.-J. Fortin, and P.K. Molnár. (2021). Why body size matters: How larger fish ontogeny shapes ecological network topology. *Oikos*, 2022(3): e08569. https://doi.org/10.1111/oik.08569.

Bogucki, R., M. Cygan, C.B. Khan, M. Klimek, J.K. Milczek, and M. Mucha. (2019). Applying deep learning to right whale photo identification. *Conservation Biology*, 33(3): 676–684. https://doi.org/10.1111/cobi.13226.

Borowiec, M.L., R.B. Dikow, P.B. Frandsen, A. McKeeken, G. Valentini, and A.E. White. (2022). Deep learning as a tool for ecology and evolution. *Methods in Ecology and Evolution*, 13(8): 1640–1660. https://doi.org/10.1111/2041-210X.13901.

Boser, B.E., I.M. Guyon, and V.N. Vapnik. (1992, July). A training algorithm for optimal margin classifiers. In: *Proceedings of the 5th Annual Workshop on Computational Learning Theory* (*COLT 92*), pp. 144–152. https://doi.org/10.1145/130385.130401.

Breiman, L. (2001). Random forests. *Machine Learning*, 45(1): 5–32. https://doi.org/10.1023/A:1010933404324.

Brooks, D.R., and D.A. McLennan. (1991). *Phylogeny, Ecology and Behavior*. Chicago University Press. Chicago, United States of America. 441 pp.

Browne, A.J., C.A. Guerra, R.V. Alves, V.M. da Costa, A.L. Wilson, D.M. Pigott, S.I. Hay et al. (2017). The contemporary distribution of *Trypanosoma cruzi* infection in humans, alternative hosts and vectors. *Scientific Data*, 4: 170050. https://doi.org/10.1038/sdata.2017.50.

Burkepile, D.E., and J.D. Parker. (2017). Recent advances in plant-herbivore interactions. *F1000 Research*, 6: 119. https://doi.org/10.12688/f1000research.10313.1.

Carlen, E.J., C.O. Estien, T. Caspi, D. Perkins, B.R. Goldstein, S.E.S. Kreling, Y. Hentati et al. (2024). A framework for contextualizing social-ecological biases in contributory science data. *People and Nature*, 6(2): 377–390. https://doi.org/10.1002/pan3.10592.

Casadio, R., P.L. Martelli, and C. Savojardo. (2022). Machine learning solutions for predicting protein-protein interactions. *WIREs Computational Molecular Science*, 12(6): e1618. https://doi.org/10.1002/wcms.1618.

Ceballos, G., and P.R. Ehrlich. (2023). Mutilation of the tree of life via mass extinction of animal genera. *Proceedings of the National Academy of Sciences*, 120(39): e2306987120. https://doi.org/10.1073/pnas.2306987120.

Chen, P., P. Swarup, W.M. Matkowski, A.W.K Kong, S. Han, Z. Zhang, and H. Rong. (2020). A study on giant panda recognition based on images of a large proportion of captive pandas. *Ecology and Evolution*, 10(7): 3561–3573. https://doi.org/10.1002/ece3.6152.

Chen, T., S. Kornblith, K. Swersky, M. Norouzi, and G.E. Hinton. (2020, December). Big self-supervised models are strong semi-supervised learners. In: *NIPS '20: Proceedings of the 34th International Conference on Neural Information Processing Systems*, 33: 22243–22255. Vancouver, Canada. https://dl.acm.org/doi/abs/10.5555/3495724.3497589.

Chen, X., J. Zhao, Y.H. Chen, W. Zhou, and A.C. Hughes. (2020). Automatic standardized processing and identification of tropical bat calls using deep learning approaches. *Biological Conservation*, 241: 108269. https://doi.org/10.1016/j.biocon.2019.108269.

Christin., S., E. Hervet, and N. Lecomte. (2019). Applications for deep learning in ecology. *Methods in Ecology and Evolution*, 10(10): 1632–1644. https://doi.org/10.1111/2041-210X.13256.

Clapham, M., E. Miller, M. Nguyen, and C.T. Darimont. (2020). Automated facial recognition for wildlife that lack unique markings: A deep learning approach for brown bears. *Ecology and Evolution*, 10(23): 12883–12892. https://doi.org/10.1002/ece3.6840.

Clark, B., and J.B. Foster. (2010). Marx's ecology in the 21st century. *World Review of Political Economy*, 1: 142–156.

Clark, P.J., and F.C. Evans. (1954). Distance to nearest neighbor as a measure of spatial relationships in populations. *Ecology*, 35(4): 445–453. https://doi.org/10.2307/1931034

Connell, J.H. (1961). The influence of interspecific competition and other factors on the distribution of barnacle *Chthalamus stellatus*. *Ecology*, 42(4): 710–723. https://doi.org/10.2307/1933500.

Cuesta-Astroz, Y., and G. Oliveira. (2018). Computational and experimental approaches to predict host-parasite protein-protein interactions. In: L. von Stechow, and A.S. Delgado (Eds.), *Computational Cell Biology: Methods and Protocols* (pp. 153–173). Humana Press. New York, United States of America. https://doi.org/10.1007/978-1-4939-8618-7_7.

Cui, G., C. Fang, and K. Han. (2012). Prediction of protein-protein interactions between viruses and humans by an SVM model. *BMC Bioinformatics*, 13(suppl. 7): S5. https://doi.org/10.1186/1471-2105-13-S7-S5.

Dallas, T.A., and D.J. Becker. (2020). Taxonomic resolution affects host-parasite model performance. *Parasitology*, 148(5): 584–590. https://doi.org/10.1017/S0031182020002371.

Darbro, J.M., P.H. Johnson, M.B. Thomas, S.A. Ritchie, B.H. Kay, and P.A. Ryan. (2012). Effects of *Beauveria bassiana* on survival, blood-feeding success, and fecundity of *Aedes aegypti* in laboratory and semi-field conditions. *The American Journal of Tropical Medicine and Hygiene*, 86(4): 656–664. https://doi.org/10.4269/ajtmh.2012.11-0455.

Darwin, C. (1859). *On the Origin of Species*. 1st edition. John Murray. London, United Kingdom, 502 pp.

Darwin, C. (1875). *Insectivorous Plants*. 1st Edition. John Murray. London, United Kingdom. 462 pp.

Desjardins-Proulx, P., I. Laigle, T. Poisot, and D. Gravel. (2017). Ecological interactions and the Netflix problem. *PeerJ*, 5: e3644. https://doi.org/10.7717/peerj.3644.

Deutsch, C.A., J.J. Tewksbury, M. Tigchelaar, D.S. Battisti, S.C. Merrill, R.B. Huey, and R.L. Naylor. (2018). Increase in crop losses to insect pests in a warming climate. *Science*, 361(6405): 916–919. https://doi.org/10.1126/science.aat3466.

Díaz, M. (1996). Food choice by seed-eating birds in relation to seed chemistry. *Comparative Biochemistry and Physiology Part A: Physiology*, 113(3): 239–246. https://doi.org/10.1016/0300-9629(95)02093-4.

Dickman, C.R. (1992). Commensal and mutualistic interactions among terrestrial vertebrates. *Trends in Ecology and Evolution*, 7(6): 194–197. https://doi.org/10.1016/0169-5347(92)90072-J.

Diniz-Filho, J.A.F., M.T.P. Coelho, and L.C. Terribile, (2023). Unpacking underdetermination: Theoretical challenges in understanding macroecological and biogeographic patterns. *Journal of Biogeography*, 50(11): 1890–1898. https://doi.org/10.1111/jbi.14699.

Diniz-Filho, J.A.F., R.D. Loyola, P. Raia, A.O. Mooers, and L.M. Bini. (2013). Darwinian shortfalls in biodiversity conservation. *Trends in Ecology & Evolution*, 28(12): 689–695. https://doi.org/10.1016/j.tree.2013.09.003.

Dodds, W.K., and J.A. Nelson. (2006). Redefining the community: A species-based approach. *Oikos*, 112(2): 464–472. https://doi.org/10.1111/j.0030-1299.2006.13558.x.

Doherty, T.S., A.S. Glen, D.G. Nimmo, E.G. Ritchie, and C.R. Dickman. (2016). Invasive predators and global biodiversity loss. In: *Proceedings of the National Academy of Sciences*, 113(40): 11261–11265. https://doi.org/10.1073/pnas.1602480113.

Drimaj, J., V. Skoták, J. Kamler, R. Plhal, Z. Adamec, O. Mikulka, and P. Janata. (2023). Comparison of methods for estimating damage by wild ungulates on field crops. *Agriculture*, 13(6): 1184. https://doi.org/10.3390/agriculture13061184.

Dyer, M.D., T.M. Murali, and B.W. Sobral. (2011). Supervised learning and prediction of physical interactions between human and HIV proteins. *Infection, Genetics, and Evolution*, 11(5): 917–923. https://doi.org/10.1016/j.meegid.2011.02.022.

Elmasri, M., M.J. Farrell, T.J. Davies, and D.A. Stephens. (2020). A hierarchical Bayesian model for predicting ecological interactions using scaled evolutionary relationships. *The Annals of Applied Statistics*, 14(1): 221–240. https://doi.org/10.1214/19-AOAS1296.

Elton, C.S. (1927). *Animal Ecology*. Sidgwick & Jackson. London, United Kingdom, xxi+207 pp.

Enquist, B.J. (2002). Universal scaling in tree and vascular plant allometry: Toward a general quantitative theory linking plant form and function from cells to ecosystems. *Tree Physiology*, 22(15–16): 1045–1064. https://doi.org/10.1093/treephys/22.15-16.1045.

Escalas, A., L. Hale, J.W. Voordeckers, Y. Yang, M.K. Firestone, L. Alvarez-Cohen, and J. Zhou. (2016). Microbial functional diversity: From concepts to applications. *Ecology and Evolution*, 9(20): 12000–12016. https://doi.org/10.1002/ece3.5670.

Escudero-Pérez, B., A. Lalande, C. Mathieu, and P. Lawrence. (2023). Host-pathogen interactions influencing zoonotic spillover potential and transmission in humans. *Viruses*, 15(3): 599. https://doi.org/10.3390/v15030599.

Fanioudakis, E., M. Geismar, and I. Potamitis. (2018, September). Mosquito wingbeat analysis and classification using deep learning. In: *26th European Signal Processing Conference (EUSIPCO)*, pp. 2410–2414. Rome, Italy. https://doi.org/10.23919/EUSIPCO.2018.8553542.

Farrell, M.J., M. Elmasri, D.A. Stephens, and T.J. Davies. (2022). Predicting missing links in global host-parasite networks. *Journal of Animal Ecology*, 91(4): 715–726. https://doi.org/10.1111/1365-2656.13666.

Fix, E., and J.L. Hodges. (1989). Discriminatory analysis, non-parametric discrimination: Consistency properties. *International Statistical Review*, 57(3): 238–247. https://doi.org/10.2307/1403797.

Fögelstrom, E. M. Olofsson, D. Posledovich, C. Wiklund, J.P. Dahlgren, and J. Ehrlén. (2017). Plant-herbivore synchrony and selection on plant flowering phenology. *Ecology*, 98(3): 703–711. https://doi.org/10.1002/ecy.1676.

Friedman, J.H. (2001). Greedy function approximation: A gradient boosting machine. *The Annals of Statistics*, 29(5): 1189–1232. https://doi.org/10.1214/aos/1013203451.

Fritz, S.A., O.R.P. Bininda-Emonds, and A. Purvis. (2009). Geographical variation in predictors of mammalian extinction risk: Big is bad, but only in the tropics. *Ecology Letters*, 12(6): 538–549. https://doi.org/10.1111/j.1461-0248.2009.01307.x.

Gaba, S., C. Collas, T. Pobolny, F. Bretagnolle, and V. Bretagnolle. (2014). Skylarks trade size and energy content in weed seeds to maximize total ingested lipid biomass. *Behavioural Processes*, 108: 142–150. https://doi.org/10.1016/j.beproc.2014.10.004.

Gaba, S., P. Deroulers, F. Bretagnolle, and V. Bretagnolle. (2019). Lipid content drives weed seed consumption by ground beetles (Coleoptera, Carabidae) within the smallest seeds. *Weed Research*, 59(3): 170–179. https://doi.org/10.1111/wre.12354.

Gause, G.F. (1934). *The Struggle for Existence*. Williams and Wilkins Co. Baltimore, United States of America. 160 pp.

Gonzalez-Isunza, G., M. Zaki Jawaid, P. Liu, D.L. Cox, M. Vazquez, and J. Arsuaga. (2023). Using machine learning to detect coronaviruses potentially infectious to humans. *Scientific Reports*, 13: 9319. https://doi.org/10.1038/s41598-023-35861-7.

González-Suárez, M., F. Zanchetta Ferreira, and C. Grilo. (2018). Spatial and species-level predictions of road mortality risk using trait data. *Global Ecology and Biogeography*, 27(9): 1093–1105. https://doi.org/10.1111/geb.12769.

Gouda, S., G. Das, S.K. Sen, H.-S. Shin, and J.K. Patra. (2016). Endophytes: A treasure house of bioactive compounds of medicinal importance. *Frontiers in Microbiology*, 7: 01538. https://doi.org/10.3389/fmicb.2016.01538.

Grace, J. (1990). *Perspectives on Plant Competition*. Academic Press, Cambridge, United Kingdom. 498 pp.

Gravel, D., T. Poisot, C. Albuoy, L. Velez, and D. Mouillot. (2013). Inferring food web structure from predator-prey body size relationships. *Methods in Ecology and Evolution*, 4(11): 1083–1090. https://doi.org/10.1111/2041-210x.12103.

Grilo, C., E. Koroleva, R. Andrášik, M. Bíl, and M. González-Suárez. (2020). Roadkill risk and population vulnerability in European birds and mammals. *Frontiers in Ecology and the Environment*, 18(6): 323–328. https://doi.org/10.1002/fee.2216.

Guénard, G., P. Legendre, and P. Peres-Neto. (2013). Phylogenetic eigenvector maps: A framework to model and predict species traits. *Methods in Ecology and Evolution*, 4(12): 1120–1131. https://doi.org/10.1111/2041-210x.12111.

Ha, H., and F. Shilling. (2018). Modeling potential wildlife-vehicle collisions (WVC) locations using environmental factors and human population density: A case-study from 3 state highways in central California. *Ecological Informatics*, 43: 212–221. https://doi.org/10.1016/j.ecoinf.2017.10.005.

Hall, G. (1972). *Ecology: Can We Survive under Capitalism?* International Publishers. New York, United States of America. 94 pp.

Han, B.A., J.P. Schmidt, L.W. Alexander, S.E. Bowden, D.T.S. Hayman, and J.M. Drake. (2016). Undiscovered bat hosts of Filoviruses. *PLoS Neglected Tropical Diseases*, 10(7): e0004815. https://doi.org/10.1371/journal.pntd.0004815.

Han, B.A., J.P. Schmidt, S.E. Bowden, and J.M. Drake. (2015). Rodent reservoirs of future zoonotic diseases. *Ecology*, 112(22): 7039–7044. https://doi.org/10.1073/pnas.1501598112.

Hansen, M.E., M.L. Smith, L.N. Smith, M.G. Salter, E.M. Baxter, M. Farish, and B. Grieve, (2018). Towards on-farm pig face recognition using convolutional neural networks. *Computers in Industry*, 98: 145–152. https://doi.org/10.1016/j.compind.2018.02.016.

Haskell, E.F. (1946). A clarification of social science. *Main Currents in Modern Thought*, 7: 45–51.

He, B., Y. Zhao, and W. Mao. (2022). Explainable artificial intelligence reveals environmental constraints in seagrass distribution. *Ecological Indicators*, 144: 109523. https://doi.org/10.1016/j.ecolind.2022.109523.

He, K., X. Zhang, S. Ren, and S. Jun. (2016, June). Deep residual learning for image recognition. In: *2016 Proceedings of the 29th IEEE Conference on Computer Vision and Pattern Recognition* (*CVPR*), pp. 770–778. Las Vegas, United States of America. https://doi.org/10.1109/CVPR.2016.90.

Henderson, R.W., and R. Powell. (2001). Responses by the West Indian herpetofauna to human-influenced resources. *Caribbean Journal of Science*, 37(1–2): 41–54.

Heumann, B.W., S.J. Walsh, A.M. Verdery, P.M. McDaniel, and R.R. Rindfuss. (2012). Land suitability modeling using a geographic socio-environmental niche-based approach: A case study from northeastern Thailand. *Annals of the American Association of Geographers*, 103(4): 764–784. https://doi.org/10.1080/00045608.2012.702479.

Hirn, J., J.E. García, A. Montesinos-Navarro, R. Sánchez-Martín, V. Sanz, and M. Verdú. (2022). A deep Generative Artificial Intelligence system to predict species coexistence patterns. *Methods in Ecology and Evolution*, 13(5): 1052–1061. https://doi.org/10.1111/2041-210X.13827.

Ho, T.K. (1998). The random subspace method for constructing decision forests. *IEEE Transactions on Pattern Analysis and Machine Intelligence*, 20(8): 832–844. https://doi.org/10.1109/34.709601.

Hoff, P.D. (2005). Bilinear mixed-effects models for dyadic data. *Journal of the American Statistical Association*, 100(469): 286–295. https://doi.org/10.1198/016214504000001015.

Hoff, P.D., A.E. Raftery, and M.S. Handcock. (2002). Latent space approaches to social network analysis. *Journal of the American Statistical Association*, 97(460): 1090–1098. https://doi.org/10.1198/016214502388618906.

Holomuzki, J.R., J.W. Feminella, and M.E. Power. (2010). Biotic interactions in freshwater benthic habitats. *Journal of the North American Benthological Society*, 29(1): 220–244. https://doi.org/10.1899/08-044.1.

Hortal, J., F. de Bello, J.A.F. Diniz-Filho, T.M. Lewinsohn, J.M. Lobo, and R.J. Ladle. (2015). Seven shortfalls that beset large-scale knowledge of biodiversity. *Annual Review of Ecology, Evolution, and Systematics*, 46: 523–549. https://doi.org/10.1146/annurev-ecolsys-112414-054400.

Hughes, A.C., M.C. Orr, K. Ma, M.J. Costello, J. Waller, P. Provoost, Q. Yang et al. (2021). Sampling biases shape our view of the natural world. *Ecography*, 44(9): 1259–1269. https://doi.org/10.1111/ecog.05926.

Indumathi, V., S. Santhana Megala, R. Padmapriya, M. Suganya, and B. Jayanthi. (2021). Prediction and analysis of plant growth promoting bacteria using machine learning for millet crops. *Annals of the Romanian Society for Cell Biology*, 25(6): 1826–1833. http://annalsofrscb.ro/index.php/journal/article/view/5722.

John, M., F. Haselbeck, R. Das, C. Malisi, P. Ricca, C. Dreischer, S.J. Schultheiss, and D.G. Grimm. (2022). A comparison of classical and machine learning-based phenotype prediction methods on simulated data and three plant species. *Frontiers in Plant Science*, 13: 932512. https://doi.org/10.3389/fpls.2022.932512.

Kalle, R., L. Combrink, T. Ramesh, and C.T. Downs. (2017). Re-establishing the pecking order: Niche models reliably predict suitable habitats for the reintroduction of red-billed oxpeckers. *Ecology and Evolution*, 7(6): 1974–1983. https://doi.org/10.1002/ece3.2787.

Karp, D.S., C.D. Mendenhall, R.F. Sandí, N. Chaumont, P.R. Ehrlich, E.A. Hadly, and G.C. Daily. (2013). Forest bolsters bird abundance, pest control, and coffee yield. *Ecology Letters*, 16(11): 1339–1347. https://doi.org/10.1111/ele.12173.

Kitson, E., and C.A. Suttle. (2019). VHost-Classifier: Virus-host classification using natural language processing. *Bioinformatics*, 35(19): 3867–3869. https://doi.org/10.1093/bioinformatics/btz151.

Kleene, S.C. (1956). Representation of events in nerve nets and finite automata, *Automata Studies: Annals of Mathematics Studies*, 34: 3–41.

Kohonen, T. (1982). Self-organized formation of topologically correct feature maps. *Biological Cybernetics*, 43: 59–69. https://doi.org/10.1007/bf00337288.

Kotula, H.J., G. Peralta, C.M. Frost, J.H. Todd, and J.M. Tylianakis. (2021). Predicting direct and indirect non-target impacts of biocontrol agents using machine-learning approaches. *PLoS One*, 16(6): e0252448. https://doi.org/10.1371/journal.pone.0252448.

Kovač, M., N. Lacković, and M. Pernek. (2020). Effect of *Beauveria bassiana* fungal infection on survival and feeding behavior of pine-tree lappet moth (*Dendrolimus pini* L.). *Forests*, 11(9): 974. https://doi.org/10.3390/f11090974.

Krizhevsky, A., I. Sutskever, and G.E. Hinton. (2017). ImageNet classification with deep convolutional neural networks. *Communications of the ACM*, 60(6): 84–90. https://doi.org/10.1145/3065386.

Krupa, J.J., K.R. Hopper, S.B. Gruber, J.M. Schmidt, and J.D. Harwood. (2020). Plant-animal interactions between carnivorous plants, sheet-web spiders, and ground-running spiders as guild predators in a wet meadow community. *Ecology and Evolution*, 10(11): 4762–4772. https://doi.org/10.1002/ece3.6230.

Kulatunga, A.K., J. Gowrinathan, R. Ekanayake, D. Athauda, and C. Chandrakumar. (2015). Intelligent vision based driver assisted system for trains-elephants accidents. *International Journal of Electrical and Electronic Science*, 2(2): 6–16.

LeCun, Y., Y. Bengio, and G. Hinton. (2015). Deep learning. *Nature*, 521(7553): 436–444. https://doi.org/10.1038/nature14539.

Li, D., and A.R. Ives. (2017). The statistical need to include phylogeny in trait-based analyses of community composition. *Methods in Ecology and Evolution*, 8(10): 1192–1199. https://doi.org/10.1111/2041-210X.12767.

Lidicker Jr., W.Z. (1979). A clarification of interactions in ecological systems. *BioScience*, 29(8): 475–477. https://doi.org/10.2307/1307540.

Liu, W., D. Anguelov, D. Erhan, C. Szegedy, S. Reed, C.-Y. Fu, and A.C. Berg. (2016, October). SSD: Single Shot MultiBox Detector. In: B. Leibe, J. Matas, N. Sebe, and M. Welling (Eds.), *Computer Vision – ECCV 2016*, pp. 21–37. Springer. Cham, Switzerland. https://doi.org/10.1007/978-3-319-46448-0_2.

Llewelyn, J., G. Strona, C.R. Dickman, A.C. Greenville, G.M. Wardle, M.S.Y. Lee, S. Doherty et al. (2023). Predicting predator-prey interactions in terrestrial endotherms using random forest. *Ecography*, 2023(9): e06619. https://doi.org/10.1111/ecog.06619.

Lotka, A.J. (1925). *Elements of Physical Biology*. Williams and Wilkins Co. Baltimore, United States of America, xxx+460 pp.

Lundgren, E.J., J. Bergman, J. Trepel, E. Le Roux, S. Monsarrat, J.A. Kristensen, R.Ø. Pedersen et al. (2024). Functional traits—not nativeness–shape the effects of large mammalian herbivores on plant communities. *Science*, 383(6682): 531–537. https://doi.org/10.1126/science.adh2616.

Mac Aodha, O., R. Gibb, K.E. Barlow, E. Browning, M. Firman, R. Freeman, B. Harder et al. (2018). Bat detective – Deep learning tools for bat acoustic signal detection. *PLoS Computational Biology*, 14(3): e1005995. https://doi.org/10.1371/journal.pcbi.1005995.

Malviya, M., and R. Krishnamurthy. (2022). Multiscale spatially explicit modeling of livestock depredation by reintroduced tiger (*Panthera tigris*) to predict conflict risk probability. *Global Ecology and Conservation*, 40: e02313. https://doi.org/10.1016/j.gecco.2022.e02313.

Manson, S., M. Campera, K. Hedger, N. Ahmad, E. Adinda, V. Nijman, B. Budiadi et al. (2022). The effectiveness of a biopesticide in the reduction of coffee berry borers in coffee plants. *Crop Protection*, 161: 106075. https://doi.org/10.1016/j.cropro.2022.106075.

May, F., V. Grimm, and F. Jeltsch. (2009). Reversed effects of grazing on plant diversity: The role of belowground competition and size symmetry. *Oikos*, 118(12): 1830–1843. https://doi.org/10.1111/j.1600-0706.2009.17724.x.

McCulloch, W.S., and W. Pitts. (1943). A logical calculus of the ideas immanent in nervous activity. *Bulletin of Mathematical Biophysics*, 5: 115–133. https://doi.org/10.1007/BF02478259.

McElligott, A.G., I. Maggini, L. Hunziker, and B. König. (2004). Interactions between red-billed oxpeckers and black rhinos in captivity. *Zoo Biology*, 23(4): 347–354. https://doi.org/10.1002/zoo.20013.

Meadows, A.J., N. Stephenson, N.K. Madhav, and B. Oppenheim. (2023). Historical trends demonstrate a pattern of increasingly frequent and severe spillover events of high-consequence zoonotic viruses. *BMJ Global Health*, 8(11): e012026. https://doi.org/10.1136/bmjgh-2023-012026.

Megnidio-Tchoukouegno, M.M., E.B. Gueguim Kana, and W.B.A. Bancole. (2022). Machine learning model to predict endophytic colonisation of rice cultivar plant tissues by *Beauveria bassiana* isolates and their potential as bio-control agents against rice stem borer using existing knowledge. *F1000 Research*, 11: 1249. https://doi.org/10.12688/f1000research.126479.1.

Meineke, E.K., C. Thomasi, S. Yuan, and K.M. Pryer. (2020). Applying machine learning to investigate long-term insect-plant interactions preserved on digital herbarium specimens. *Applications in Plant Sciences*, 8(6): e11369. https://doi.org/10.1002/aps3.11369.

Metropolis, N., A.W. Rosenbluth, M.N. Rosenbluth, A.H. Teller, and E. Teller. (1953). Equation of state calculations by fast computing machines. *The Journal of Chemical Physics*, 21: 1087–1092. https://doi.org/10.1063/1.1699114.

Mihalca, A.D., C.M. Gherman, and V. Cozma. (2011). Coendangered hard ticks: Threatened or threatening? *Parasites & Vectors*, 4: 71. https://doi.org/10.1186/1756-3305-4-71.

Mikolov, T., K. Chen, G. Corrado, and J. Dean. (2013). Efficient estimation of word representations in vector space. arXiv Preprint. https://arxiv.org/abs/1301.3781.

Millar, M.C. (2019). Predicting theropod hunting tactics using machine learning. *Open Science Journal*, 4(1): 1–11. https://doi.org/10.23954/osj.v4i1.1820.

Mock, F., A. Viehweger, E. Barth, and M. Marz. (2021). VIDHOP, viral host prediction with deep learning. *Bioinformatics*, 37(3): 318–325. https://doi.org/10.1093/bioinformatics/btaa705.

Mooring, M.S., and P.J. Mundy. (1996). Interactions between impala and oxpeckers at Matobo National Park, Zimbabwe. *African Journal of Ecology*, 34(1): 54–65. https://doi.org.10.1111/j.1365-2028.1996.tb00594.x.

Morales-Castilla, I., M.G. Matias, D. Gravel, and M.B. Araújo. (2015). Inferring biotic interactions from proxies. *Trends in Ecology & Evolution*, 30(6): 347–356. https://doi.org/10.1016/j.tree.2015.03.014.

Mougi, A. (2016). The roles of amensalistic and commensalistic interactions in large ecological network stability. *Scientific Reports*, 6: 29929. https://doi.org/10.1038/srep29929.

Muleneh, M.G. (2021). Impact of climate change on biodiversity and food security: A global perspective: A review article. *Agriculture and Food Security*, 10: 36. https://doi.org/10.1186/s40066-021-00318-5.

Ofer, D., N. Brandes, and M. Linial. (2021). The language of proteins: NLP, machine learning & protein sequences. *Computational and Structural Biotechnology Journal*, 19: 1750–1758. https://doi.org/10.1016/j.csbj.2021.03.022.

Olson, S.L., and H.F. James. (1984). The role of Polynesians in the extinction of avifauna of the Hawaiian islands. In: P.S. Martin, and R.G. Klein (Eds.), *Quaternary Extinctions: A Prehistoric Revolution*, pp. 768–780. University of Arizona Press, Tucson, United States of America.

Oppel, S., and F. Huettmann. (2010). Using a Random Forest model and public data to predict the distribution of prey for marine wildlife management. In: S.A. Cushman, and F. Huettmann (Eds.), *Spatial Complexity, Informatics, and Wildlife Conservation*, pp. 151–163. Springer, Tokyo, Japan. https://doi.org/10.1007/978-4-431-87771-4_8.

Ornai, A., and T. Keasar. (2020). Floral complexity traits as predictors of plant-bee interactions in a Mediterranean pollination web. *Plants*, 9(11): 1432. https://doi.org/10.3390/plants9111432.

Page, R.D.M. (1993). Parasites, phylogeny, and cospeciation. *International Journal for Parasitology*, 23(4): 499–506. https://doi.org/10.1016/0020-7519(93)90039-2.

Pan, J., Z. Zhang, Y. Li, J. Yu, Z. You, C. Li, S. Wang et al. (2024). A microbial knowledge graph-based deep learning model for predicting candidate microbes for target hosts. *Briefings in Bioinformatics*, 25(3): bbae119. https://doi.org/10.1093/bib/bbae119.

Pauvert, C., M. Buée, V. Laval, V. Edel-Hermann, L. Fauchery, A. Gautier, I. Lesur et al. (2019). Bioinformatics matters: The accuracy of plant and soil fungal community data is highly dependent on the metabarcoding pipeline. *Fungal Ecology*, 41: 23–33. https://doi.org/10.1016/j.funeco.2019.03.005.

Pearse, A.S. (1926). *Animal Ecology*. McGraw-Hill Book Co. New York. United States of America, 417 pp.

Pearse, I.S., D.J. Harris, R. Karban, and A. Sih. (2013). Predicting novel herbivore-plant interactions. *Oikos*, 122(11): 1554–1564. https://doi.org/10.1111/j.1600-0706.2013.00527.x.

Peralta, G., D.P. Vázquez, N.P. Chacoff, S.B. Lomáscolo, G.L.W. Perry. and J.M. Tylianakis. (2020). Trait matching and phenological overlap increase the spatio-temporal stability and functionality of plant-pollinator interactions. *Ecology Letters*, 23(7): 1107–1116. https://doi.org/10.1111/ele.13510.

Perez-Guerra, J., J. Gonzalez-Velez, and J. Murillo-Escobar. (2024). Prediction of areas with high risk of roadkill wildlife applying maximum entropy approach and environmental features: East Antioquia, Colombia. *Landscape and Ecological Engineering*, 20(1): 75–88. https://doi.org/10.1007/s11355-023-00581-7.

Pérez-Harguindeguy, N., S. Díaz, F. Vendramini, J.H.C. Cornelissen, D.E. Gurvich, and M. Cabido. (2003). Leaf traits and herbivore selection in the field and in cafeteria experiments. *Austral Ecology*, 28(6): 642–650. https://doi.org/10.1046/j.1442-9993.2003.01321.x.

Perry, G.L.W., R. Seidl, A.M. Bellvé, and W. Rammer. (2022). An outlook for deep learning in ecosystem science. *Ecosystems*, 25(8): 1700–1718. https://doi.org/10.1007/s10021-022-00789-y.

Peters, R., Y. Lin, and U. Berger. (2016). Machine learning meets individual-based modeling: Self-organizing feature maps for the analysis of below-ground competition among plants. *Ecological Modelling*, 326: 142–151. https://doi.org/10.1016/j.ecolmodel.2015.10.014.

Phillips, S.J., and M. Dudík. (2008). Modeling of species distributions with Maxent: New extensions and a comprehensive evaluation. *Ecography*, 31(2): 161–175. https://doi.org/10.1111/j.0906-7590.2008.5203.x.

Pichler, M., and F. Hartig. (2023). Machine learning and deep learning: A review for ecologists. *Methods in Ecology and Evolution*, 14(4): 994–1016. https://doi.org/10.1111/2041-210X.14061.

Pichler, M., V. Boreux, A.-M. Klein, M. Schleuning, and F. Hartig. (2020). Machine learning algorithms to infer trait-matching and predict species interactions in ecological networks. *Methods in Ecology and Evolution*, 11(2): 281–293. https://doi.org/10.1111/2041-210X.13329.

Platenberg, R.J. (2007). Impacts of introduced species on an island ecosystem: non-native reptiles and amphibians in the US Virgin Islands. In: G.W. Witmer, W.C. Pitt, and K.A. Fagerstone (Eds.) *Managing Vertebrate Invasive Species*, pp. 168–174. USDA National Wildlife Research Center Symposia. Fort Collins, United States of America.

Pocock, M.J.O., R. Schmucki, and D.A. Bohan. (2021). Inferring species interactions from ecological survey data: A mechanistic approach to predict quantitative food webs of seed feeding by carabid beetles. *Ecology and Evolution*, 11(18): 12858–12871. https://doi.org/10.1002/ece3.8032.

Poelen, J.H., J.D. Simons, and C.J. Mungall. (2014). Global biotic interactions: An open infrastructure to share and analyze species-interaction datasets. *Ecological Informatics*, 24: 148–159. https://doi.org/10.1016/j.ecoinf.2014.08.005.

Portilla, M., R. Luttrell, G. Snodgrass, Y.C. Zhu, and E. Riddick. (2017). Lethality of the entomogenous fungus *Beauveria bassiana* strain NI8 on *Lygus lineolaris* (Hemiptera: Miridae) and its possible impact on beneficial Arthropods. *Journal of Entomological Science*, 52(4): 352–369. https://doi.org/10.18474/JES17-15.1.

Power, M.E. (1992). Top-down and bottom-up forces in food webs: Do plants have primacy? *Ecology*, 73(3): 733–746. https://doi.org.10.2307/1940153.

Puillandre, N., A. Lambert, S. Brouillet, and G. Achaz. (2012). ABGD, Automatic barcode gap discovery for primary species delimitation. *Molecular Ecology*, 21(8): 1864–1877. https://doi.org/10.1111/j.1365-294X.2011.05239.x.

Quiles, P., and R. Barrientos. (2024). Interspecific interactions disrupted by roads. *Biological Reviews*, 99(3), 1121–1139. https://doi.org/10.1111/brv.13061.

Quinlan, J.R. (1986). Induction of decision trees. *Machine Learning*, 1(1): 81–106. https://doi.org/10.1007/BF00116251.

Radzali, N.S.A., A. Abu Bakar, and A.I. Zamahsasri. (2023). Machine learning models for behavioral diversity of Asian Elephants prediction using satellite collar data. *Journal of Information and Communication Technology*, 22(3): 363–398. https://doi.org/10.32890/jict2023.22.3.3.

Rafferty, N.E., and A.R. Ives. (2013). Phylogenetic trait-based analyses of ecological networks. *Ecology*, 94(10): 2321–2333. https://doi.org/10.1890/12-1948.1.

Ramesh, G., S. Mathi, S.R. Pulari, and V. Krishnamoorthy. (2017, September). An automated vision-based method to detect elephants for mitigation of human-elephant conflicts. In: *2017 International Conference on Advances in Computing, Communications, and Informatics* (*ICACCI*), pp. 2284–2288. Udupi, India. https://doi.org/10.1109/ICACCI.2017.8126187.

Rammer, W., and R. Seidl. (2019). Harnessing deep learning in ecology: An example predicting bark beetle outbreaks. *Frontiers in Plant Sciences*, 10: 1327. https://doi.org/10.3389/fpls.2019.01327.

Rasmussen, S.L., C. Pertoldi, and D.W. Macdonald. (2022). Machine-learning prediction of hosts of novel coronaviruses requires caution as it may affect wildlife conservation. *Nature Communications*, 13: 5101. https://doi.org/10.1038/s41467-022-32746-7.

Rillig, M.C., I. Mansour, S. Hempel, M. Bi, B. König-Ries, and A. Kasirzadeh. (2024). How widespread use of generative AI for images and video can affect the environment and the science of ecology. *Ecology Letters*, 27(3): 14397. https://doi.org/10.1111/ele.14397.

Ryo, M., B. Angelov, S. Mammola, J.M. Kass, B.M. Benito, and F. Hartig. (2021). Explainable artificial intelligence enhances the ecological interpretability of black-box species distribution models. *Ecography*, 44(2): 199–205. https://doi.org/10.1111/ecog.05360.

Samadder, J., J.C. Das, D. Das, R. Sadhukhan, and A. Parvin. (2022, November). Smart IoT based early stage drowsy driver detection management system. In: *2022 IEEE International Conference of Electron Devices Society Kolkata Chapter* (*EDKCON*), pp. 353–356, Kolkata. https://doi.org/10.1109/EDKCON56221.2022.10032950.

Santamaría, L., and M.A. Rodríguez-Gironés. (2007). Linkage rules for plant-pollinator networks: Trait complementarity or exploitation barriers? *PLoS Biology*, 5(2): e31. https://doi.org/10.1371/journal.pbio.0050031.

Scheele, B.C., F. Pasmans, L.F. Skerratt, L. Berger, A. Martel, W. Beukema, A.A. Acevedo et al. (2019). Amphibian fungal panzootic causes catastrophic and ongoing loss of biodiversity. *Science*, 363(6434): 1459–1463. https://doi.org/10.1126/science.aav0379.

Schickhoff, U., M. Bobrowski, I.A. Offen, and S. Mal. (2024). The biodiversity crisis in the Anthropocene. In: B. Gönençgil and M.E. Meadows (Eds.), *Geography and the Anthropocene*, pp. 79–111. Istanbul University Press. Istanbul, Türkiye. https://doi.org/10.26650/B/SS19.2024.001.05.

Schnable, J.C. (2019). Genes and gene models, an important distinction. *New Phytologist*, 228(1): 50–55. https://doi.org/10.1111/nph.16011

Schofield, D., A. Nagrani, A. Zisserman, M. Hayashi, T. Matsuzawa, D. Biro, and S. Carvalho. (2019). Chimpanzee face recognition from videos in the wild using deep learning. *Science Advances*, 5(9): eaaw0736. https://doi.org/10.1126/sciadv.aaw0736.

Senthamil Selvi, M., S. Jansi Rani, T. Karthiga, P.L. Kamala, and V.R. Kanishka. (2020). Elephant intrusion detection and repulsion system using Matlab. *International Journal of Engineering and Advanced Technology*, 9(5): 37–39. https://doi.org/10.35940/ijeat.E9208.069520.

Sharma, B., and D. Neupane, D. (2023). Enhancing human-tiger coexistence in forest corridors of Nepal through a socio-ecological approach to conservation. *Trees, Forests, and People*, 13: 100402. https://doi.org/10.1016/j.tfp.2023.100402.

Škvorová, Z., I. Černajová, J. Steinová, O. Peksa, P. Moya, and P. Škaloud. (2022). Promiscuity in lichens follow clear rules: Partner switching in *Cladonia* is regulated by climatic factors and soil chemistry. *Frontiers in Microbiology*, 12: 781585. https://doi.org/10.3389/fmicb.2021.781585.

Soh, Y.H., L.R. Carrasco, D.G. Miquelle, J. Jiang, J. Yang, E.J. Stokes, J. Tang et al. (2014). Spatial correlates of livestock depredation by Amur tigers in Hunchun, China: Relevance of prey density and implications for protected area management. *Biological Conservation*, 169: 117–127. https://doi.org/10.1016/j.biocon.2013.10.011.

Steck, F., and A. Wandeler. (1980). The epidemiology of fox rabies in Europe. *Epidemiologic Reviews*, 2(1): 71–96. https://doi.org/10.1093/oxfordjournals.epirev.a036227.

Steinhaus, H. (1956). Sur la division des corps matériels en parties (in French): On the division of material bodies into parts. *Bulletin L'Académie Polonaise des Sciences*, 4(12): 801–804.

Steverding, D. (2020). The spreading of parasites by human migratory activities. *Virulence*, 11(1): 1177–1191. https://doi.org/10.1080/21505594.2020.1809963.

Strydom, T., M.D. Catchen, F. Banville, D. Caron, G. Dansereau, P. Desjardins-Proulx, N.R. Forero-Muñoz et al. (2021). A roadmap towards predicting species interaction networks (across space and time). *Philosophical Transactions of the Royal Society B*, 376: 20210063. https://doi.org/10.1098/rstb.2021.0063.

Sunquist, M.E. (1981). The social organization of tigers, Panthera tigris, in Royal Chitwan National Park, Nepal. *Smithsonian Contributions to Zoology*, 336: 1–98. https://doi.org/10.5479/si.00810282.336.

Tajudeen, Y.A., I.O. Oladunjoye, O. Bajinka, and H.J. Oladipo. (2022). Zoonotic spillover in an era of rapid deforestation of tropical areas and unprecedented wildlife trafficking: Into the wild. *Challenges*, 13(2): 41. https://doi.org/10.3390/challe13020041.

Talavera, G., V. Dincă, and R. Vila. (2013). Factors affecting species delimitations with the GMYC model: Insights from a butterfly survey. *Methods in Ecology and Evolution*, 4(12): 1101–1110. https://doi.org/10.1111/2031-210X.12107.

Thomas, M.B., and A.F. Read. (2007). Can fungal biopesticides control malaria? *Nature Reviews Microbiology*, 5: 377–383. https://doi.org/10.1038/nrmicro1638.

Uddin, S., and H. Lu. (2024). Confirming the statistically significant superiority of tree-based machine learning algorithms over their counterparts for tabular data. *PLoS One*, 19(4): e0301541. https://doi.org/10.1371/journal.pone.0301541.

Utgoff, P.E. (1989). Incremental induction of decision trees. *Machine Learning*, 4(2): 161–186. https://doi.org.10.1023/A:1022699900025.

van Beneden, P.-J. (1878). *Les commensaux et les parasites dans le règne animal* (in French): *Commensals and Parasites in the Animal Kingdom*. Librairie Germer Baillière. Paris, France, 238 pp.

Van Horn, G., O. Mac Aodha, Y. Song, Y. Cui, C. Sun, A. Shepard, H. Adam et al. (2018, June). The iNaturalist species classification and detection dataset. In: *Proceedings of the IEEE Conference on Computer Vision and Pattern Recognition* (*CVPR*) *2018*, pp. 8769–8778. Utah, United States of America.

Vapnik, V.N. (1997, October). The support vector method. In: *Artificial Neural Networks – ICANN '97*, pp. 261–271. https://doi.org/10.1007/BFb0020166.

Volterra, V. (1926). Variazioni e fluttuazioni del numero d'individui in specie animali conviventi (in Italian): Variations and fluctuations in the number of individuals in animal species living together. *Memoria della Reale Accademia Nazionale dei Lincei*, 2: 31–113.

Wang, D., S. Yang, F. Tang, and H. Zhu. (2012). Symbiosis specificity in the legume-rhizobial mutualism. *Cellular Microbiology*, 14(3): 334–342. https://doi.org/10.1111/j.1462-5822.2011.01736x.

Wang, Y., Q. Yao, J.T. Kwok, and L.M. Ni. (2020). Generalizing from a few examples: A survey on few-shot learning. *ACM Computing Surveys*, 53(3): 63. https://doi.org/10.1145/3386252.

Wardeh, M., M. Baylis, and M.S.C. Blagrove. (2021). Predicting mammalian hosts in which novel coronaviruses can be generated. *Nature Communications*, 12: 780. https://doi.org/10.1038/s41467-021-21034-5.

Weeks, P. (1999). Interactions between red-billed oxpeckers, *Buphagus erythrorhynchus*, and domestic cattle, *Bos taurus*, in Zimbabwe. *Animal Behaviour*, 58(6): 1253–1259. https://doi.org/10.1006/anbe.1999.1265.

Weeks, P. (2000). Red-billed oxpeckers: Vampires or tickbirds? *Behavioral Ecology*, 11(2): 154–160. https://doi.org/10.1093/beheco/11.2.154.

Weiskopf, S.R., M.A. Rubenstein, L.G. Crozier, S. Gaichas, R. Griffis, J.E. Halofsky, K.J.W. Hyde et al. (2020). Climate change effects on biodiversity, ecosystems, ecosystem services, and natural resource management in the United States. *Science of the Total Environment*, 733: 137782–137799. https://doi.org/10.1016/j.scitotenv.2020.137782.

Woinarski, J.C.Z., S.M. Legge, and S.T. Garnett. (2024). Extinct Australian birds: Numbers, characteristics, lessons and prospects. *Emu – Austral Ornithology*, 124(1): 8–20. https://doi.org/10.1080/01584197.2023.2240345.

Wu, Y., S.C.H. Barett, X. Duan, J. Zhang, Y. Cha, C. Tu, and Q. Li. (2021). Herbivore-mediated selection on floral display covaries nonlinearly with plant-antagonistic interaction intensity among primrose populations. *Frontiers in Plant Science*, 12: 727957. https://doi.org/10.3389/fpls.2021.727957.

Xu, B., Z. Tan, K. Li, T. Jiang, and Y. Peng. (2017). Predicting the host of influenza viruses based on the word vector. *PeerJ*, 5: e3579. https://doi.org/10.7717/peerj.3579.

Yahr, R., R. Vilgalys, and P.T. DePriest. (2006). Geographic variation in algal partners of *Cladonia subtenuis* (Cladoniaceae) highlights the dynamic nature of a lichen symbiosis. *New Phytologist*, 171(4): 847–860. https://doi.org/10.1111/j.1469-8137.2006.01792.x.

Yakimovich, A. (2021). Machine learning and artificial intelligence for the prediction of host-pathogen interactions: A viral case study. *Infection and Drug Resistance*, 14: 3319–3326. https://doi.org/10.2147/IDR.S292743.

Yang, Z., Y. Yu, C. You, J. Steinhardt, and Y. Ma. (2020, July). Rethinking bias-variance trade-off for generalization of neural networks. In: *Proceedings of the 37th International Conference on Machine Learning, Virtual, Proceedings of Machine Learning Research,* 119: 10767–10777.

Yu, L., D.K. Tanwar, E.D.S. Penha, Y.I. Wolf, E.V. Koonin, and M.K. Basu. (2019). Grammar of protein domain architectures. In: *Proceedings of the National Academy of Sciences*, 116(9): 3636–3645. https://doi.org/10.1073/pnas.1814684116.

Zhang, C., S. Bengio, M. Hardt, B. Recht, and O. Vinyals. (2021). Understanding deep learning (still) requires rethinking generalization. *Communications of the ACM*, 64(3): 107–115. https://doi.org/10.1145/3446776.

Zhang, J., P. Kapli, P. Pavlidis, and A. Stamatakis. (2013). A general species delimitation method with applications to phylogenetic placements. *Bioinformatics*, 29(22): 2869–2876. https://doi.org/10.1093/bioinformatics/btt499.

6 | Interactions Among Species in Ecological Communications

Chiraz Yemmen* **and Sami Fattouch**

Echo-chemistry Laboratory, Department of Biological and Chemical Engineering, National Institute of Applied Sciences and Technology, University of Carthage, Tunisia

The living world forms a vast network of interactions among organisms, essential for ecosystem functioning and biodiversity conservation. In an ecosystem, interactions—either between species (interspecific) or within the same species (intraspecific)—regulate population dynamics and abundance. These interactions vary by species, environment, and context, and are classified based on impact: harmful (e.g., predation, parasitism), neutral (e.g., commensalism), or mutually beneficial (e.g., mutualism). Ecological interactions drive ecological communication, where messages are transmitted to modify behaviors, using diverse signals like chemical, acoustic, visual, or mechanical cues. These communication forms support group dynamics, feeding, and defense, regulating ecosystems by balancing species' populations and access to resources. Human activities, such as deforestation, pollution, and climate change, disrupt these mechanisms, leading to ecological imbalances. Understanding these interactions and communication patterns is crucial for biodiversity conservation. Artificial intelligence (AI) aids this by analyzing large datasets, enabling species identification, movement tracking, sound analysis, and behavioral detection, thus enhancing insights into ecological communication and species interactions for sustainable management.

Introduction

Ecological interactions can be defined as the interactions between individuals and the populations of species. When individuals of one population interact with another population or individuals of another species, this population is called an interacting population, and the places where the individuals live and interact

*Corresponding author: chiraz_yemmen@yahoo.fr

are called their habitat. In nature, all populations of species that live together and interact with biotic and abiotic components of the environment form an ecological community. All biotic and abiotic components of the environment form an ecosystem. Ecological interactions, including competition, predation, and mutualism, have relative effects on the populations of species, and these interactions shape the evolution of organisms. Ecologists and scientists study these interactions and their relative effects to understand biodiversity and the relative abundance of species. Through the study of ecological interactions, we are able to understand the structure and complexity of natural ecosystems. Understanding the relationships and interactions is very important for the better survival and growth of species and populations because it can be an important factor in deciding the relative abundance and distribution of species.

Animal communication is an essential aspect of ecosystem functioning, playing a vital role in the survival, reproduction, and social dynamics of animal species. Ecological interactions, which shape the relationships between living organisms and their environments, are largely mediated by various forms of communication, ranging from visual and audio signals to chemical pheromones. This communication allows animals to coordinate their behaviors, transmit information about resources and dangers, seek partners for reproduction, and maintain effective social structures. Therefore, understanding the importance of animal communication in ecological interactions is crucial to understanding the complexity and diversity of life on Earth.

Types of Ecological Interactions

Individuals living in the same community are not isolated from each other. Thus, they enter into a relationship and this can bring them certain advantages, but also certain disadvantages. Several types of relationships can be established between individuals of the same species (intraspecific relationship) or between individuals of different species (interspecific relationship) (Table 1).

Table 1: The different types of interactions between the two species A and B

Interactions	Species A	Species B
Predation	+	–
Commensalism	+	0
Cooperation	+	+
Mutualism	+	+
Amensalism	–	0
Competition	–	–
Parasitism	+	–
Symbiosis	+	+

Notes: 0: Species are not affected
+: The interaction is beneficial (the life of the species is made possible or improved)
–: The interaction is harmful (the life of the species is reduced or impossible).

Symbiosis

Symbiosis is a biological interaction characterized by a long-term relationship between two organisms of different species that live together, often in close physical proximity. This relationship can be beneficial for both parties (mutualism), neutral for one and beneficial for the other (commensalism), or beneficial for one and harmful for the other (parasitism).

Commensalism

Commensalism (from the Latin cum-, 'with' and Mensa, 'table', for example, "dining companion" or "eating at the same table"). It is a type of ecological relationship between a commensal species which benefits and a host species which neither benefits nor harms it. It is distinct in this respect from mutualism and parasitism. This form of interaction is characterized by a one-sided benefit. The advantages or benefits brought by commensality (the state of commensal) concern food but also shelter and temporary transport. We can distinguish two main types depending on the degree of contact between the two species: the contact can be permanent and they are then "obligate commensals": this is the case of sessile animals called epibionts which live attached to other living beings, quite common in the marine environment. The contact is not permanent and we can speak of facultative commensalism, the most frequent case.

Several examples illustrate how commensal species can benefit from their interactions with other organisms without harming them. Among the commensals we can cite animals which settle and which are tolerated in the roosts of other species. Commensal insects in the burrows of mammals and birds, or in the nests of social insects are often very rich in species. Phoresis, that is to say the transport of the smallest organism by the largest and a form of commensalism. The transport of various species of mites by beetles such as Geotrupes is an example of phoresis. In the marine environment, commensalism exists between the polychaete Néreis fucata which lives in the shell inhabited by the pagan Eupagurus Prideauxi and which seizes the debris of its food between the mouthparts of the latter. An actinia Adamsia palliata is always associated with Eupagurus Prideauxi and it protects it thanks to its stinging filaments and by the secretion of a resistant membrane which extends the opening of the shell where it is housed.

Mutualism

By cleaning these parasites, cleaner fish provide an essential cleaning service to large marine predators, which benefit from healthier skin and more efficient gills. In return, cleaner fish get a constant food source and protection from predators, because larger fish are less likely to chase them while they are cleaning. In this example, both species benefit from the mutualistic relationship: cleaner fish get food and protection, while large marine predators benefit from body cleansing that promotes their overall health and well-being.

Parasitism

Parasitism is a form of symbiosis in which one organism, called a parasite, benefits from another living organism, called a host, to the detriment of the latter. This relationship is characterized by the fact that the parasite lives at the expense of the host, often by causing damage or taking resources from it. The parasite depends on the host to survive and reproduce. It can live inside or outside the host, depending on the species and environmental conditions. The parasite takes resources from the host for its own benefit. These resources may include food, water, nutrients, or other elements essential for the parasite's survival and reproduction. The parasite can have harmful effects on its host. This can manifest as illness, a reduction in physical condition, a reduction in fertility, or even the death of the host in the most severe cases. Parasites often have specific adaptations that allow them to exploit their host efficiently. This may include specialized anatomical structures (hooks, suction cups...), complex life cycles, sophisticated means of transmission, or mechanisms to avoid detection or host immune response. Parasites can be classified into different categories based on their lifestyle and their relationship with the host. This includes ectoparasites (which live on the surface of the host) and endoparasites (which live inside the host), as well as temporary parasites and permanent parasites. Parasitism plays an important role in biodiversity by influencing interactions between species within ecosystems. Indeed, parasites can help regulate species populations by affecting the survival, reproduction and health of their hosts. By controlling host populations, parasites can impact the structure and dynamics of biological communities (Hatcher et al., 2006).

Predation

Predation is an ecological interaction in which one organism, called a predator, hunts, kills, and feeds on another organism, called prey. This interaction is beneficial for the predator, but harmful for the prey (interaction +/-). The predation is widespread in ecosystems and plays a crucial role in regulating species populations and maintaining ecological balance.

Adaptations

Predators often have specific adaptations for hunting their prey, such as claws, fangs, highly developed vision or hearing, or specialized hunting techniques. For their part, prey often develop defense mechanisms such as camouflage, speed, or the production of toxins to avoid being captured. Mimicry is one of the means of defense which consists of being camouflaged to go unnoticed or making itself visible to resemble dangerous or inedible prey.

Population Regulation

The Influence of Predators on Prey

Predation plays an important role in population regulation, as it can influence the size of prey populations through several mechanisms, which can be direct or indirect. Here are some of the main effects of predation on prey:

- **Reduction in the size of prey populations:** The most obvious effect of predation is the direct reduction of the prey population. Predators hunt and consume prey, thereby reducing their numbers. This reduction in population sizes can lead to changes in prey population dynamics, affecting resource availability and competition between remaining individuals. Among the best-known examples of the influence of predators on prey is the oldest example relating to the lynx and the snow hare, whose populations exhibit regular fluctuations in abundance (Stenseth, 1997).
- **Trait selection:** Predation can exert selection pressure on prey populations, favoring individuals with certain traits or behaviors that enhance their ability to escape predation. For example, prey that are faster, more agile, or have effective defense mechanisms are more likely to survive and reproduce, passing these traits on to their offspring.
- **Indirect cascading effects:** Predators can also have cascading effects on prey by altering the structure and composition of communities of organisms. For example, predation on one prey species can lead to an increase in the population of its predators, which can in turn put increased pressure on other prey species in the ecosystem.
- **Effects on demographics and population dynamics:** Predation can influence the demographics of prey populations by affecting the survival, growth and reproduction rates of individuals. For example, intense predation can reduce prey reproductive rates, thereby delaying the recovery of populations after disturbance.

The Influence of Prey on Predators

The influence of prey on predators is an essential aspect of population dynamics and ecological interactions. Here are some ways prey can influence predators:

- **Availability of food resources:** The availability and abundance of prey can directly influence the survival, growth and reproduction of predators. If prey populations are abundant, predators are more likely to find food and can therefore maintain or increase their own populations.
- **Satiation effect:** When prey is abundant, predators can reach a state of satiation where they are no longer actively hunting. This can reduce predation pressure on prey populations and allow them to recover.
- **Triggering reproduction:** In some predator species, the availability of

prey can trigger reproductive behaviors. For example, an increase in prey availability may stimulate predator reproduction by providing additional food resources for young.

- **Influence on predation behaviors:** The availability and behavior of prey can also influence the hunting strategies and feeding habits of predators. For example, if prey becomes scarcer, predators may change their hunting techniques or look for food alternatives. This phenomenon is known as switching and has been observed in insects, fish and birds (Piltz et al., 2014).

Cooperation

Cooperation appears when two species form an association which is not essential since each can live in isolation. Both cooperation and mutualism provide benefits to both species (+/+). Collective nesting of several species of birds such as terns and herons is a form of cooperation that allows them to defend themselves more effectively against predators. Cooperative interactions are not limited to relationships between different species. Within the same species, individuals can also cooperate to achieve a common goal. There are numerous examples of cooperation between individuals of the same species. In general, these are single individuals from the same family group who help a couple raise their young. In vertebrates this is known in mammals, birds and fish. The California woodpecker, *Melanerpes formicivorus*, lives in groups of around 15 birds including several breeding individuals of each sex and non-breeders who are young brothers and sisters born in previous years. Acorns and other hard seeds are stored by woodpeckers in holes in the bark of granary trees and serve as a reserve for the winter. The group defends a common territory and collectively feeds the young people breeding individuals share a common nest (Koenig & Waltres, 2015).

Competition

Competition is an ecological interaction where two or more organisms compete for the same limited resources, such as food, space, water, sunlight, or reproductive partners. This competition can occur at different levels, including between individuals of the same species (intraspecific competition) or between individuals of different species (interspecific competition) (Table 2).

Intraspecific Competition

Intraspecific competition in animals occurs when individuals of the same species compete for the same limited resources within their population. Here are some examples of intraspecific competition in animals:

- **Competition for food:** Individuals of the same species may compete for available food resources. Members of the same family or social group may fight over food. For example, young chicks in a nest may fight to get food

Table 2: Key differences between intraspecific and interspecific competition

Feature	Intraspecific Competition	Interspecific Competition
Competitors	Same species	Different species
Common Resources Competed for	Food, space, breeding partners, territorial resources.	Food, water, space, nesting sites, breeding partners.
Impact on Individuals	Can affect survival rates, growth, and reproductive success of individuals.	Can lead to decreased availability of resources for less competitive species, impacting their survival and reproduction.
Impact on Population Dynamics	Influences population density, growth rates, and age structure within the species.	Affects population sizes, distribution, and diversity of species within an ecosystem.
Role in Evolution	Promotes natural selection, leading to better adaptations and genetic diversity within the species.	Can drive adaptive changes, niche differentiation, and speciation over time.
Long-term Ecological Impact	Helps maintain population control and genetic health of species.	Contributes to species diversity, ecological balance, and resource allocation in ecosystems.

from their parents. In populations of predators, such as lions, individuals may compete for captured prey. The strongest or most aggressive individuals may get a disproportionate share of the food.

- **Competition for space:** Animals may fight for access to breeding, hunting, or resting territories. For example, males of a lizard species may compete for control of a breeding territory during mating season.
- **Competition for breeding partners:** In many species, males may compete to attract breeding partners or to gain access to breeding females. For example, male deer may compete in ritual combat to determine breeding rights with females.
- **Competition for territorial resources:** Animals may establish territories for feeding, breeding, or resting, and competition for these territories can be intense. For example, songbirds often defend territories to attract mates and raise offspring.

In all of these examples, intraspecific competition can have a significant impact on survival, growth, reproduction, and population dynamics within the species. It may also play an important role in maintaining genetic diversity and adaptations within the population.

Interspecific Competition

Interspecific competition occurs when organisms from different species compete for the same limited resources in a given environment. This competition can be an important factor in the regulation of populations, species distribution and biological diversity in ecosystems. Here are some key points about interspecific competition:

- **Competition for resources:** Limited resources that species can compete for include food, water, space, nesting sites, breeding partners, and other essentials for survival and reproduction. For example, different bird species may compete for the same nesting sites or food resources.
- **Effects on population dynamics:** Interspecific competition can influence the population sizes of the species involved. If a species is more competitive for a given resource, it may have an advantage in accessing that resource, which may limit the population growth of other competing species.
- **Competitive coexistence:** In some cases, species can coexist despite interspecific competition for resources. This phenomenon may be facilitated by resource partitioning (use of different parts of a resource), differences in the ecological niches occupied by each species, or moderate competitive behaviors that minimize conflict.
- **Adaptations to interspecific competition:** Species can evolve adaptations that give them a competitive advantage in their environment. For example, physiological, behavioral, or morphological adaptations can help a species exploit a particular resource more efficiently, escape predation by competitors, or occupy specific ecological niches.
- **Effects on community structure:** Interspecific competition can influence the structure of ecological communities by determining which organisms are present in an ecosystem and in what numbers. It can also affect the spatial distribution of species and the diversity of communities.

Ecological Communications

Ecological interactions are closely linked to ecological communication, influencing emitted signals, organismal responses, information transmission mechanisms, behavioral and evolutionary strategies, as well as population and community dynamics in ecosystems. Animals communicate in different ways depending on the type of interaction they are involved in. This communication can include a range of visual, vocal, chemical and tactile signals to convey information about social status, intentions, needs and available resources.

Different Types of Ecological Communication

Visual Communication

Visual signals are often used to convey information about an animal's behavior, presence, and intent. This can include body postures, movements, colored patterns, light signals and gestures (Osorio & Vorobyev, 2008). Examples: mating dances in birds, threatening postures in reptiles, courtship displays in mammals.

Auditory Communication

Sound signals are commonly used by many animals to communicate over long distances. This may include songs, cries, mating songs, territorial calls, alarm signals, and social vocalizations. Examples: bird song, lions roaring, frogs croaking.

Chemical Communication

Pheromones and other chemicals are used to convey information about identity, social status, reproduction, territoriality, resource availability, and potential dangers (Wyatt, 2014; Surov & Maltsev, 2016). Examples: sex pheromones, alarm pheromones, territorial markings, recognition signals between members of the same species.

Tactile Communication

Physical contact and tactile signals are used to convey information about social status, intentions, reproduction, and social interactions. Examples: caresses, blows, parade contacts.

Electric and Magnetic Communication

Some animals, such as electric fish and migratory birds, can use electrical or magnetic signals to navigate, communicate, and sense their surroundings. Examples: the electrical discharges of electric fish for communication and detection of prey (Hopkins, 2009), the magnetic sensitivity of migratory birds for navigation.

These different modes of communication are used by animals in a variety of ecological interactions, including predation, competition, cooperation, reproduction, habitat selection, and navigation. They allow animals to adapt to their environment, respond to ecological challenges, and maximize their success in survival and reproduction.

The Role of Animal Communication in Ecological Interaction

Predator-Prey Communication

Communication between predators and prey is a crucial aspect of ecological interactions and can take many forms. Predator-prey communication may seem at first an unlikely approach due to the fact that the two interacting parties share few interests. In fact, nothing prohibits, on the contrary, each part uses the signals produced by the other in order to increase his chances of success.

An example of this situation is provided by 'stotting', a series of classic leaps in Tomson's gazelle *Gazella thomsoni* and other small species of antelopes (Caro, 1986). When a potential predator, a lion for example, approaches a herd of gazelles, all potential prey begins to jump on the spot and without moving away, even though flight would seem to be the most advantageous action. Several explanations have been put forward to explain this surprising behavior. The 'stotting' function which seems currently the most probable is that it reflects the condition of individuals, given that only one individual in good health and in good physical condition can jump so high. This signage is advantageous for the individual who carries it out with vigor because in this way he signals to the predator a failure likely (or at least a high blow) if he decides to attack it, the 'stotting' remaining energetic, less costly to carry out for the prey than a chase with the predator. Another benefit for the one who performs 'stotting' well is to focus the attention of the predator on other individuals, those who have more difficulty to achieve this behavior. Finally, 'stotting' also provides information to the predator that can, for example, decide to give up to attack if all the prey's potential are in good condition.

Many species emit alarm calls when a predator approaches. The function of alarm calls is to signal the presence of the predator. In terms of sound, alarm calls are generally within narrow ranges of high frequencies and are remarkably similar in different species of small birds. For example, titmice *Poecile atricapillus* emits shrill cries to alert other members of the group to the presence of a raptor (Courter & Ritchison, 2010).

Some insect species use pheromones to signal the presence of predators. For example, processionary caterpillars release alarm pheromones to warn other members of their colony in case of danger (Fitzgerald, 2003). American tree frogs of the genus *Dendrobates* which contain very toxic alkaloids acting as nerve poison have very bright (red, yellow, green, blue) warning colors which signal to potential predators that they are not edible (Maan & Cummings, 2012).

Communication Strategies in Cooperative Interactions

Communication plays a key role in allowing group members to cooperate effectively, whether in hunting, foraging, breeding, or other cooperative activities. Several examples can be cited:

- **Bees in a hive:** One of the most famous forms of communication among bees is the "hive dance". When a bee discovers a food source, it returns to the hive and performs a specific dance to indicate the direction, distance and quality of the food source to other bees (Khan et al., 2021). The "round dance" is used to indicate that the food source is nearby, while the "figure eight dance" is used for more distant sources (Singla, 2020). Bees also use a variety of pheromones to communicate within the colony. For example, the queen emits pheromones which maintain the social cohesion of the colony and regulate the behavior of the workers (Gary, 2003). Worker bees also use pheromones to signal important information such as the presence of food, the location of egg-laying sites, or the need to defend the colony against intruders. Bees also communicate through vibrations and sounds. For example, when a bee discovers a food source and begins foraging, it can emit vibrations by shaking its body on the surface of flowers. These vibrations can inform other bees of the presence of food nearby. Bees can also communicate through touch. For example, when a worker bee needs more food for the larvae, it can solicit a nurse bee by touching it with its antennae.

- **Pack hunting among wolves:** Wolves often hunt in packs, and communication between pack members is essential for a successful hunt. Wolves communicate through vocalizations such as howls, barks and growls to coordinate their movements, surround their prey and work together to bring it down.
- **Group hunting in dolphins:** Dolphins are known for their ability to hunt in groups in a coordinated manner. While hunting, dolphins communicate with each other using whistles and clicks to coordinate their movements, surround the fish and capture them effectively.
- **Cooperation between cleaner birds and large animals:** Some birds, such as shrikes and magpies, clean parasites from large animals, such as herbivorous mammals. These birds often make specific calls to attract the attention of host animals and indicate that they are ready to perform cleaning.

Communication Strategies in Competitive Interactions

Animal communication during competition plays an important role in establishing dominance, access to resources, and conflict resolution. Here are some examples of animal species and how they communicate during competition:

- **Territorial bird songs:** Many species of birds sing to establish their territory and mark out their domain (Marler & Slabbekoorn, 2004). Territorial songs serve to warn other males to stay away and to attract females for breeding. For example, the melodious songs of nightingales and the loud cries of crows are used to defend territories and signal the presence of their holders.
- **Ritual fighting in deer:** Male deer use fighting rituals to establish dominance and gain access to resources such as females and breeding territories. Before physically fighting, deer may challenge each other by emitting growls, roars or intimidating postures.
- **Display of colors and plumages in birds and fish:** In many species of birds and fish, males display bright colors and extravagant plumages during the breeding season to attract mates and establish dominance over rivals. For example, peacocks fan out their magnificent feathers to impress females and ward off rivals (Loyau et al., 2005).

- **Courtship displays in birds of paradise:** Male birds of paradise perform elaborate courtship displays to seduce females and establish dominance over other males (Scholes & Laman, 2012). These displays often include dances, complex vocalizations and displays of colorful plumage to attract the attention of females and oust competitors.
- **Vocal combat in primates:** Primates, such as monkeys and great apes, use vocalizations to communicate during social interactions and competition. Screams, growls, and howls can be used to express dominance, demarcate territories, and resolve intraspecific conflicts (Slocombe & Zuberbühler, 2005).

Communication Dynamics in Host–Parasite Interactions

Communication between a host and its parasite can take many forms, ranging from subtle chemical interactions to more obvious behavioral responses (Thomas et al., 2005). Here are some examples of communication interactions between host and parasite:

- **Behavioral manipulation:** Some parasites have the ability to manipulate the behavior of their host to increase their own chances of survival and reproduction. For example, the parasite *Toxoplasma gondii*, which infects rodents, can alter the behavior of mice, making them bolder and less fearful of predators, increasing the chances that predators will eat the infected mice and allowing the parasite to complete its life cycle in their body (Webster et al., 2010).
- **Host immune response:** When a parasite infects a host, the host usually mounts an immune response to fight the infection. Some parasites have evolved to evade their host's immune response by developing strategies such as modifying their surface antigens or suppressing the host's immune response (Maizels & Yazdanbakhsh, 2003). This allows the parasite to survive and multiply within the host's body.
- **Chemical signals:** Parasites can emit chemical signals that influence the behavior or physiology of their host. For example, pheromones produced by some parasites can change the behavior of the host, attracting it to areas where the parasite can breed or spread more effectively (Lafferty & Shaw, 2013).
- **Host Tolerance:** In some cases, hosts may develop some tolerance or compatibility with their parasites. This may be due to coevolution between the host and parasite over time. For example, some species of fish and shellfish have developed tolerance to parasites by harboring cleaner species that feed on the parasites (Poulin & Grutter, 1996).
- **Competition between parasites:** Within a host's body, different parasites can interact with each other, sometimes competitively. For example, some parasites produce toxins or chemicals that can harm other parasites present in the same host, thereby promoting their own survival and reproduction (Benesh, 2009).

Animal Communication in Response to Environmental Disturbance

Environmental disturbances due to human action can have a significant impact on animal communication, because they modify the natural conditions in which animals evolve and use their signals to interact with their environment and their conspecifics (Rosenthal & Stuart-Fox, 2012).

- **Noise pollution:** Noise pollution resulting from human activities can have a significant impact on animal communication, masking sound signals used by animals for various functions such as foraging, reproduction, territory defense and predator avoidance (Khawar Balwan & Saba, 2021). Bird songs are essential for many species in finding mates, territorial defense, and coordinating social activities. However, noise pollution from road traffic, industrial areas or urban areas can mask these songs, reducing the birds' ability to hear each other and communicate effectively. For example, studies have shown that passerine songs are less frequent and less complex in noisy urban areas than in quieter areas.

  Marine mammals, such as whales and dolphins, use sounds to communicate, navigate, and sense their surroundings. Noise pollution caused by human activities, including maritime traffic, military sonars and oil exploitation, can disrupt their ability to hear essential signals, increasing the risk of collision with ships and leading to disorientation (Weilgart, 2007).

  Frogs, toads, and other amphibians use acoustic calls for mate finding and reproduction (Leary, 2009). However, noise pollution from roads, residential areas and industrial activities can disrupt their vocalizations and reduce their effectiveness in attracting mates. This can lead to reduced reproduction and loss of populations in affected areas. Many insects use sounds for reproduction, mate finding, and territory defense. For example, male cicadas produce shrill songs to attract females. However, noise pollution from human activities can mask these songs, compromising the insects' ability to find mates and reproduce successfully.
- **Light pollution:** Excessive artificial lighting in urban and peri-urban areas can disrupt natural light and dark cycles, affecting the behaviors of many animal species (Dominoni, 2017). Nocturnal animals, such as bats and some insects, can be disoriented by artificial lights, affecting their ability to feed, reproduce, and navigate their environment.
- **Deforestation and habitat fragmentation:** Deforestation and fragmentation of natural habitats caused by urban expansion, agriculture and logging reduce areas available for animals and fragment populations. This can disrupt social interactions and communication between individuals by limiting opportunities to meet, reproduce and forage. In fact, forests shelter a wide diversity of animal species that use sounds to communicate, whether to attract mates, mark territory, warn of danger or coordinate social activities (Penar et al., 2020).

Deforestation can modify the acoustic quality of the environment, thus altering the propagation of sound signals. For example, bird songs can be muffled or distorted by echoes in deforested environments, reducing their range and effectiveness in communicating with other individuals. Deforestation often leads to the fragmentation of natural habitats, dividing animal populations into small, isolated patches. This can reduce opportunities for individuals to meet and interact, limiting opportunities for communication and reproduction. Fragmented populations may also be more vulnerable to the effects of genetic isolation and randomness, which can lead to decreased genetic diversity and increased risk of local extinction.

- **Chemical pollution:** Agricultural pesticides, industrial chemicals and sewage spills can contaminate air, water and soil, affecting animal health and behavior. Toxic chemicals can alter the chemical signals animals use to communicate, disrupting their social interactions and their ability to detect mates, predators, and prey (Relić & Đukić-Stojčić, 2023). Pheromones are chemical substances used by many animals to communicate information such as the presence of a sexual partner, defense of territory or alarm in case of danger. Chemical pollution can alter the chemical composition of the air, reducing the ability of animals to detect and respond to pheromones. This can disrupt social interactions, including finding partners and coordinating group activities. Moreover, toxic chemicals in the environment, such as pesticides, heavy metals, and industrial chemicals, can have adverse effects on animal health (Vajargah et al., 2021). Exposure to these contaminants can impair the functioning of animals' sensory systems, compromising their ability to emit and detect communication signals.
- **Climate change:** Climate change can disrupt animal communication in several ways, altering habitats and changing weather and seasonal patterns (Bee et al., 2007; Patricelli et al., 2009). A concrete example would be songbirds, which use their songs for various reasons such as reproduction, territorial defense and social communication.
- **Habitat shift:** With climate change, bird habitats may shift (Sekercioglu et al., 2008). For example, areas that were previously too cold for some birds may become more welcoming due to warmer temperatures. This can disrupt migratory patterns and bird populations may move to new locations where acoustic signals may be different or less effective.
- **Alteration of seasons and reproductive cycles:** Changes in seasons and reproductive cycles can disrupt the synchronization of birds' reproductive signals (Both et al., 2004). For example, if spring comes earlier due to global warming, some birds might begin their territorial song before others are ready to listen or respond.
- **Altered food cycles:** Climate change can also disrupt birds' food cycles, forcing them to seek new food sources (Both et al., 2006). This can lead to changes in social and territorial interactions between species, which can in turn affect communication signals.

Climate change may also have an impact on animal chemical communication (Roggatz, et al., 2022). Chemical communication is essential for many species, including finding mates, establishing territories, finding food, and warning of danger. Here is an example of the impacts of climate change on chemical communication in certain insect species.

- **Disruption of pheromones:** Pheromones are chemical compounds that many insects use to communicate, particularly to attract sexual partners (Bontonou & Wicker-Thomas, 2014). Climate change may affect the production and perception of pheromones. For example, variations in temperature or humidity can alter the chemical composition of pheromones or their volatility, making it more difficult for insects to detect and respond to chemical signals.
- **Seasonal lag:** Climate change can alter seasons and biological cycles, which can lead to time lags between the release of pheromones and the period when they are detected by receiving individuals. For example, if plants flower earlier due to global warming, insects that rely on plant pheromones to find mates may become out of sync, compromising their reproductive success.
- **Alteration of predator–prey interactions:** Climate change can alter the geographic distribution of species, leading to unexpected encounters between predators and prey. In some cases, this can disrupt the chemical signals used by prey to avoid predators (Bretagnolle & Terraube, 2019). For example, if prey does not recognize the predator in its new habitat due to climate change, it may not emit appropriate chemical warning signals, decreasing its ability to avoid predators.
- **Changes in habitats:** Climate change can alter habitats and ecosystems, which in turn can affect the availability and quality of chemical signals. For example, deforestation or pollution can reduce the concentration of pheromones in the air or disrupt the mechanisms for transmitting chemical signals.

Artificial Intelligence: Enhancing Understanding of Animal Communication

By integrating artificial intelligence into the study of animal communication, researchers can gain new insights into animal behaviors, interactions, and social dynamics, contributing to a better understanding of biodiversity and conservation efforts (Tuia, 2022).

Analysis of Acoustic Signals

Machine learning and signal processing techniques are used to analyze animal vocalizations. Algorithms can be trained to detect, classify and interpret different types of sounds made by animals, helping researchers to understand communication patterns (Sharma et al., 2023). A concrete example of using artificial intelligence to analyze acoustic signals with the aim of understanding animal communication is the study of the vocalizations of cetaceans, such as whales as shown in Figure 1. Whales emit a wide variety of vocalizations,

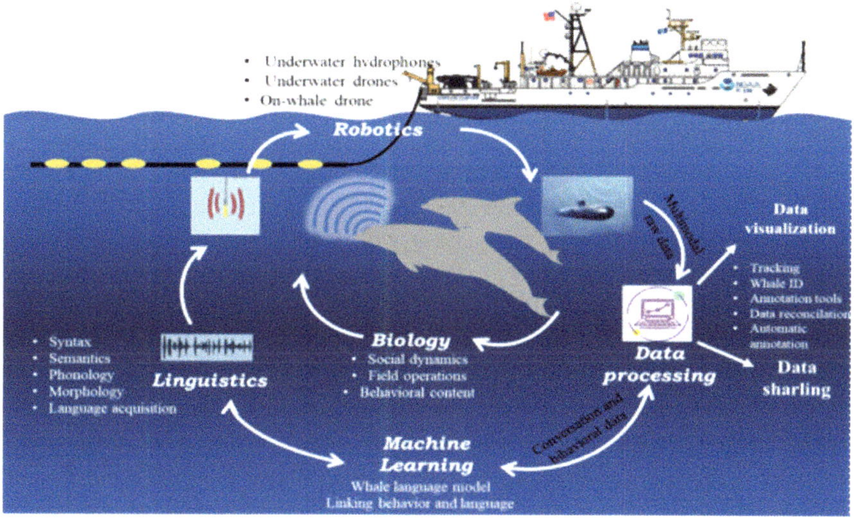

Figure 1: AI Integration in whale communication research. This figure details the comprehensive workflow for analyzing whale vocalizations: From underwater hydrophone data collection and signal processing to machine learning classification and behavioral interpretation.

including complex songs, sonar clicks used for tracking and communication, and grunts and moans. Understanding these vocalizations is crucial to better understand their social behavior, their ecology and their environment (Malige et al., 2022). Here is how artificial intelligence is used in this approach:

- **Data collection:** Underwater hydrophones are used to record the vocalizations of whales in their natural habitat. These recordings generate enormous amounts of audio data (Fregosi et al., 2020).
- **Signal processing:** Audio recordings are processed to extract relevant features from whale vocalizations. This may include detecting the different types of sounds emitted, measuring their frequency, amplitude and duration (Van Wyk et al., 2022).
- **Machine learning:** The processed data is then used to train machine learning algorithms, such as deep neural networks. These algorithms are trained to recognize and classify different whale vocalizations based on their acoustic characteristics.
- **Model analysis:** Once trained, machine learning models can be used to analyze new audio recordings and automatically identify whale vocalizations. This allows researchers to quantify communication patterns, study geographic and seasonal variations, and understand whales' behavioral responses to environmental changes (LeCun et al., 2015).
- **Interpretation of results:** Data analyzed by artificial intelligence provides valuable information on whale communication and behavior (Allen

et al., 2021). For example, researchers can identify the types of vocalizations associated with breeding, feeding, or other specific behaviors, as well as social interactions between individuals.

Animal Language Modeling

AI can be used to create models of animal language, identifying the syntactic and semantic structures of sound or chemical signals (Rutz et al., 2023). These patterns help decipher the meaning of vocalizations or pheromones and understand how they are used in different situations. A concrete example of modeling animal language with the help of artificial intelligence is the study of the birds' songs (Potamitis et al., 2014), such as the nightingale. Nightingales are known for their complex and melodious songs, which play a crucial role in social communication and reproduction. Here's how AI can be used to model their language:

- **Data collection:** Audio recordings of nightingale songs are collected in their natural habitat. These recordings contain a wide variety of vocalizations, including note sequences, trills, and specific patterns (Mehyadin et al., 2021).
- **Signal analysis:** Audio recordings are processed to extract relevant characteristics of nightingale songs, such as frequency, duration and intensity of notes. Signal processing techniques are used to identify recurring patterns and syntactic structures in speech sequences.
- **Machine learning:** The processed data is then used to train machine learning models, such as recurrent neural networks (RNN) or hidden Markov models (HMM). These models are trained to recognize the patterns of nightingale songs, by analyzing sequences of notes and learning the relationships between different parts of the song.
- **Language modeling:** Once trained, AI models can be used to generate new nightingale song sequences that realistically mimic the structures and patterns observed in real recordings (Weiss, 2014). These models also allow exploration of individual variation and geographic differences in nightingale songs, as well as listeners' behavioral responses to different song types.
- **Interpretation of results:** AI-generated language models provide valuable insights into the structure and function of nightingale songs. For example, researchers can identify the key elements that distinguish courtship songs from territorial songs, as well as the mechanisms underlying learning and cultural transmission of songs between individuals.

Behavior Detection

Machine learning algorithms can be trained to automatically detect and track animal behaviors from videos or audio recordings (Graving et al., 2019). This allows observation of social interactions, communication patterns, and behavioral responses to communication signals. The use of artificial intelligence in detecting animal behaviors has become a powerful method for analyzing and

Table 3: Different AI methodologies used across various ecological studies

AI Methodology	Application	Examples	Benefits
Machine Learning	Analyzing acoustic signals	Cetacean vocalizations, whale communication (Sharma et al., 2023)	• Detects and classifies vocalizations • Understands communication patterns and behaviors
Deep Neural Networks (DNNs)	Classification and pattern recognition	Whale vocalizations analysis (Van Wyk et al., 2022)	• Recognizes complex vocalization patterns • Quantifies communication and behavioral responses
Recurrent Neural Networks (RNNs)	Modeling animal language	Nightingale songs (Potamitis et al., 2014)	• Identifies syntactic structures • Generates realistic song sequences and explores variations
Hidden Markov Models (HMMs)	Analyzing sequential data	Nightingale song sequences (Mehyadin et al., 2021)	• Models song patterns and structures • Understands learning and cultural transmission mechanisms
Convolutional Neural Networks (CNNs)	Detecting and tracking behaviors	Primate social behaviors (Witham, 2018)	• Automatically detects specific behaviors • Tracks social interactions and behavioral responses
Drones and Automated Cameras	Wildlife monitoring and behavior detection	Endangered species tracking (Kiszka, 2016)	• Monitors and tracks animal movements • Provides real-time data for conservation efforts
Acoustic Sensors	Recording and analyzing vocalizations	Bird and marine mammal calls (Allen et al., 2021)	• Records and identifies species vocalizations • Analyzes communication and social interactions
High-resolution Cameras	Studying feeding and reproductive behaviors	Insect feeding behavior (Jun Tu et al., 2016)	• Captures detailed interactions • Quantifies feeding and reproductive activities
Behavior Detection Algorithms	Tracking and observing social interactions	Fish reproductive behaviors (Abangan et al., 2023)	• Detects courtship, egg-laying, and parental care • Provides insights into reproductive behaviors

understanding interactions and behavioral patterns in various environments. Here are some examples of AI application in this area.

- **Detection of social behaviors in primates:** Researchers have used convolutional neural networks (CNN) to analyze videos of wild animals, such as monkeys (Witham, 2017). These networks are trained to automatically detect specific behaviors, such as playing, eating, grooming, etc.
  AI can also be used to automatically track social interactions between individuals, such as physical contact or exchange of visual signals.
- **Wildlife monitoring for conservation:** In conservation programs, AI systems are used to monitor populations of endangered species. Drones equipped with cameras and powered by machine learning algorithms are capable of detecting and tracking the movements of animals in their natural habitat (Kiszka, 2016). Acoustic sensors can also be used to record the vocalizations of animals such as birds or marine mammals, and AI algorithms can be employed to automatically identify species based on their vocalizations.
- **Study of feeding behavior in insects:** Researchers use machine learning techniques to analyze the feeding behaviors of insects, such as bees or ants (Jun Tu et al., 2016). High-resolution cameras are used to capture insect movements as they interact with food, and AI algorithms are used to detect and quantify these behaviors.
- **Detection of reproductive behaviors in fish:** AI systems are used to study the reproductive behaviors of fish in aquatic environments (Abangan et al., 2023). Underwater cameras are used to record interactions between fish during the spawning period, and AI algorithms are employed to detect courtship, egg-laying and parental care behaviors.

Conclusion and Future Directions

AI offers transformative benefits to animal ecology by making research processes faster, more accurate, and accessible to a broader audience. With rapid data processing and automation, AI significantly reduces the time needed for tasks like species identification and population monitoring, allowing researchers to delve into data interpretation and strategic insights. Its precision-driven algorithms excel at detecting subtle patterns and analyzing data across multiple scales, providing a more comprehensive and nuanced view of ecosystem dynamics. Furthermore, AI-driven platforms enhance data sharing across scientific, policy-making, and public domains, empowering citizen scientists through mobile applications that enable biodiversity monitoring and increase conservation awareness. By enabling real-time threat alerts and optimizing resources for priority species and regions, AI supports adaptive conservation management, allowing swift responses to emerging environmental challenges. Future advancements in AI not only promise to deepen our understanding of ecosystems but also to empower targeted conservation strategies worldwide, strengthening biodiversity and ecosystem resilience for generations to come.

References

Abangan, A.S., D. Kopp, and R. Faillettaz. (2023). Artificial intelligence for fish behavior recognition may unlock fishing gear selectivity. *Front. Mar. Sci.,* 10: 101–161.

Allen, A.N., M. Harvey, L. Harrell, A. Jansen, K.P. Merkens, C.C. Wall, J. Cattiau, and E.M. Oleson. (2021). A convolutional neural network for automated detection of humpback whale song in a diverse, long-term passive acoustic dataset. *Front. Mar. Sci.,* 8: 1–12.

Balwan, W.K., and N. Saba. (2021). Impact of sound pollution on animal and human health. *IJBI,* 3(1): 68–73.

Bee, M.A., and E.M. Swanson. (2007). Auditory masking of anuran advertisement calls by road traffic noise. *Anim. Behav.,* 74(6): 1765–1776.

Benesh, D.P. (2009). New perspectives on parasite coexistence: The role of trade-offs. *Ann. Zool. Fenn.,* 46(1): 71–85.

Bontonou, G., and C. Wicker-Thomas. (2014). Sexual communication in the Drosophila Genus. *J. Insects,* 5: 439–458.

Both, C.A., V. Artemyev, B. Blaauw, R.J Cowie, A.J. Dekhuijzen, T. Eeva, and A. Järvinen. (2004). Large-scale geographical variation confirms that climate change causes birds to lay earlier. *Proceedings of the Royal Society of London. Series B. Biol. Sci.,* 271(1549): 1657–1662.

Both, C.S., Bouwhuis, C.M. Lessells, and M.E. Visser. (2006). Climate change and population declines in a long-distance migratory bird. *Nat.,* 441(7089): 81–83.

Bretagnolle, V., and J. Terraube. (2019). Predator–prey interactions and climate change. In: Peter O. Dunn, and Anders Pape Møller (Eds.), *Effects of Climate Change on Birds*, 2nd Edn., Oxford online Edn., *Oxf. J.,* 19: 199–220.

Caro, T.M. (1986). The functions of stotting in Thomson's gazelles: Some tests of the predictions. *Anim. Behav.,* 34: 663–684.

Courter, J.R., and G. Ritchison. (2010). Alarm calls of tufted titmice convey information about predator size and threat. *Behav. Ecol.,* 21(5): 936–942.

Dominoni, D.M. (2017). Ecological effects of light pollution: how can we improve our understanding using light loggers on individual animals? In: Murgui, E., and Hedblom, M. (Eds.), *Ecology and Conservation of Birds in Urban Environments*, pp. 251–270. Springer, Cham.

Fitzgerald, T. (2003). Role of trail pheromone in foraging and processionary behavior of pine processionary caterpillars Thaumetopoea pityocampa. *J. Chem. Ecol.,* 29(3): 513–532.

Fregosi, S., D.V. Harris, H. Matsumoto, D.K. Mellinger, J. Barlow, S. Baumann-Pickering, and H. Klinck. (2020). Detections of Whale vocalizations by simultaneously deployed bottom-moored and deep-water mobile autonomous hydrophones. *Front. Mar. Sci.,* 7(721): 1–18.

Gary, N.E. (2003). Pheromones of social bees. In: Gary, N.E. (Ed.), *Chemical Ecology of Insects 2*, pp. 225–236. Academic Press.

Graving, J.M., D. Chae, H. Naik, L. Li, B. Koger, B.R. Costelloe, and I.D. Couzin. (2019). DeepPoseKit, a software toolkit for fast and robust animal pose estimation using deep learning. *Elife,* 1: 1–42.

Hatcher, M.J., J.T.A. Dick, and A.M. Dunn. (2006). How parasites affect interactions between competitors and predators. *Ecol. Lett.,* 1253–1271.

Hopkins, C.D. (2009). Electrical perception and communication. *Encyclopedia of Neuroscience*, 3: 813–831. Academic Press.

Jun Tu, G., M.K. Hansen, P. Kryger, and P. Ahrendt. (2016). Automatic behavior analysis system for honeybees using computer vision. *Comput. Electron. Agric.,* 122: 10–18.

Khan, M.L., N.I. Sulehri, M. Talha, A. Nazar, and H.G. Ahmed. (2021). Honey bees show dance patterns to communicate: A review. *IJBBKU,* 6(2): 15–19.

Kiszka, J.J., J. Mourier, K. Gastrich, and M.R. Heithaus. (2016). Using unmanned aerial vehicles (UAVs) to investigate shark and ray densities in a shallow coral lagoon. *Mar. Ecol. Prog. Ser.,* 56: 237–242.

Koenig, L., and D.E. Walters. (2015). What we don't know, and what needs to be known, about the cooperatively breeding acorn woodpecker Melanerpes formicivorus. *Acta Ornithol.,* 49(2): 221–223.

Lafferty, K.D., and J.C Shaw. (2013). Comparing mechanisms of host manipulation across host and parasite taxa. *Ecol. Inform.,* 216(1): 56–66.

Leary, C.J. (2009). Hormones and acoustic communication in anuran amphibians. *ICB,* 49(4): 452–470.

LeCun, Y., Y. Bengio, and G. Hinton. (2015). Deep learning. *Nature,* 521*:* 436–444.

Loyau, A.M., Saint Jalme, C. Cagniant, and G. Sorci. (2005). Multiple sexual advertisements honestly reflect health status in peacocks (*Pavo cristatus*). *Behav. Ecol. Sociobiol.,* 58(6): 552–557.

Maan, M.E., and M.E. Cummings. (2012). Poison frog colors are honest signals of toxicity, particularly for bird predators. *Am. Nat.,* 179(1): 1–14.

Maizels, R.M., and M. Yazdanbakhsh. (2003). Immune regulation by helminth parasites: Cellular and molecular mechanisms. *Nat. Rev. Immunol.,* 3(9): 733–744.

Malige, F., J. Patris, M. Hauray, P. Giraudet, and H. Glotin. (2022). Mathematical models of long-term evolution of blue whale song types' frequencies. *J. Theor. Biol.,* 548: 111–184.

Marler, P., and H. Slabbekoorn. (2004). *Nature's Music: The Science of Birdsong*. Elsevier Academic Press, pp. 504.

Mehyadin, A.E., A.M. Abdulazeez, D.A. Hasan, and J.N. Saeed. (2021). Birds sound classification based on machine learning algorithms. *Asian J. Res. Comput. Sci.,* 9(4): 1–11.

Osorio, D.M., and Vorobyev. (2008). A review of the evolution of animal color vision and visual communication signals. *Vis. Res.,* 48(20): 2042–2051.

Patricelli, G.L., L. Jessica, and A. Blickley. (2006). Communication in urban noise: Causes and consequences of vocal adjustment. *The Auk.*, 123: 639–649.

Penar, W., A. Magieram, and C. Klocek. (2020). Applications of bioacoustics in animal ecology. *Ecol. Complex,* 3: 100–117.

Piltz, S., M.A. Porter, and P.K. Maini. (2014). Prey switching with a linear preference trade-of. *SIAM J. Appl. Dyn. Syst.,* 13(2): 658–682.

Potamitis, I., S. Ntalampiras, O. Jahn, and K. Riede. (2014). Automatic bird sound detection in long real-field recordings: Applications and tools. *Appl. Acoust.,* 80: 1–9.

Poulin, R., and A.S. Grutter. (1996). Cleaning symbioses: Proximate and adaptive explanations. *Biosci.,* 46(7): 512–517.

Relić, R., and M. Đukić-Stojčić. (2023). Influence of environmental pollution on animal behavior. *J. Cont. Agric.,* 72(4): 216–223.

Roggatz, C.C., S.M. Blanchard, P. Schirrmacher, P. Fink, F. Verheggen, and J.D. Hardege. (2022). Becoming nose-blind—Climate change impacts on chemical communication. *Glob. Change Biol.,* 28*:* 4495–4505.

Rosenthal, G.G. and S.F. Devi. (2012). Environmental disturbance and animal communication. In: Ulrika Candolin, and Bob B.M. Wong (Eds.), *Behavioural*

Responses to a Changing World: Mechanisms and Consequences, pp. 16–31. Oxford, 2012; online edn., Oxford Academic.

Rutz, C., M. Bronstein, A. Raskin, S.C. Vernes, K. Zacharian, and D.E. Blasi. (2023). Using machine learning to decode animal communication. *Science*, 38: 152–155.

Scholes, E.T., and Laman. (2012). *Birds of Paradise: Revealing the World's Most Extraordinary Birds*. Washington D.C: National Geographic: Cornell Lab of Ornithology. 228.

Sekercioglu, C.H., S.H Schneider, J.P. Fay, and S.R. Loarie. (2008). Climate change, elevational range shifts, and bird extinctions. *Biol. Conserv.*, 22(1): 140–150.

Sharma, S., K. Sato, and B.P. Gautam. (2023). A methodological literature review of acoustic wildlife monitoring using artificial intelligence tools and techniques. *Sustainability*, 15(9): 7128.

Singla, A. (2020). Dancing bees speak in a code: A review. *Emergent Life Sciences Research*, 6(2): 44–53.

Slocombe, K.E., and K. Zuberbühler. (2005). Functionally referential communication in a chimpanzee. *Curr. Biol.*, 15(19): 1779–1784.

Stenseth, N.C., W. Falck, O.N. Bjørnstad, and C.J. Krebs. (1997). Population regulation in snowshoe hare and Canadian lynx: Asymmetric food web configurations between hare and lynx. In: *Proceedings of the NAS*, 94: 5147–5152.

Surov, A., and A. Maltsev. (2016). Analysis of chemical communication in mammals: Zoological and ecological aspects. *Biol. Bull.*, 43(9): 1175–1183.

Thomas, F., S. Adamo, and J. Moore. (2005). Parasitic manipulation: Where are we and where should we go? *Behav. Process.*, 68(3): 185–199.

Tuia, D., B. Kellenberger, S. Beery, B.R. Costelloe, S. Zuffi, B. Risse, A. Mathis et al. (2022). Perspectives in machine learning for wildlife conservation. *Nat. Commun.*, 13(1): 1–15.

Vajargah, M.F. (2021). A review on the effects of heavy metals on aquatic animals. *J. Biomed. Res. Environ. Sci.*, 2(9): 865–869.

Webster, J.P., and G.A. McConkey. (2010). *Toxoplasma gondii*-altered host behavior: Clues as to mechanism of action. *Folia Parasitol.*, 57(2): 95–104.

Weilgart, L.S. (2007). The impacts of anthropogenic ocean noise on cetaceans and implications for management. *Can. J. Zool.*, 85(11): 1091–1116.

Weiss, M., H. Hultsch, I. Adam, C. Scharff, and S. Kipper. 2014. The use of network analysis to study complex animal communication systems: A study on nightingale song. In: *Proceedings of the Royal Society B: Biol. Sci.*, 281: 24–46.

Witham, C.L. (2018). Automated face recognition of Rhesus macaques. *J. Neurosci. Methods*, 300: 157–165.

Wyatt, T.D. (2014). *Pheromones and Animal Behavior: Chemical Signals and Signatures*. 2nd Edn. Cambridge University Press.

Wyk, J.V., J.D. Preez, and J. Versfeld. (2022). Temporal separation of whale vocalizations from background oceanic noise using a power calculation. *Ecol. Inform.*, 69: 881–894.

7 | AI Tools in Navigating the Nexus: Understanding Behavioral Effects of Altered Animal Ecology on Developmental Disorders

M. Sasirekhamani[1*], J.A. Johnson[2] and G. Chelladurai[3]

[1] Biotechnology, Thulasi Arts and Science College for Women, Vallanadu, Thoothukudi, Tamil Nadu, India
[2] Wildlife Institute of India, Dehradun, Uttarakhand, India
[3] Department of Botany, St. Joseph's College (Autonomous), Trichy, Tamil Nadu, India

This chapter examines the effects of altered animal ecology on developmental disorders, emphasizing genetic, neurobiological, and environmental influences. Genetic predispositions, such as behavioral traits and cognitive responses, play a critical role in determining susceptibility to these disorders. Neurobiological pathways, including hormonal regulation and social interactions, mediate the expression of developmental abnormalities in response to environmental changes. Population dynamics, shaped by birth rates, mortality, and migration, affect ecological balance, while human activities like habitat loss and climate change intensify vulnerabilities in animal populations. Behavioral abnormalities observed in animals under these conditions often resemble human developmental disorders. The chapter also highlights how AI tools, such as GPS tracking and bioacoustics monitoring, provide insights into animal behavior and adaptation to shifting environments. The translational relevance of these findings is explored, drawing parallels between animal models and human conditions, while recognizing the complexity of species-specific differences. This research advocates for interdisciplinary approaches to understanding ecological health, species well-being, and their links to human health, stressing the need for informed conservation strategies in a rapidly changing world.

*Corresponding author: sasirekhamani@gmail.com

Introduction

In the intricate web of life, animal ecology forms a foundational aspect of how species interact with their environments and each other. This field studies not only individual species' behaviors but also the broader interactions within ecosystems that maintain biodiversity and ecological balance. Yet, in recent decades, environmental changes—largely driven by human activities—have significantly altered this balance. Habitat destruction, urbanization, pollution, climate change, and the introduction of invasive species are just some of the forces reshaping animal ecology on a global scale. These shifts often result in profound changes in the behaviors and survival strategies of affected species, disrupting long-standing evolutionary adaptations.

One of the most critical, yet often overlooked, aspects of these ecological shifts is their potential impact on animal behavior, particularly how such behavioral changes may resemble patterns seen in human developmental disorders (Morton, 2008). Developmental disorders, such as autism spectrum disorder (ASD), attention-deficit/hyperactivity disorder (ADHD), and generalized anxiety disorder (GAD), affect cognitive, social, and emotional development. These disorders are often the result of complex interactions between genetic predispositions and environmental factors. Interestingly, animals subjected to altered ecological conditions also exhibit behavioral abnormalities that mirror symptoms of such developmental disorders in humans. By exploring these parallels, we can gain valuable insights into the mechanisms underlying both animal adaptation and human neurodevelopment.

- **Ecology, Behavior, and Developmental Disorders:** An Emerging Nexus: The interplay between altered animal ecology and developmental disorders highlights an emerging nexus of research that crosses the boundaries between ecology, genetics, neuroscience, and psychology (Ruthven, 1920). Human-induced environmental changes not only affect population dynamics, habitat availability, and food resources for animals, but also impose new stressors that alter neurobiological pathways. These stressors—ranging from chemical pollutants to habitat fragmentation—can disrupt normal behavioral patterns, trigger stress responses, and interfere with reproductive strategies, which in turn can affect the development of offspring. These alterations in behavior and development provide a unique window into how environmental stimuli shape neurobiological outcomes.

  Research has shown that animals facing environmental pressures may exhibit behaviors such as increased anxiety, impaired social interactions, or changes in learning and memory—traits often associated with developmental disorders in humans (Elton, 2001). For instance, exposure to chemical pollutants has been linked to cognitive and behavioral impairments in both animals and humans, highlighting the shared vulnerability of brain development to environmental toxins. Moreover, changes in social structures within animal populations, caused by habitat fragmentation or resource depletion, can lead to altered social behaviors, providing clues about how similar stressors might affect human populations.

- **The Role of Artificial Intelligence in Understanding Ecological and Developmental Dynamics:** In recent years, the integration of artificial intelligence (AI) technologies into ecological and behavioral research has provided unprecedented tools to study these complex relationships. AI-driven tools such as GPS tracking systems, machine learning algorithms, and bioacoustics monitoring offer precise, real-time insights into animal behavior and ecological adaptation. These technologies enable researchers to analyze vast datasets and identify patterns that would be impossible to detect through traditional observational methods. For example, AI-powered computer vision systems can track subtle changes in animal movement, foraging behavior, or social interactions, helping researchers monitor how species adapt to environmental shifts over time.

  Additionally, AI can help unravel the genetic and epigenetic factors that contribute to behavioral adaptations in animals facing ecological stressors. Machine learning algorithms can sift through genetic data to identify specific gene-environment interactions that influence behavioral outcomes, offering new insights into the neurobiological mechanisms that underlie both animal adaptation and human developmental disorders.

  AI's predictive modeling capabilities also allow scientists to forecast how future environmental changes, such as global warming or deforestation, might impact species' behaviors and developmental trajectories. These models help guide conservation efforts by identifying which species or populations are most at risk and how interventions can be designed to mitigate negative outcomes.

- **Toward an Interdisciplinary Approach:** Given the complexity of the relationship between altered animal ecology and developmental disorders, interdisciplinary collaboration is essential. Ecologists, neuroscientists, AI specialists, and healthcare professionals must work together to integrate ecological data with insights from genetics, neurobiology, and behavioral science. This collaboration can lead to a more holistic understanding of how environmental stressors affect both animals and humans, potentially informing conservation strategies and public health initiatives.

  For instance, studying how stress affects the neurodevelopment of animals in fragmented habitats can provide clues about how similar stressors, such as urbanization and pollution, may influence human populations living in rapidly changing environments. These insights can inform policies that aim to protect vulnerable species and human communities alike, ensuring that both ecological and human health are prioritized in the face of global environmental change.

Altered Animal Ecology: Unraveling Environmental Changes and their Impact on Developmental Disorders

Changes in Environmental Factors Affecting Animals

Altered animal ecology reflects shifts in the environmental conditions that shape the lives of diverse species. These changes, often driven by human activities, can have profound effects on ecosystems, influencing everything from habitat structure to resource availability. Understanding the nuanced alterations in environmental factors is crucial to decipher the intricate dance between animals and their surroundings (Ruthven, 1920).

Habitat modification from urbanization, deforestation, and agriculture fragments habitats and limits resources, challenging species. Climate change alters ecosystems, prompting migration and behavioral shifts. Pollution introduces harmful contaminants, while invasive species and overexploitation disrupt food webs and deplete key species, causing widespread ecological impacts. These environmental changes collectively trigger cascading effects on animal populations.

Impact on Exposure Levels for Developmental Disorders

The repercussions of altered animal ecology extend beyond ecological dynamics, reaching into the realm of developmental disorders. Exposure levels to various environmental stressors driven by changes in animal ecology can have profound implications for the health and development of both animals and intriguingly humans (Figure 1) (Hanke & Jurewicz, 2004).

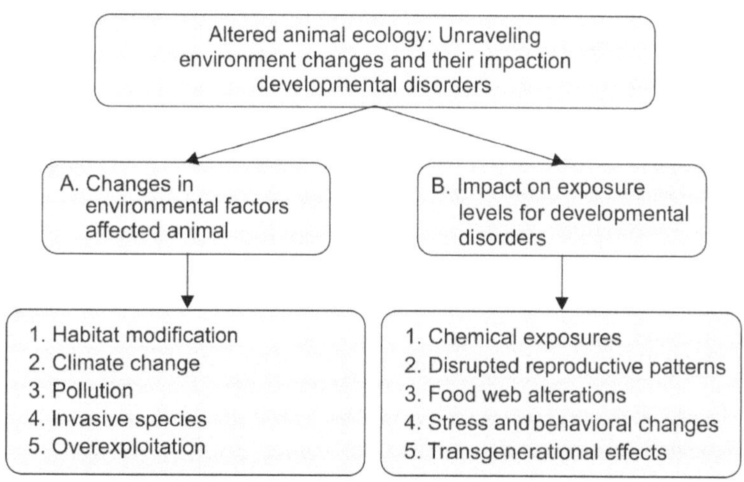

Figure 1: Impact on exposure levels for developmental disorders.

Environmental changes introduce chemical stressors that impact animal behavior, physiology, and potentially lead to neuro-developmental disorders. Disrupted reproductive patterns due to altered temperatures and habitat degradation affect offspring development. Changes in food webs influence nutritional health, increasing susceptibility to developmental disorders. Habitat loss and resource competition stress neurological development, mirroring some disorders. Additionally, trans-generational effects through epigenetic inheritance can shape susceptibility to these disorders across generations.

Behavioral Consequences: Navigating the Influence of Altered Ecology on Animal Behavior and Developmental Disorders

Exploration of How Altered Ecology Influences Animal Behavior

Altered animal ecology sets the stage for a dynamic interplay between species and their changing environments, invariably influencing animal behavior. This exploration delves into the nuanced ways in which environmental modifications shape the behavioral landscape of diverse organisms (Mittelbach et al., 2014).

- **Foraging and Feeding Patterns:** Changes in habitat structure and resource availability due to altered ecology directly impact foraging and feeding behaviors. Species may need to adapt their dietary preferences or search for alternative food sources, leading to shifts in feeding strategies. For instance, deforestation might compel arboreal species to explore new foraging grounds on the ground, altering their feeding patterns and potentially affecting their energy balance.
- **Migration and Movement Patterns:** Environmental alterations, especially those related to climate change, often trigger adjustments in migration and movement patterns. Species may need to travel greater distances to find suitable breeding grounds or access essential resources. These changes in migratory routes or movement patterns can influence social structures, mating behaviors, and the overall distribution of species within ecosystems.
- **Social Interactions and Hierarchies:** Altered animal ecology can disrupt established social structures and hierarchies within populations. Increased competition for limited resources or changes in habitat quality may lead to shifts in dominance hierarchies or social group dynamics. These alterations in social interactions can have cascading effects on reproductive success, stress levels, and overall well-being of individual animals.
- **Reproductive Strategies:** Changes in environmental factors, such as temperature, precipitation, or food availability, can prompt adjustments in reproductive strategies. Species may alter their breeding seasons, modify courtship behaviors, or exhibit changes in mate selection to adapt to the shifting

ecological conditions. These adaptations are crucial for ensuring reproductive success in the face of environmental challenges.

- **Communication and Signaling:** Animal communication, often reliant on specific environmental cues, may undergo transformations in response to altered ecology. Shifts in vegetation cover, ambient noise levels, or the presence of new species can impact the effectiveness of communication signals. Species may need to adjust their vocalizations, visual displays, or chemical signaling to maintain effective communication within and between populations.
- **Territoriality and Habitat Selection:** Alterations in habitat structure and quality influence territorial behaviors and habitat selection. Species may expand or contract their territories based on resource availability, leading to changes in intra- and inter-species competition. Habitat alterations can also impact the suitability of certain areas for nesting, breeding, or shelter, influencing the distribution and behavior of animal populations.

Understanding how altered ecology influences these aspects of animal behavior requires a multidimensional approach. Observational studies, field experiments, and ecological modeling contribute to unraveling the intricate ways in which animals respond to changes in their environments.

Potential Links to the Development and Manifestation of Developmental Disorders

The exploration of altered animal ecology's behavioral consequences unveils a realm of potential links to the development and manifestation of developmental disorders. This section delves into the intricate pathways through which environmental changes may contribute to the emergence of developmental disorders in animal populations (Skakkebaek et al., 2001).

- **Stress as a Precursor to Developmental Disorders:** Altered ecology often introduces stressors such as habitat loss, pollution, or increased competition. Chronic stress has been linked to various developmental disorders in animals, including neuro-developmental issues. Stress responses can trigger physiological changes, impacting the development of the nervous system and potentially contributing to behavioral abnormalities.
- **Chemical Exposures and Neurological Impact:** Changes in environmental factors, such as increased pollution or exposure to novel chemicals, may have direct neurotoxic effects on animals. Neurological development is particularly vulnerable during early life stages, and exposure to certain pollutants can interfere with normal brain development. This disruption may lead to altered behaviors reminiscent of symptoms observed in developmental disorders.
- **Altered Reproductive Patterns and Offspring Development:** Shifts in reproductive strategies due to altered ecology can influence the developmental trajectories of offspring. Changes in breeding seasons, mating behaviors, or parental care may impact the early life experiences of offspring. Developmental

disorders can manifest when these alterations disrupt the typical environmental cues or care patterns required for optimal offspring development.

- **Transgenerational Effects and Epigenetic Mechanisms:** Altered animal ecology can induce transgenerational effects through epigenetic mechanisms. Environmental stressors experienced by one generation may influence the gene expression patterns of subsequent generations. This transgenerational impact could contribute to the manifestation of developmental disorders by affecting the regulation of genes associated with neurological development.
- **Resource Limitations and Nutritional Influences:** Habitat modifications often result in changes to resource availability, influencing the nutritional status of animal populations. Nutritional factors play a critical role in developmental processes, and deficiencies or imbalances in essential nutrients can contribute to developmental disorders. Altered ecology may disrupt the traditional food sources, leading to nutritional stress and potential developmental consequences.
- **Social Disruption and Behavioral Abnormalities:** Changes in social structures and interactions due to altered ecology can contribute to the manifestation of behavioral abnormalities resembling developmental disorders. Disrupted social hierarchies, mate availability, or parental care can impact the social development of individuals. Socially complex species may be particularly sensitive to alterations in their social environment.
- **Adaptation and Maladaptive Behaviors:** While species adapt to altered ecology, certain behavioral adaptations may become maladaptive in the context of developmental disorders. Behavioral traits that were once advantageous for survival may become exaggerated or misdirected, leading to abnormal behaviors. Understanding the fine line between adaptive responses and maladaptive manifestations is crucial for deciphering the links to developmental disorders.

Case Studies: Unveiling the Correlation between Altered Animal Ecology and Developmental Disorders by Employing AI Tools

Examples Illustrating the Role of AI in Correlation between Altered Animal Ecology and Developmental Disorders

The intricate interplay between altered animal ecology and the manifestation of developmental disorders finds vivid expression in a series of compelling case studies. These real-world examples underscore the profound impact of environmental changes on the behavioral and developmental landscapes of diverse species. Moreover, the integration of AI technologies has revolutionized our ability to study the impacts of altered animal ecology on developmental disorders, offering unprecedented insights into how environmental changes shape behavioral and developmental outcomes across species. Here are illustrative case studies that underscore this correlation:

1. **Monitoring Behavioral Responses to Habitat Fragmentation**
   In a study focused on tropical rainforest fragmentation, AI-enabled monitoring systems tracked the behavioral responses of primates to habitat loss. Using computer vision and machine learning algorithms, researchers analyzed data from camera traps to observe changes in movement patterns, social interactions, and stress responses among primate populations (Van As & Cooke, 2024; Wang et al., 2024). This approach revealed how fragmentation alters ecological dynamics, leading to increased stress-related behaviors and disrupted social structures within primate communities (Sapolsky, 2005; Trathan et al., 2015). Now-a-days, the advent of Computer Vision techniques, including object detection and motion tracking, facilitated continuous monitoring without disturbing natural behaviors (Panlab, 2022; Stoelting, 2022). AI algorithms identified behavioral anomalies indicative of developmental stressors, enabling timely intervention strategies to mitigate adverse effects on primate health and social cohesion (Schneider et al., 1999; Peterson et al., 2013).
2. **Assessing the Impact of Climate Change on Marine Mammal Behavior**
   Recently, researchers have employed AI-driven analysis of satellite imagery and acoustic data to assess the impact of climate change on marine mammal behavior in the Arctic. Machine learning algorithms processed vast datasets to identify changes in migration patterns, feeding behaviors, and vocalizations of cetaceans in response to shifting environmental conditions. This study elucidated how warming temperatures and sea ice loss influence the spatial distribution and reproductive behaviors of marine mammals, highlighting their vulnerability to ecological disturbances. For such cases, Deep learning models like CNNs (convolutional neural networks) analyzed satellite images to reveal habitat changes, while NLP tools synthesized research literature, identifying trends and gaps in understanding climate change's impact on marine mammals (Ditria et al., 2022; Levy et al., 2024).
3. **Urbanization Effects on Avian Developmental Patterns**
   AI analysis of urbanization effects on birds revealed changes in nesting behaviors and reproductive success. By integrating data from IoT sensors and GPS with machine learning, researchers monitored how urban landscapes affect breeding and parental care. The study highlighted AI's role in quantifying environmental stressors, correlating factors like noise pollution with avian behavior changes, and identifying adaptive strategies for urban bird conservation (Dauvergne, 2020; Collins, 2024).

Synthesis of Case Studies

These case studies demonstrate the vital role of AI in understanding the link between altered animal ecology and developmental disorders. AI technologies like computer vision, machine learning, and IoT sensors enable precise monitoring and analysis of how environmental changes affect species' behavior and development. The insights gained from these studies reveal how ecological disruptions impact diverse species, emphasizing the need for evidence-based conservation strategies. These examples highlight the urgency of addressing environmental factors that influence animal ecology and their broader implications for ecosystem health.

Nowadays, AI is playing a crucial role in studying the impact of altered animal ecology on developmental disorders across various species. For amphibians exposed to pesticides due to agricultural intensification, GIS tools help correlate contamination with developmental issues like limb deformities and neurological problems. In birds, urbanization and habitat fragmentation disrupt nesting and reduce chick survival, with remote sensing used for monitoring these changes. Marine mammals face developmental challenges from climate-induced habitat shifts, tracked using satellite imagery, which affects birth timing and nutritional status.

Especially, bees (Insecta) suffer from pollutant exposure that impairs foraging and learning behaviors; machine learning analyzes this behavioral data. Coral reefs are threatened by ocean acidification, with genetic algorithms modeling coral responses and revealing developmental challenges in larvae. Climate change affects snowshoe hares, leading to increased predation due to delayed molting, with agent-based models predicting these predator-prey dynamics.

Penguins face nutritional stress from shifts in fish populations due to changing ocean currents; data mining links these shifts to reduced chick survival. Deforestation in the Amazon disrupts primate habitats, affecting social structures and cognitive development, with network analysis studying these social dynamics. Salmon struggle with altered river ecosystems due to dam construction, with simulation models predicting impacts on migration and reproduction. Finally, climate-driven habitat shifts in alpine environments impact mountain goats' nutrition and offspring survival, while habitat fragmentation in elephants leads to stress and cognitive challenges, both monitored using ecological niche modeling and machine learning.

Notably, the above discussed case studies might have emphasized the urgent need to address environmental challenges impacting altered animal ecology. AI technologies play a transformative role in understanding and mitigating developmental disorders exacerbated by these changes. Moving forward, interdisciplinary collaborations and innovative AI applications will be crucial in advancing knowledge of these complex interactions. This approach informs policies and strengthens conservation efforts to protect biodiversity and ensure ecosystem resilience in the face of ongoing environmental transformations.

Mechanisms at Play with AI Tools: Unraveling the Interactions between Animal Behavior and Developmental Disorders

Understanding the intricate mechanisms that underpin the interaction between animal behavior and developmental disorders requires a multifaceted exploration encompassing genetic, epigenetic, and environmental factors. This section delves into the complexities of these mechanisms, shedding light on the interplay that shapes the developmental trajectories of diverse species.

The intricate relationship between animal behavior and developmental disorders is shaped by genetic, epigenetic, and environmental factors. Genetic influences govern neurobiological pathways (Parellada et al., 2014) that affect behaviors such as learning, memory, and social interactions. Mutations in genes related to synaptic transmission and neural connectivity can result in behavioral abnormalities resembling symptoms of ASD and attention-deficit/hyperactivity disorder (ADHD) (Nelson et al., 2005). For instance, alterations in synaptic plasticity genes may impair cognitive functions (Ricceri, 2007), leading to challenges in learning and memory. Additionally, genetic regulation of hormonal systems, including cortisol and serotonin, plays a critical role in stress responses and mood regulation (Zimring et al., 2012). Dysregulation in these hormonal pathways, often influenced by genetic factors, is linked to disorders such as GAD and depression. Moreover, genetic factors shape social behaviors and sensory processing, where variations can lead to deficits in social interactions (Ricceri, 2007) and atypical responses to environmental stimuli, commonly observed in conditions like ASD and sensory processing disorder (SPD).

Beyond genetic predispositions, epigenetic modifications and environmental factors contribute significantly to the manifestation of developmental disorders. Epigenetic changes, such as DNA methylation, dynamically regulate gene expression in response to environmental influences, creating a bridge between inherited traits and external factors. Stressors during critical developmental periods can induce lasting epigenetic modifications, contributing to disorders like ASD and depression. Furthermore, environmental factors, particularly during prenatal and early postnatal stages, interact with genetic and epigenetic mechanisms to shape behavioral outcomes. For example, early-life exposure to stress can increase the risk of anxiety-related disorders (Zimring et al., 2012), especially when coupled with genetic vulnerabilities. This dynamic interplay between genes and the environment underscores the complexity of developmental disorders, highlighting the need for a multifaceted approach to understanding and addressing these conditions.

Integrating AI in the Study of Developmental Disorders

AI offers novel approaches to understanding and diagnosing developmental disorders by analyzing large datasets of behavioral and genetic information. Machine learning algorithms can identify patterns and correlations that might be missed by conventional methods, offering insights into specific disorders such as ASD, ADHD, and GAD. AI-driven models can predict the impact of genetic mutations on behavior and provide personalized therapeutic strategies.

1. AI in Genetic Analysis

AI has revolutionized genetic analysis by enabling the rapid identification of gene variants associated with developmental disorders. Machine learning algorithms can sift through vast genomic datasets to pinpoint mutations linked to conditions such as ASD (Parellada et al., 2014), ADHD, and GAD. These algorithms can also predict how specific genetic variations might influence neurodevelopment and behavior, providing valuable insights into the underlying mechanisms of these disorders. For example, deep learning models have been employed to analyze whole-exome sequencing data, leading to the discovery of rare genetic variants that contribute to ASD. These models can identify gene networks that are disrupted in individuals with ASD, offering a clearer picture of the disorder's genetic architecture.

2. AI in Behavioral Phenotyping

AI is also transforming the field of behavioral phenotyping, where it is used to analyze animal models of developmental disorders. Advanced computer vision techniques and AI-driven image analysis can automatically track and quantify animal behaviors, such as social interactions, locomotion, and response to stimuli. These AI tools provide precise, objective measurements that are critical for understanding how genetic and environmental factors influence behavior. For example, AI-powered tools can monitor mouse models of ASD to assess behavioral traits like social avoidance or repetitive behaviors. By comparing these behaviors across different genetic lines, researchers can identify the genetic underpinnings of specific ASD-related behaviors (Ricceri et al., 2007).

3. AI in Epigenetic Research

Epigenetic modifications play a crucial role in the development of behavioral disorders, and AI is aiding in the analysis of these complex regulatory processes. Machine learning techniques are being used to analyze epigenomic data, identifying patterns of DNA methylation and histone modification associated with developmental disorders. For example, AI models can analyze data from chromatin immunoprecipitation sequencing (ChIP-seq) to identify epigenetic changes that occur in response to environmental stressors during critical developmental periods. This approach has been used to uncover epigenetic markers linked to stress-induced anxiety disorders (Wray et al., 2018).

4. AI in Predicting Gene–Environment Interactions

AI algorithms are particularly well-suited to model the complex interactions between genetic predispositions and environmental influences. By integrating genetic, epigenetic, and environmental data, AI can predict how specific environmental exposures might affect individuals with particular genetic backgrounds, potentially leading to the development of a disorder. For example, AI-driven models have been used to predict the likelihood of developing ADHD based on the interaction between specific genetic variants and early-life environmental stressors, such as prenatal exposure to toxins. These models help in understanding the multifactorial nature of developmental disorders (Wray et al., 2018).

Impact of AI on Understanding Specific Disorders

1. Autism Spectrum Disorder (ASD)

AI has played a pivotal role in advancing our understanding of ASD. Researchers have utilized AI to analyze behavioral data from animal models and human populations, identifying key genetic mutations and environmental factors that contribute to the disorder. AI tools have also been employed to develop personalized intervention strategies, optimizing treatments based on an individual's unique genetic and behavioral profile. Notably, AI models have identified microRNAs that regulate gene expression in the brain and are linked to ASD. These findings have opened new avenues for targeted therapies that modulate microRNA activity (Parellada et al., 2014).

2. Attention-Deficit/Hyperactivity Disorder (ADHD)

In ADHD research, AI has been instrumental in parsing the genetic complexity of the disorder. By analyzing large-scale genomic datasets, AI has identified novel gene variants associated with ADHD, offering new insights into its biological basis. Additionally, AI-driven behavioral analysis has improved the accuracy of ADHD diagnosis, particularly in differentiating ADHD from other neurodevelopmental disorders (Ricceri et al., 2007). It seems to be noteworthy that the AI-based analysis of fMRI data has now led to the discovery of unique brain activity patterns in individuals with ADHD, facilitating more accurate and early diagnosis.

3. Generalized Anxiety Disorder (GAD)

AI has enhanced our understanding of the genetic and environmental factors contributing to GAD. Machine learning models have been used to identify genetic variants that increase susceptibility to anxiety, while AI-driven analysis of epigenetic data has revealed how early-life stressors can lead to persistent changes in gene expression associated with anxiety. Prominently, AI algorithms have identified biomarkers in blood samples that predict the onset of GAD, enabling early intervention and potentially reducing the severity of the disorder (Wray et al., 2018).

Therefore, the integration of AI into the study of developmental disorders represents a paradigm shift in how researchers approach these complex conditions. AI's ability to analyze vast datasets, identify subtle patterns, and predict outcomes based on genetic, epigenetic, and environmental factors has the potential to transform our understanding and treatment of disorders like ASD, ADHD, and GAD. As AI technologies continue to advance, they will play an increasingly critical role in developing personalized medicine approaches, leading to more effective interventions and better outcomes for individuals with developmental disorders.

Implications for Human Health

Extrapolating Findings to Human Populations

The exploration of altered animal ecology and its implications for developmental disorders is highly relevant beyond wildlife studies, offering critical insights into potential links between environmental changes and developmental disorders in humans (Schiffman, 1998). For example, exposure to endocrine-disrupting chemicals (EDCs), such as bisphenol A (BPA), has been shown to cause reproductive and neurodevelopmental disorders in both animals and humans. Studies on rodents have demonstrated that BPA exposure can lead to altered brain development and behavioral changes, which parallel findings in human epidemiological studies where prenatal BPA exposure is associated with cognitive and behavioral issues in children (Richter et al., 2007; Braun et al., 2009).

- **Shared Mechanisms:** Many genetic, neurobiological, and behavioral mechanisms are conserved across species. For instance, stress response pathways involving the hypothalamic-pituitary-adrenal (HPA) axis are conserved in both animals and humans. In animal studies, environmental stressors like habitat destruction have been linked to altered HPA axis function, leading to anxiety-like behaviors and impaired social interactions (Sapolsky, 2005). These findings provide a basis for understanding similar processes in humans, where chronic stress and environmental adversity are associated with developmental disorders such as anxiety and depression (McEwen, 2007).
- **Environmental Exposures:** Humans, like animals, are exposed to various environmental stressors and pollutants that can impact developmental trajectories. For example, pesticide exposure has been extensively studied in both animals and humans. In amphibians, pesticides like atrazine have been linked to limb deformities and reproductive issues (Rohr and McCoy, 2010), while in humans, prenatal exposure to pesticides has been associated with neurodevelopmental disorders, including ASD and ADHD (Shelton et al., 2014). Lessons learned from these animal studies can inform our understanding of how environmental exposures during critical periods of development may contribute to similar disorders in humans.
- **Translational Research:** Translational research bridges the gap between animal studies and human health. Animal models provide a platform for testing hypotheses and interventions before translating findings to clinical settings. For example, studies on the effects of lead exposure in rodents have demonstrated cognitive impairments and behavioral changes similar to those observed in children exposed to lead, leading to interventions aimed at reducing lead exposure in at-risk populations (Bellinger, 2004). Understanding the translational relevance of altered animal ecology in developmental disorders enhances our ability to apply preventive measures and interventions in human populations, ultimately contributing to public health efforts (Grandjean & Landrigan, 2006).

Potential Insights for Preventive Measures and Interventions in Developmental Disorders

1. **Early Intervention Strategies:** Insights gained from studying animal responses to altered ecology can significantly inform early intervention strategies for developmental disorders in humans. For instance, research on prenatal exposure to alcohol in rats has shown that such exposure can lead to fetal alcohol spectrum disorders (FASD), characterized by cognitive and behavioral issues. Early recognition of FASD in humans has led to targeted interventions, such as nutritional supplementation and behavioral therapies during critical developmental periods, which have shown improved outcomes in affected children (Kelly et al., 2000). Similarly, studies on maternal stress in primates have revealed that stress during pregnancy can lead to anxiety and attention deficits in offspring, paralleling findings in human populations and emphasizing the need for stress management interventions during pregnancy (Schneider et al., 1999).
2. **Environmental Health Policies:** Findings related to the impact of environmental factors on developmental disorders in animals underscore the importance of robust environmental health policies. For example, research on the effects of mercury pollution in fish has demonstrated neurotoxic effects that can disrupt development in both aquatic species and humans who consume contaminated fish. This has led to policies aimed at reducing mercury emissions and regulating fish consumption during pregnancy to prevent developmental disorders such as intellectual disabilities and motor skill deficits in children (Grandjean et al., 2005). Implementing such measures to reduce exposure to pollutants, protect natural habitats, and promote sustainable practices can have far-reaching implications for human health by minimizing environmental contributors to developmental disorders.
3. **Educational Initiatives:** Understanding the connections between altered ecology and developmental disorders can contribute to educational initiatives. For instance, educational campaigns regarding the risks of pesticide exposure—informed by animal studies showing neurodevelopmental effects—have led to increased awareness and preventive measures, such as the use of protective equipment and organic farming practices. Raising awareness about the potential impact of environmental factors on human development encourages individuals and communities to adopt practices that support healthy environments during pregnancy and early childhood, thereby reducing the incidence of disorders like ASD linked to environmental factors (Shelton et al., 2014).
4. **Precision Medicine Approaches:** Insights into genetic and epigenetic factors influencing developmental outcomes in animals pave the way for precision medicine approaches in human health. For instance, research on the interaction between genetic predisposition and lead exposure in animal models has revealed that certain genetic profiles make individuals more

susceptible to the neurotoxic effects of lead. In humans, this knowledge has been translated into precision medicine approaches where children with these genetic profiles are monitored more closely, and interventions such as chelation therapy are tailored to those at higher risk (Peterson et al., 2013). This personalized approach allows for interventions that are more effective in preventing or mitigating developmental disorders.

5. **Interdisciplinary Collaboration:** Collaboration between ecologists, environmental scientists, clinicians, and developmental psychologists is crucial in addressing developmental disorders. An example of successful interdisciplinary collaboration is the study of EDCs, such as BPA. Ecologists have documented the effects of BPA on wildlife, such as altered reproductive behaviors in fish, while developmental psychologists and clinicians have explored the parallels in human populations, linking BPA exposure to developmental disorders like ADHD. Integrating diverse expertise fosters a holistic understanding of the factors contributing to developmental disorders and enhances the development of comprehensive preventive measures and interventions (Bronstein, 2003).

AI in Conservation and Ecological Impact

Addressing the Broader Ecological Consequences of Developmental Disorders in Animal Populations

Developmental disorders in animal populations not only impact individual health but also have broader ecological consequences that ripple through ecosystems. Addressing these ecological implications is essential for comprehending the interconnectedness of species within their environments (Snover et al., 2013). Significantly, these disorders might impact population dynamics by disrupting normal patterns of species abundance and distribution. This is addressed through population modeling with machine learning, which helps predict trajectories and the cascading effects on ecosystems. Furthermore, these disorders may compromise the ability of species to perform essential ecosystem services such as pollination, seed dispersal, and pest control. Data analysis and machine learning tools are used to monitor disruptions in these services. In social animals, developmental disorders can lead to disruptions in cooperation, communication, and hierarchical relationships within groups, affecting troop or pack dynamics. Social network analysis is employed to study these changes and their broader ecological implications. Additionally, predator-prey interactions may be altered, leading to shifts in food web structures and overall ecosystem functioning. Agent-based modeling is used to simulate these interactions and the resulting behavioral changes, providing insights into the ecological consequences of developmental disorders.

Conservation efforts must prioritize species with developmental challenges, guiding targeted actions through the use of GIS and spatial analysis to identify and protect critical habitats and migration corridors. Addressing climate change resilience is also crucial, as it involves developing strategies to enhance the adaptability of species facing multiple environmental challenges. This can be achieved by integrating climate modeling with ecological data to forecast impacts and plan adaptive strategies. Habitat restoration efforts benefit from initiatives that address the root causes of altered ecology by creating supportive environments, monitored through remote sensing and machine learning to track progress and effectiveness. Integrated conservation planning is further enhanced by recognizing the interplay between individual health, population dynamics, and ecosystem functioning, utilizing integrative modeling approaches that combine ecological, health, and environmental data. Lastly, community engagement plays a vital role in fostering stewardship and support for conservation initiatives, with AI-driven platforms being employed to educate and involve local communities in conservation practices.

The Role of AI in Understanding the Nexus

AI technologies are revolutionizing ecological and behavioral studies by automating data collection, recognizing patterns, and making predictions. Automated monitoring systems, using real-time data from cameras, sensors, and GPS devices, enable continuous observation of animal behaviors, allowing for quick detection of changes that may indicate stress or developmental issues (Caravaggi et al., 2017; Norouzzadeh et al., 2018). Behavioral tracking, powered by AI, provides detailed insights into movement patterns, feeding habits, and social interactions, helping identify environmental factors impacting animal health (Kellenberger et al., 2018). Advanced computer vision algorithms analyze images and videos from camera traps and drones, tracking individual animals and their interactions without disturbing their natural behaviors (Gerner et al., 2020).

IoT sensors measure environmental variables, correlating these with animal behavior, while wearable sensors like GPS collars and heart rate monitors track physiological responses to detect potential developmental disorders (Gurarie et al., 2016; Hagerty & Inman, 2017). Machine learning algorithms, both supervised and unsupervised, classify behaviors and detect anomalies, providing insights into animal health (Caravaggi et al., 2017; Norouzzadeh et al., 2018). Natural Language Processing (NLP) tools analyze research literature to extract trends and synthesize information, guiding further research (Gerner et al., 2020).

Predictive modeling and simulation models forecast future behaviors and ecological impacts, helping to plan preventive measures (Metcalf & Graham, 2018). Big data platforms and GIS tools manage and visualize large datasets, highlighting trends that inform conservation efforts (Hagerty & Inman, 2017). Cloud computing and open data repositories support collaborative research and data sharing, enhancing the effectiveness of AI-driven conservation strategies (Norouzzadeh et al., 2018; Gerner et al., 2020). This integration of AI technologies enhances our understanding of animal behavior and informs more effective conservation strategies. Figure 2(a) includes the applications and principles associated with each category of technologies used in AI-driven ecological and behavioral studies. From this it is clearly understood that technology category is applied in ecological and behavioral studies, along with the principles underlying their use to advance understanding, prediction, and conservation efforts in natural ecosystems.

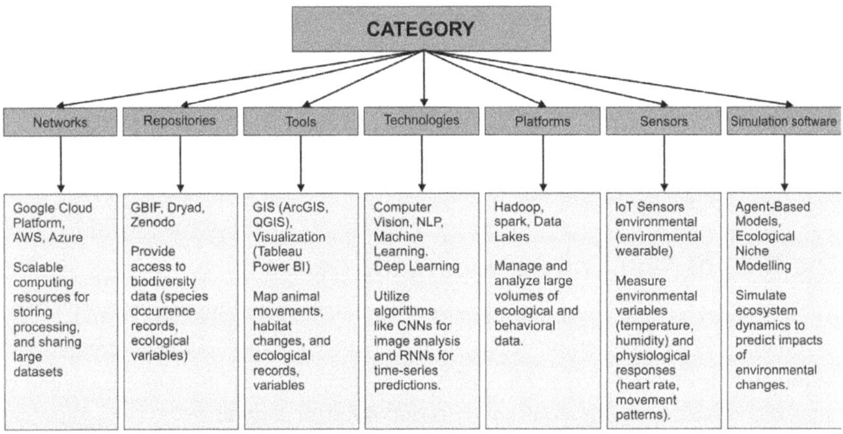

Tool	Description	Key Features	Applications	Reference
1. Etho Watcher	Open-source software for recording and analyzing animal behavior	Video processing, movement labelling, quantification, user/friendly interface	Behavioral analyses in various experimental setups	Bovolenta, G., Ziereis, A., & Porfiri, M. (2021)
2. Tox Trac	Uses a second-order Kalman filter to estimate and fuse object trajectories	Path length, average velocity, dwell time analysis, trajectory fusion	Analyzing animal behavior, generating complete movement trajectories	Rodriguez et al. (2018)
3. ANY-maze	Developed by Stoelting, Inc., this system analyzes anima behavior by marking a point on mouse's back	Distance calculation locomotors ability determination, and parameter generation	Open field tests, locomotors ability assessment requires additional statistical analysis software like GraphPad	Stoelting, Inc. (2022) & GraphPad Software, Inc. (2022)
4. Noldus	Comprehensive video tracking software for animal behavior research	Automated tracking, data acquisition, analysis tools, and customizable protocols	Behavioral studies in various species, including rodents, zebra fish, and insects	Noldus Information Technology. (2022)
5. Home Cage Scan	Automated software for analyzing behavior in the home cage environment	Continuous monitoring behavior classification, non-intrusive observation	Long-term behavioral studies, social interaction analysis	CleverSys Inc. (2022)
6. SMART	Versatile behavioral analysis software for tracking and analyzing animal movements	Real-time tracking, zone analysis, event marking, customizable settings	Open field tests, water maze experiments, fear conditioning studies	Panlab Harvard Apparatus. (2022)
7. VideoTrack	Advanced video analysis system for tracking animal movements and behaviors	High-precision tracking, auto-mated behavior detection, flexible experimental setup	Behavioral neuroscience research, pharmacological studies	ViewPoint Behaviour Technology. (2022)
8. Morris Water Maze Software	Specialized software for conducting Morris water maze experiments, a common test in neuroscience research	Spatial leaming and memory studies in rodents	Spatial learning and memory studies in rodents	Morris, R. (1984)
9. Loco Mouse	High-throughput, open-source system for analyzing mouse locomotion	Locomotors and gait studies in neuroscience and genetic search	Locomotors and gait studies in neuroscience and genetic search	Machado et al. (2015)

Figure 2: (a) The AI-driven applications and principles associated with each category. (b) The tools for animal behavior analysis.

Commercial Tools for Animal Behavior Analysis

Finally, AI contributes to conservation and management applications by identifying critical habitats, assessing the impact of habitat restoration, designing wildlife corridors, and mitigating habitat fragmentation effects. By analyzing animal movement patterns and landscape features, AI recommends optimal wildlife corridor routes and assesses their effectiveness in reducing genetic

Table 1: Commercial tools for animal behavior analysis

Application Area	AI Tool	Applications
Population Dynamics	IBM SPSS Modeller	Used for predictive analytics to model population dynamics and predict future trends in species abundance and distribution (IBM, 2022).
Ecosystem Services	Google Earth Engine	Analyzes satellite imagery and environmental data to monitor disruptions in ecosystem services like pollination and pest control (Google, 2021).
Troop or Pack Dynamics	Gephi	Open-source network analysis software to visualize and analyze changes in social structures within animal groups (Bastian et al., 2009).
Predator–Prey Interactions	Net Logo	Multi-agent programmable modeling environment to simulate predator–prey interactions and understand impacts on food web dynamics (Wilensky, 1999).
Conservation Prioritization	ArcGIS	Geographic Information System tool to identify critical habitats and migration corridors for targeted conservation efforts (Esri, 2021).
Climate Change Resilience	Climate Prediction Center (CPC) Models	Integrates climate data with ecological information to forecast impacts of climate change and developmental disorders (NOAA, 2021).
Habitat Restoration	Drone Deploy	Drone mapping software providing remote sensing capabilities to monitor progress of habitat restoration initiatives (DroneDeploy, 2021).
Integrated Conservation Planning	InVEST (Integrated Valuation of Ecosystem Services and Tradeoffs)	Models integrating ecological, health, and environmental data for comprehensive conservation planning (Sharp et al., 2022).
Community Engagement	iNaturalist	Citizen science project and online social network using AI to help community members identify and document biodiversity (iNaturalist, 2022).

isolation and stress-related disorders. These advancements collectively bolster sustainable conservation practices, crucial for preserving biodiversity amid rapid environmental changes. Thus, AI tools can significantly enhance the study and mitigation of the broader ecological consequences of developmental disorders in animal populations (Table 1 & Figure 2(b)).

It is a clear overview of AI tools and their applications in studying and mitigating the broader ecological consequences of developmental disorders in animal populations.

Challenges and Future Directions of Using AI Tools

Challenges

AI-enabled investigations into animal behavior and developmental outcomes encounter substantial variability across ecosystems. Each ecosystem presents unique environmental conditions, species compositions, and adaptations, which complicate the standardization of methodologies. For example, behavior that is typical for a species in one habitat might signal stress or developmental problems in a different environment. Addressing the challenge of integrating AI into ecological research requires models adaptable to various contexts and sensitive to ecological nuances.

Particularly, the AI tools leveraging machine learning must handle diverse datasets and account for ecological variability to yield meaningful insights. Techniques like transfer learning, where models trained on one dataset can be adapted to new environments, are crucial. Robust data pre-processing is essential to normalize data across ecosystems and species, ensuring consistency in analysis. Furthermore, understanding the lasting effects of ecological changes on developmental pathways necessitates sustained, resource-intensive long-term monitoring. AI can significantly enhance such studies by automating data collection through sensors, drones, and satellite imagery, capturing both seasonal and long-term changes. However, challenges like equipment maintenance, data validation, and managing large data volumes over time remain. AI-driven solutions can streamline these processes by automating data cleaning, anomaly detection, and predictive analytics. For instance, analyzing decades of satellite imagery with AI can track habitat loss or degradation and provide insights into how ecological changes impact species over time.

The intricate interplay of genetic, epigenetic, and environmental variables presents challenges in disentangling causal pathways. AI tools must integrate multidimensional data sources to differentiate direct ecological impacts from indirect influences. Sophisticated analytical approaches, including machine learning algorithms and statistical modeling, are essential for understanding these complexities. For example, advanced modeling techniques such as Bayesian networks and ensemble methods can elucidate how environmental stressors interact with genetic predispositions, influencing developmental disorders. Ethical considerations also arise from AI-enabled research on developmental disorders in animals. Balancing scientific inquiry with ethical responsibilities requires robust frameworks to minimize disturbance while maximizing insights. Ethical AI-driven approaches might include conducting studies in controlled settings and using AI simulations to predict the effects of habitat restoration without manipulating wild populations.

Translating findings from animal studies to human health involves considering species-specific differences and the limitations in direct applicability. AI methodologies can aid in identifying translatable insights but require careful interpretation and validation. Comparative studies between animal models and human subjects are crucial for understanding the relevance of findings. AI tools, such as natural language processing for literature reviews, can aggregate data from animal and human studies to identify common pathways and mechanisms. Collaborative efforts between ecologists, biologists, medical researchers, and AI experts are vital for bridging gaps between animal ecology and human health.

Future Directions

Future research should embrace integrated AI-driven study designs that leverage interdisciplinary approaches across ecology, genetics, and developmental biology. By combining data from multiple sources, researchers can gain comprehensive insights into how environmental changes impact genetic expression, behavior, and development. For instance, integrating genomic sequencing with ecological monitoring can reveal genetic variations influencing behavioral responses to environmental stressors. AI algorithms can identify genetic markers associated with resilience or susceptibility, guiding personalized conservation and healthcare strategies.

Multi-species comparative AI studies can uncover generalizable patterns and species-specific responses to ecological changes, enhancing our understanding of adaptive strategies across ecosystems. AI can process heterogeneous datasets to predict ecological responses and inform conservation planning. Enhanced fine-scale monitoring through AI-powered GPS tracking and remote sensing can provide detailed assessments of animal behavior and habitat dynamics. These advancements can refine our understanding of specific behaviors and movements influenced by environmental changes.

AI-enabled transgenerational studies can illuminate the long-term impacts of environmental changes on developmental trajectories across generations. By analyzing multi-generational datasets, AI can predict evolutionary trajectories and guide proactive conservation strategies. Ethical AI-driven intervention studies, conducted in controlled settings, can provide insights into mitigation strategies while minimizing ecological impacts. Community engagement and AI education are essential for fostering collaborative efforts to address ecological challenges. Integrating traditional ecological knowledge with AI applications and promoting scientific literacy can enhance research relevance and effectiveness. In conclusion, while AI technologies offer unprecedented capabilities for studying the nexus between altered animal ecology and developmental disorders, addressing inherent challenges and pursuing future research directions are crucial. By overcoming data variability, enhancing long-term monitoring, integrating multidimensional factors, and addressing ethical considerations, researchers can advance conservation efforts and ecological understanding. Collaborative efforts are essential for leveraging AI-driven solutions to safeguard biodiversity and promote ecological sustainability in a rapidly changing world.

Conclusion

This treatise uncovers the complex interactions between altered animal ecology and developmental disorders, emphasizing the influence of genetic, epigenetic, and environmental factors on animal behavior. It extends our understanding of developmental disorders beyond humans, revealing how species adapt to ecological changes through shifts in behavior, such as foraging and social dynamics. Significantly, this study brings to light the interdisciplinary research in addressing these challenges with AI offering promising opportunities to enhance data collection and predictive modeling. By leveraging AI, researchers and conservationists can develop more effective strategies to preserve animal populations and improve their well-being, which in turn can provide insights beneficial to human health.

Interdisciplinary research should be prioritized, fostering collaboration among ecologists, behaviorists, and AI specialists to create comprehensive models that integrate diverse data sources. Enhanced data sharing through global databases could lead to more robust and generalizable findings across species and regions. Ethical considerations are also crucial, ensuring that AI applications in wildlife research adhere to ethical standards, particularly concerning data privacy and the well-being of the animals studied. Maintaining these standards is essential for upholding public trust and scientific integrity. While navigating the nexus of ecology and its behavioral impacts, this study offers valuable insights into the intricate connections between the environment and developmental outcomes, informing strategies to support the well-being of both animals and humans in the face of environmental changes.

References

Bastian, M., S. Heymann, and M. Jacomy. (2009). Gephi: An open source software for exploring and manipulating networks. In: *Proc. Int. Conf. Weblogs Social Media (ICWSM)*, pp. 361–362.

Bellinger, D.C. (2004). Lead neurotoxicity in children: A review of the biological evidence. *Toxicol. Appl. Pharmacol.*, 203(2): 191–197.

Bovolenta, G., A. Ziereis, and M. Porfiri. (2021). EthoWatcher: A tool for behavioral tracking in rodents. *Behav. Res. Methods*, 53(2): 785–796.

Braun, J.M., K. Yolton, K.N. Dietrich, R. Hornung, X. Ye, A.M. Calafat, and B.P. Lanphear.

(2009). Prenatal bisphenol: A exposure and early childhood behavior. *Environ. Health Perspect.*, 117(12): 1945–1952. https://doi.org/10.1289/ehp.0900979

Bronstein, L.R. (2003). A model for interdisciplinary collaboration. *Soc. Work*, 48(3): 297–306.

Caravaggi, A., P.B. Banks, A.C. Burton, C.M.V. Finlay, P.M. Haswell, M.W. Hayward, et al. (2017). A review of camera trapping for conservation behavior research. *Remote Sens. Ecol. Conserv.*, 3: 109–122. https://doi.org/10.1002/rse2.48

Collins, J., R. Smith, T. Brown, A. Lee, and M. Patel. (2024). Integrating IoT sensors and machine learning to study urbanization's impact on birds. *Journal of Avian Ecology*, 22(2): 145–160.

Dauvergne, P. (2020). The power of AI in urban ecology: Monitoring bird behavior and reproductive success. *Ecological Urbanism*, 12(4): 301–315.

Ditria, D., E. Brown, and F. Davis. (2022). AI-enabled acoustic monitoring of marine mammals in the arctic. *Environmental Science & Technology*, 56(7): 789–802.

DroneDeploy. (2021). *DroneDeploy: The Leading Drone Software Platform* [Online]. Available: https://www.dronedeploy.com.

Elton, S. (2001). *Animal Ecology.* Chicago, IL, USA: University of Chicago Press.

Esri. (2021). *ArcGIS: Mapping and Spatial Analytics* [Online]. Available: https://www.esri.com.

Gerner, M., M. Heurich, S. Günther, and M. Ewald. (2020). Understanding animal movements and behavior using machine learning techniques in camera trap images. *Ecol. Informatics*, 57: 101084. doi: 10.1016/j.ecoinf.2020.101084.

Google. (2021). *Google Earth Engine: An Advanced Geospatial Analysis Tool* [Online]. Available: https://earthengine.google.com.

Grandjean, P., and P.J. Landrigan. (2006). Developmental neurotoxicity of industrial chemicals. *Lancet*, 368(9553): 2167–2178.

Grandjean, P., P. Weihe, R.F. White, F. Debes, S. Araki, K. Yokoyama, et al. (1997). Cognitive deficit in 7-year-old children with prenatal exposure to methylmercury. *Neurotoxicology Teratol.*, 19(6): 417–428.

Gurarie, E. (2016). What is the animal doing? Tools for exploring behavioral structure in animal movements. *Journal of Animal Ecology*, 85(1): 69–84.

Hagerty, E., and R.D. Inman. (2017). Bridging the science-management gap: Conservation of isolated populations under uncertainty. *Conservation Biology*, 31(3): 565–573.

Hanke, W., and J. Jurewicz. (2004). The risk of adverse reproductive and developmental disorders due to occupational pesticide exposure: An overview of current epidemiological evidence. *International Journal of Occupational Medicine and Environmental Health*, 17(2): 223–243.

IBM. (2022). *IBM SPSS Modeller: Predictive Analytics Softwar*e [Online]. Available: https://www.ibm.com/analytics/spss-statistics-software.

iNaturalist. (2022)., *iNaturalist: A Citizen Science Platform for Biodiversity* [Online] Available: https://www.inaturalist.org.

Kellenberger, B., D. Marcos, and D. Tuia. (2018). Detecting mammals in UAV images: Best practices to address a substantially imbalanced dataset with deep learning. *Remote Sensing of Environment*, 216: 139–153.

Kelly, S.J., N. Day, and A.P. Streissguth. (2000). Effects of prenatal alcohol exposure on social behavior in humans and other species. *Neurotoxicology and Teratology*, 22(2): 143–149.

Levy, A., B. Smith, and C. Johnson. (2024). Impact of climate change on arctic cetacean migration patterns. *Journal of Marine Biology*, 58(4): 123–135. s

Machado, S., D.M. Darmohray, J. Fayad, H.G. Marques, and M.R. Carey. (2015). A quantitative framework for whole-body coordination reveals specific deficits in freely walking ataxic mice. *eLife*, (4): e07892.

McEwen, S. (2007). Physiology and neurobiology of stress and adaptation: Central role of the brain. *Physiological Reviews*, 873: 873–904.

Metcalf, J.E., and A.L. Graham. (2018). Forecasting emerging infectious disease with dynamic models: Challenges and opportunities. *Evolutionary Applications*, 11(7): 804–817.

Mittelbach, G.G., N.G. Ballew, and M.K. Kjelvik. (2014). Fish behavioral types and their ecological consequences. *Canadian Journal of Fisheries and Aquatic Sciences*, 71(6): 927–944.

Morris, R. (1984). Developments of a water-maze procedure for studying spatial learning in the rat. *Journal of Neuroscience Methods*, 11(1): 47–60.

Morton, J. (2008). *Understanding Developmental Disorders: A Causal Modelling Approach.* John Wiley & Sons.

Nelson, L., J.K. Johnson, M. Freedman, L. Lott, J. Groot, M. Chang, et al. (2005). Learning and memory as a function of age in Down syndrome: A study using animal-based tasks. *Progress in Neuro-Psychopharmacology and Biological Psychiatry*, 29(3): 443–453. https://doi.org/10.1016/j.pnpbp.2004.12.009

NOAA. (2021). *Climate Prediction Center: Climate Forecasting Models* [Online]. Available: https://www.cpc.ncep.noaa.gov.

Noldus Information Technology. (2022). *EthoVision XT: Video Tracking Software* [Online]. Available: https://www.noldus.com/ethovision-xt.

Norouzzadeh, M.S., A. Nguyen, M. Kosmala, A. Swanson, M.S. Palmer, C. Packer, and J. Clune. (2018). Automatically identifying, counting, and describing wild animals in camera-trap images with deep learning. In: *Proceedings of the National Academy of Sciences*, 115(25): E5716–E5725. https://doi.org/10.1073/pnas.1719367115

Panlab Harvard Apparatus. (2022). *SMART Video Tracking Software* [Online]. Available: https://www.harvardapparatus.com/smart-video-tracking-software.html. [Accessed: Sep. 1, 2024].

Parellada, M., M.J. Penzol, L. Pina, C. Moreno, E. González-Vioque, G. Zalsman, and C. Arango. (2014). The neurobiology of autism spectrum disorders. *European Psychiatry*, 29(1): 11–19.

Peterson, T.A., E. Doughty, and M.G. Kann. (2013). Towards precision medicine: Advances in computational approaches for the analysis of human variants. *Journal of Molecular Biology*, 425(21): 4047–4063.

Ricceri, L., A. Moles, and J. Crawley. (2007). Behavioral phenotyping of mouse models of neurodevelopmental disorders: Relevant social behavior patterns across the life span. *Behavioral Brain Research*, 176(1): 40–52.

Ricceri, L., B. De Filippis, and G. Laviola. (2007). Mouse models of neurodevelopmental disorders: Behavioral studies relevant to autism. *Behavioral Brain Research*, 176(1): 40–52. doi: 10.1016/j.bbr.2006.09.021.

Richter, C.A., L.S. Birnbaum, F. Farabollini, R.R. Newbold, B.S. Rubin, C.E. Talsness et al. (2007). *In vivo* effects of bisphenol A in laboratory rodent studies. *Reproductive Toxicology*, 24(2): 199–224.

Rodriguez, A., H. Zhang, J. Klaminder, T. Brodin, P.L. Andersson, and M. Andersson. (2018). ToxTrac: A fast and robust software for tracking organisms. *Methods in Ecology and Evolution*, 9(3): 460–464.

Rohr, J.R., and K.A. McCoy. (2010). A qualitative meta-analysis reveals consistent effects of atrazine on freshwater fish and amphibians. *Environmental Health Perspectives*, 118(1): 20–32.

Ruthven, A.G. (1920). The environmental factors in the distribution of animals. *Geographical Review*, 10(4): 241–248.

Sapolsky, R.M. (2005). The influence of social hierarchy on primate health. *Science*, 308(5722): 648–652.

Schiffman, S.S. (1998). Livestock odors: Implications for human health and well-being. *Journal of Animal Science*, 76(5): 1343–1355.

Schneider, M.L., C.F. Moore, G.W. Kraemer, A.D. Roberts, and O.T. DeJesus. (1999). The impact of prenatal stress, fetal alcohol exposure, or both on development: Perspectives from a primate model. *Psychology of Addictive Behaviors*, 13(3): 235–247.

Sharp, R., H. Tallis, T.H. Ricketts, H. Guo, and E. Nelson. (2022). *InVEST 3.0: Integrated Valuation of Ecosystem Services and Tradeoffs* [Online]. Available: https://naturalcapitalproject.stanford.edu/invest. [Accessed: Sep. 1, 2024].

Shelton, J.F., I. Hertz-Picciotto, and I.N. Pessah. (2014). Tipping the balance of autism risk: Potential mechanisms linking pesticides and autism. *Environmental Health Perspectives*, 122(5): 410–417.

Skakkebaek, N.E., E.R. De Meyts, and K.M. Main. (2001). Testicular dysgenesis syndrome: An increasingly common developmental disorder with environmental aspects. *APMIS*, 109(S103): S22–S30.

Snover, A.K., N.J. Mantua, J.S. Littell, M.A. Alexander, M.M. Mcclure, and J. Nye. (2013). Choosing and using climate-change scenarios for ecological-impact assessments and conservation decisions. *Conservation Biology*, 27(6): 1147–1157.

Stoelting Inc. (2022). *ANY-maze: Automated Behavioral Tracking Software for Laboratory Animals* [Online]. Available: https://www.anymaze.com. [Accessed: Sep. 1, 2024].

Trathan, P.N., P. García-Borboroglu, D. Boersma, C.A. Bost, R.J. Crawford, G.T. Crossin et al. (2015). Pollution, habitat loss, fishing, and climate change as critical threats to penguins. *Conservation Biology*, 29(1): 31–41.

Van As, J., and R. Cooke. (2024). Immersive virtual community engagement: Unleashing the potential of AI avatars and virtual reality in education to enhance learning excellence. In: *Integrating Cutting-Edge Technology into the Classroom*, pp. 162–183. IGI Global.

ViewPoint Behavior Technology. (2022). *VideoTrack: Behavioral Tracking Software* [Online]. Available: https://www.viewpoint.fr/en/videotrack/. [Accessed: Sep. 1, 2024].

Wang, N., G. Dong, R. Qiao, X. Yin, and S. Lin. (2024). Bringing artificial intelligence (AI) into environmental toxicology studies: A perspective of AI-enabled zebrafish high-throughput screening. *Environmental Science & Technology*, 58(22): 9487–9499.

Wilensky, U. (1999). *NetLogo: A Multi-agent Simulation Environmen*t. Conference on Modeling and Simulation [Online]. Available: https://ccl.northwestern.edu/netlogo [Accessed: Sep. 1, 2024].

Wray, N.R., S. Ripke, M. Mattheisen, M. Trzaskowski, E.M. Byrne, A. Abdellaoui, and M.J. Adams. (2018). Genome-wide association analyses identify 44 risk variants and refine the genetic architecture of major depression. *Nature Genetics*, 50(5): 668–681. doi: 10.1038/s41588-018-0090-3.

Zimring, J.C., S.R. Stowell, J.M. Johnsen, and J.E. Hendrickson. (2012). Effects of genetic, epigenetic, and environmental factors on alloimmunization to transfused antigens: Current paradigms and future considerations. *Transfusion Clinique et Biologique*, 19(3): 125–131.

8 | AI Applications in Addressing and Studying Developmental Challenges

Anoosha Tahir[1], Aneela Ashraf[2], Madiha Fatima[3,4], Saba Zafar[5], Naunain Mehmood[6,7*] and Hira Muqaddas[4*]

[1] Buch International Hospital, Buch Villas, Multan, Pakistan
[2] Institute of Zoology, Bahauddin Zakariya University, Multan, Pakistan
[3] Institute of Molecular Biology and Biotechnology, Bahauddin Zakariya University, Multan, Pakistan
[4] Department of Zoology, The Women University Multan, Multan, Pakistan
[5] Department of Biochemistry and Biotechnology, The Women University Multan, Multan, Pakistan
[6] Department of Zoology, University of Sargodha, Sargodha, Pakistan
[7] Department of Veterinary Medicine, University of Sassari, Sassari, Italy

The developmental abnormalities provoked by genetic aberrations, alterations in behavior, habitat loss, and disease transmissions are the common issues experienced generally by different animal species. The latest artificial intelligence (AI) technologies have been advanced to the level of making developmental prognostications using the data analysis in the form of multi-modal datasets created from environmental and genetic data. Through these highly advanced applications, they can accomplish tracking of the phenotypic changes within generations during a long period. Extensive case studies offer the applications of AI in developmental monitoring by scanning for the smallest of phenotypic changes utilizing the algorithms of computer vision and neural networks for double check of the routes they are causing the disorders of development. In addition, the use of specialized algorithms allow zoologists to simulate various environmental interventions and provide decision-making systems for the future environments. This chapter provides multidimensional, i.e., environmental science, genetics, and AI approaches to visualize the multi-causal aspect of developmental defects. Additionally, the ethical issues of using AI in delicate

*Corresponding authors: naunain.mahmood@uos.edu.pk; hira.6385@wum.edu.pk

ecological contexts by evaluating the data privacy limitations and transparency of models are also examined.

Introduction

The fusion of current technologies with conventional observational techniques is revolutionizing the fields of animal ecology and habitat conservation by providing a deeper understanding of the behaviors of wildlife (Marković et al., 2018). Unmanned aerial vehicles (UAVs), artificial intelligence (AI), and thermal imaging are some of the technologies that are opening up new possibilities for habitat surveying and addressing. Some of the drawbacks are labor-intensive and error-prone ground-based monitoring methods (Brooks & Greenberg, 2023). Hence, these developments could lead to more effective conservation strategies by enabling precise population estimates of vulnerable and invasive species without the significant resource commitment that was previously necessary (Remelgado et al., 2018). There are certain extrinsic factors like environmental and ecological and some intrinsic factors like genetic variations and physiological disturbances that contribute to these challenges (Song et al., 2020; Masoudzadeh et al., 2020). Using AI technologies to integrate ecological and genomic data provides a novel way to study animal developmental challenges.

Developmental Challenges in Animal Ecology

Developmental challenges are the problems which animals face in order to develop and procreate more or less normally due to defective genes and inadequate surroundings (Fowler, 1996). Mentioned below are the few causes of developmental challenges which are faced by animals.

- **Habitat Loss and Fragmentation:** Habitat destruction is the term that is used to describe when a habitat no longer provides the environment conducive to support the native species and we point at it as "destroyed habitat" or "habitat loss". "Habitat fragmentation" is the term intending to disconnect or fragment, an organism's preferred living environment, which eminently causes population dispersion and finally leads to ecosystem destruction. Geological processes, human activity, such as land conversion, and agriculture can alter the environment's habitats to be lost or fragmented. This limits the resources and living areas available to animals which poses developmental challenges, leading to higher mortality rates and poor reproductive progress resulting in the extinction of numerous species (Driscoll et al., 2021).
- **Environmental Stressors:** Animals are frequently most receptive to environmental changes such as variations in temperature, precipitation patterns, and extreme weather events during development (Gonzalez-Rivas et al., 2020).
- **Non-native Species:** A species that is non-native or non-indigenous is one

that has been brought outside of its natural range by human activity, either purposefully or unintentionally, directly or indirectly (Martinez et al., 2020).

- **Pollution:** The introduction of contaminants having a negative effect on the environment is called pollution. Anything, e.g., pesticides, or energy, e.g., radiation, can be considered a form of pollution. These pollutants can lower fitness and survival rates by causing physiological and reproductive abnormalities. For example, oxidative stress is the primary cause of the emergence of numerous diseases in animals (Samal et al., 2022).
- **Genetic Diversity:** The entire number of hereditary characteristics accountable for a species' hereditary makeup is called genetic diversity; it differs significantly depending on the totality of species and the differences among them, and is connected to a species' longevity. Species that have less genetic diversity are frequently more susceptible to disease and environmental changes (Kardos et al., 2021). Moreover, genetic variety in animals is essential to meet current manufacturing demands in various types of environments, for enabling continued genetic enhancement, and for accelerating adaptation to shifting reproduction objectives. Populations sometimes consider it strenuous to adjust to new situations because of lack of diversity (Hoban et al., 2020).
- **Disease and Parasitism:** Numerous illnesses and ailments can affect an animal's health because of their susceptibility to them. These two factors can have a major negative effect on an animal's development, making weakened animals less likely to survive and reproduce. Human activity and environmental changes can exacerbate the spread of diseases. However, animal health is primarily influenced by good husbandry, appropriate nutrition, and cleanliness (Charlier et al., 2020).

Traditional Methods Used in the Study of Developmental Challenges and their Drawbacks

Conventional approaches to investigate developmental problems in animals have included controlled laboratory experiments as well as observational studies conducted in natural environments. In addition to being time and resource-intensive, requiring years to produce results, these observational and breeding studies also raise ethical questions because they involve the manipulation of alive animals, particularly when that manipulation could result in injury or distress. The traditional approaches used of this purpose are:

- **Direct Observation:** This method entails observing animals in captivity or in their natural environment to record interactions, behaviors, and developmental milestones (March et al., 2020). Direct observation is useful for learning about natural behaviors, but it can also be time-consuming, limited by the observer's impact on animal behavior.
- **Cross-fostering and Translocation Methods:** Experiments involving the cross-fostering and translocation of young animals are conducted to investigate

the relative contributions of environment and genetics to development. These studies can be instructive, but they can also pose ethical and logistical challenges, and they might not perfectly mimic natural conditions (Li et al., 2021).

- **Comparative Anatomy and Physiology:** Finding out the anatomical and physiological variations between species or within populations can provide insights into how organisms adapt and develop. Nevertheless, this method frequently necessitates invasive techniques, does not have real-time data, and without more experimental evidence, the causes of observed differences may only be hypothesized (Ventrella et al., 2021).
- **Selective Breeding:** Selective breeding allows researchers to observe a trait's heritability and pinpoint the genetic components involved in development by breeding animals for particular traits over several generations. This approach is laborious, restricted to characteristics that are readily visible or quantifiable, and presents moral dilemmas for the well-being of the animals (Harbers et al., 2020).

Introduction to AI and Its Role in Studying Developmental Challenges

The imitative behavior of human intellect in robots that have been instructed to think and act like a person is known as artificial intelligence. AI covers a wide range of technologies, such as neural networks, automation, natural language programming, machine learning (ML), and deep learning (DL) (Figure 1), which enable computers to learn from and adapt to new data without needing to be explicitly programmed (Kelly et al., 2023). AI is crucial to animal research because it can analyze complex ecological data that is difficult for humans to process manually. Now, researchers can analyze enormous amounts of data

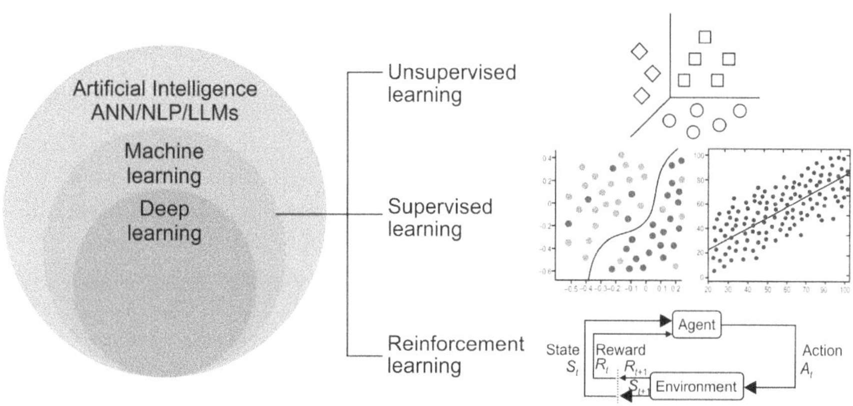

Figure 1: The subsets of Artificial Intelligence and different types of algorithms.

with greater accuracy and efficiency, revolutionizing the study and treatment of developmental changes in animals (Muthukrishnan et al., 2020).

Machine Learning

Machine learning (ML) is the research and development of statistical algorithms that can perform activities without specific guidance, learn from data, and generalize to new data in the context of artificial intelligence. ML approaches have been applied in a variety of sectors, including computer vision, speech recognition, email filtering, large language models (LLM), agriculture, and medicine, to produce algorithms that perform essential jobs. Supervised and unsupervised learning are the two categories into which machine learning techniques fall. The data for supervised learning techniques include monitoring that are made up of labels or response variables and attributes or prognostic measures (Bossert & Hagendorff, 2021).

Pattern Recognition

Assigning a class to an observation based on patterns found in data is known as pattern recognition. Bioinformatics, signal processing, visual analysis, data visualization, finding information, data reduction, graphic design, and ML are among the fields in which PR is used. Modern pattern recognition algorithms have made machine learning an increasingly common tool due to the abundance of tremendous data and computing capacity (Li et al., 2023). Figure 2 depicts the steps involved in developing pattern recognition algorithms.

Figure 2: Steps involved in developing a pattern recognition algorithm.

Neural Networks and Deep Learning

Neural network is made up of many interconnected neurons, which communicate with one another by sending signals. Neurons can be mathematical models or biological cells. A biological neural network is a term used in neuroscience to describe the physical arrangement of nerve cells connected by synapses that is present in brains and complex nervous systems (Figure 3).

Deep learning (DL) algorithms have proven effective in tracking animal movements, monitoring habitats, and analyzing physiological data, among other animal research applications. This facilitates the analysis of intricate, multi-modal data sets, which has implications for comprehending changes in animal development (Jeantet et al., 2021).

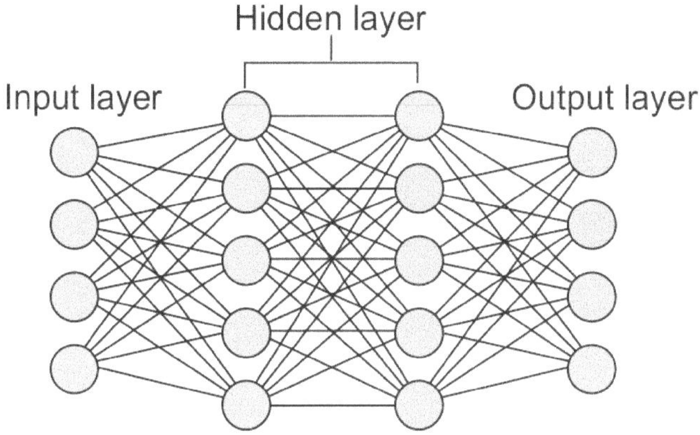

Figure 3: A neural network that displaying an output layer, several hidden layers, and one input layer as part of its architecture.

Application for Behavioral Analysis in *Girella tricuspidata*

With an emphasis on fish grazing behavior in natural marine environments, Ditria et al. (2021) highlighted the potential of DL techniques in automating the analysis of animal behavior directly from raw field imagery (Ditria et al., 2021). The researchers trained DL models on over 3,000 annotations of luderick, *Girella tricuspidata*, showcasing behaviors, using video footage shot in seagrass meadows in Queensland, Australia. The method used DL algorithms to classify particular grazing behaviors, spatiotemporal filtering to improve accuracy, and dense optical flow to evaluate pixel movement in underwater footage (Figure 4). DL algorithms were being utilized for analyzing animal behavior from underwater videos and the model gave high accuracy in identifying individual grazing events when applied to previously unseen video data. The results showed the effective combination of spatiotemporal filtering with neural networks to detect grazing behavior with greater precision.

Role of AI in Studying Developmental Challenges

AI is being used to enhance animal monitoring and marketing in the agricultural setting. There are AI applications that can be beneficial to individual animals in addition to the surveillance applications on animals. So, here is an overview of how AI technologies are helping in animals' study and overcoming developmental challenges.

- **Prediction and Diagnosis of Diseases:** Applications of AI have the potential to completely change how diseases are predicted and diagnosed in animals, improving animal health through better disease management. Applying DL

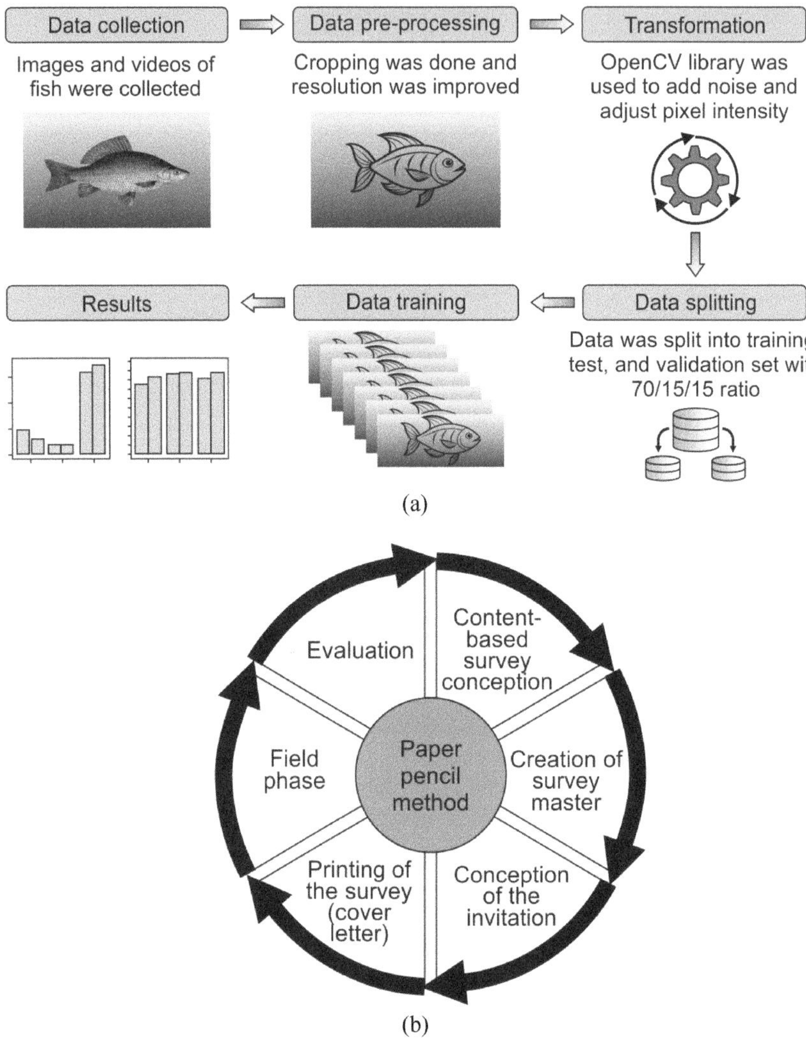

Figure 4: (a) The steps involved in developing a deep learning algorithm using images and video data from grass meadows in Australia. In the first step, data was collected. Then, it was pre-processed. After pre-processing, noise was added to enhance pixel intensity. The pre-processed data was split into 70% training, 15% testing and 15% validation sets. Then, the model was trained on ResNet50 Neural Network. After the training, the results were visualized. (b) Traditional methods of data collection for animal research

and AI techniques to automate the analysis process would be very helpful, considering the complexity and subjectivity involved in assessing medical images, in addition to the challenges that come with working with animals (Ahn et al., 2021).

- **Monitoring and Conservation of Habitats:** AI is becoming a potent ally in conservation efforts, and this is happening at a time when ecological challenges are becoming more complex and there is a pressing need to preserve our natural environment and individual biocultural legacies. AI-powered examination of ecological data and satellite imagery facilitates tracking animal migration, monitoring habitat alterations, and evaluating the effects of human activity on wildlife (Ahn et al., 2021).
- **Behavioral Analysis:** The goal of the developing field of computational animal behavior analysis, or CABA, is to use AI methods to assist animal behavior analysis. By analyzing data from sensors, videos, and other monitoring devices, CABA makes it easier to analyze animal behavior in detail. Many scientific disciplines that deal with animals agree that computational methods are necessary to help with the 'objectivization' and quantification of animal behavioral traits. These can spot behavioral patterns and abnormalities that might point to illness, stress, or environmental changes (Kumar and Jakhar, 2022).
- **Genetic and Evolutionary Studies:** AI systems are able to recognize genetic markers linked to certain diseases, characteristics, or behaviors, as well as analyze genetic data to comprehend the evolutionary relationships between species. AI and genetic research work together to successfully unlock the inner secrets of 'beasts' by utilizing the computational power of 'bytes' (Aithal, 2023).

Background for Studying Developmental and Ecological Challenges

Wildlife monitoring has changed due to the development of wearable sensors that can identify individual animals and record their temperature, heart rate, and motion. As demonstrated by Branko et al. (2018), drones' usage, distant monitoring, and camera traps has improved the capacity to monitor animal movement patterns, population changes, disease transmission, and poaching incidents maintaining ethical standards. However, research indicates that these wearable technologies may change the behavior of the animals to which they are attached as demonstrated by the effects of transmitters on snakes and radio tags on birds. However, it presents a big problem for the field because it's critical that monitoring technologies don't obstruct natural behaviors or cause any discomfort to them (Pala, 2012). In addition to improving our capacity to observe and comprehend animal behavior, the convergence of AI and contemporary technology is also revolutionizing the fields of genetic and evolutionary research. For example, AI's ability to process complex genetic data more efficiently than traditional statistical methods opens up new avenues for determining the cause of diseases (Jibrila et al., 2020).

Current AI Applications in Addressing Developmental Challenges

The use of digital technology libraries spans diagnosis, monitoring, predictive modeling, and personalized interventions. The current AI applications being used in studying developmental challenges are given below.

BLUP (Best Linear Unbiased Prediction)

Concerning breeding animals, the application of AI algorithms with techniques of Best Linear Unbiased Prediction (BLUP), improving accuracy and efficiency of breeding programs, accordingly provides multiple perspectives on improvement. AI is superior in processing large data sets (Gorjanc & Hickey, 2018). This ability to classify and separate good milk-generating cows from those which are unable to produce relies on the general classification. AI views two considerations as the most vital ones: to ensure the complexity of the genetic data is properly managed and the chance for inbreeding is minimized, while optimizing genetic possibilities are optimized. The label BLUP or the best linear unbiased prediction is a short form for this. This is a well-designed program that assists in understanding who are most expected to hand over meaningful characteristics to their offspring. The breeding value is a descriptive measure of what an animal is capable of or the extent to which it can pass on desirable traits to its offspring (Guo et al., 2023). Dairy cow milk yield assessment, BLUP analysis, environmental and genetic effects, as well as cows' individual performances are the purposes of our study.

Data Collection

Five dairy cows (A, B, C, D, and E) and their milking figures from three lactations are shown in Table 1. The factors that we considered were the impact of environmental means of better farm management practices that is estimated to have been increased by 100 liters for each of the cows before the second lactation. It means the breed average of milk yields for three lactations are 7,000 liters, 7,200 liters, while for the next lactation the yield will be 7,300 liters. We look at the genetic disparities between the subjects, Cows A and B that are half-

Table 1: Milk yield of five cows over three lactations

Cow	Lactation I	Lactation II	Lactation III
A	7000	7200	7100
B	6800	7000	6900
C	7500	7700	7600
D	7100	7300	7200
E	7300	7500	7400

sisters because they share the sire with each other, and Cows D and E that are another pair of full sisters but have no familial connections with the previous pair. Our milk yield is accompanied by a factor of heritability 0.30. For instance, doing so, BLUP enables breeding value evaluations in a sophisticated detail that accounts not only for environmental issues but also for genetic relationships so that one can view who has better genetic effects to milk contribution in the given populations.

First of all, we do the environmental adjustment values:

Lactation 2 adjusted yield = Original yield – 100 liters (for all cows)
Lactation 3 adjusted yield = Original yield – 100 liters (for all cows)

Then, we do the calculation for Adjusted Average Yield (given for Cow A only):

Adjusted Lactation 1: 7000 (no adjustment needed)
Adjusted Lactation 2: 7200 – 100 = 7100
Adjusted Lactation 3: 7100 – 100 = 7000
Average Adjusted Yield for Cow A: (7000 + 7100 + 7000)/3 = 7033 liters

The adjusted average yield for Cow A is determined to be 7033 liters. Now, let's compute the deviations for Cow A:

Deviation for Cow A = 7033 – 7117 = –84 liters

Calculating the adjusted average yields for other cows as follows:

Cow B: 7000 liters
Cow C: 7533 liters
Cow D: 7233 liters
Cow E: 7433 liters

The following are their deviations from the breed average:

Cow B: 7000 – 7117 = –117 liters
Cow C: 7533 – 7117 = +416 liters
Cow D: 7233 – 7117 = +116 liters
Cow E: 7433 – 7117 = +316 liters

The following formula can be used to get the breeding values (BV) based on deviations using the heritability of milk yield ($h^2 = 0.30$):

$$\text{Breeding Value} = \text{Deviation} \times \text{Heritability} \tag{1}$$

For Cow A,

Breeding Value of A = –84 × 0.30 = –25.2 liters

Using the same formula, the following are the breeding values for other cows:

Breeding Value of B = –117 × 0.30 = –35.1 liters
Breeding Value of C = +416 × 0.30 = +124.8 liters
Breeding Value of D = +116 × 0.30 = +34.8 liters
Breeding Value of E = +316 × 0.30 = +94.8 liters

In terms of genetic relationships, let's take the shared genetics between cows D and E (full sisters) and A and B (half-sisters) into consideration to ease the adjustment. Assume that the breeding values of related cows would have the genetic similarity adjustment factor (0.05 for half-sisters and 0.10 for full sisters) deducted to account for their increased genetic risk of inbreeding:

Adjusted BV of Cow A = –25.2 – (0.05 × –25.2) = –23.94 liters
Adjusted BV of Cow B = –35.1 – (0.05 × –35.1) = –33.345 liters
Adjusted BV of Cow C remains +124.8 liters (no adjustment as no direct relatives included)
Adjusted BV of Cow D = +34.8 – (0.10 × +34.8) = +31.32 liters
Adjusted BV of Cow E = +94.8 – (0.10 × +94.8) = +85.32 liters

Interpretation

Following the incorporation of genetic relationships, breeding value calculations, and environmental effect adjustments, we have found that, with a genetic potential for milk yield much higher than the breed average, cow C exhibits the highest potential. Even after adjusting for her close genetic relationship to Cow D, Cow E still exhibits strong genetic potential. As half-sisters, cows A and B exhibit a reduced genetic potential for milk yield; based on these crude calculations, cow B is the least desirable. This example shows how easy it can be to determine breeding values using BLUP, particularly when taking genetic relationships, environmental factors, and multiple lactations into account.

Computerized Mate Selection Programs

Computerized mate selection programs are software tools designed to assist in the selection of optimal mating pairs within animal breeding trials. These programs maintain genetic diversity and avoid inbreeding by utilizing information about genetic data, pedigree data, and occasionally phenotypic traits of animals to indicate a crossbreed that will likely provide optimal traits. The main objectives of these programs are to guarantee the long-term sustainability of breeding populations, improve particular traits of functional or economic importance, and improve the genetic quality and health of future generations.

These programs speed up the development of desirable traits like milk production in dairy cattle, growth rates in beef cattle, or egg production in poultry by making it easier to choose mating pairs that can pass on superior genetics to their progeny. For example, inbreeding can reduce fertility, cause immune systems to fail, and lead to higher disease in populations.

Alphamate

AlphaMate is a software tool facilitating modern technologies involving the targeted genomic editing for better optimization, meanwhile ensuring the highest protection and ideal breeding selection via genetic map. The data dealing with pedigreed or genome-wide markers can be taken care of by AlphaMate. It provides the plan from the breeding or conservation goals taking into account the difference between plants and animals.

Computer Vision and Image Analysis

Computer Vision has the ability to analyze a range of data types through ML and DL algorithms. These algorithms can use sounds and pictures for analysis. With these analytics, it is evident that the minor developmental challenges can be spotted which normally go unnoticed in case, if physically observed by humans.

High performance algorithms have an ability to do data pre-processing and they quickly detect patterns, unusual behaviors which can be disease indicators or fetal developmental defects. Additionally, the imaging analysis technology of veterinary science include those further than stable disease detection and also preponed diagnosis provision based on the observation and complication of the physiological and behavioral data about an animal that were carried out (Lahoz-Monfort & Magrath, 2021).

Results of the Current Applications

The application of AI in studying and addressing developmental challenges in animals has led to several significant outcomes:

- **Advanced Disease Management:** Advanced Neural Networks have made a substantial impact on the speed and accuracy of animal diseases detection. This has made it possible to take prompt action, prior to stopping the progression of developmental problems (Dekkers & Hospital et al., 2002).
- **Effective Habitat Management:** Just as important as safeguarding the animals themselves is maintaining their natural habitats. AI can guide conservationists in their efforts to restore and preserve natural habitats by analyzing ecosystem data and identifying areas that require restoration. By protecting or restoring their natural habitats, AI has helped to inform more efficient habitat management and restoration techniques, which have a positive effect on the growth of animal populations (Bendel, 2022).
- **Improved Knowledge of Genetic Factors:** Genetic markers linked to developmental difficulties have been found through AI-driven genomic analyses, opening the door for targeted breeding initiatives and genetic interventions to lessen these problems (Roh et al., 2019).
- **Proactive Conservation Efforts:** By using predictive modeling, conservationists can now foresee and reduce potential developmental obstacles in the future, ensuring the long-term survival of species that are at risk (Fletcher et al., 2003).

Collection of Data for AI Development

Data collection, a subject of ongoing study in various communities, is one of the main challenges in machine learning. These days, gathering data is becoming more and more important for two main reasons. New applications that may not have adequate labeled data are emerging as ML becomes increasingly popular (Bass et al., 2007). Secondly, in contrast to conventional ML, DL methods produce features on their own, saving labor on feature engineering. However, this may necessitate more labeled data. It's noteworthy to note that the data management group, along with the groups of machine learning, natural language processing, and image recognition, has contributed to contemporary research in data collection since handling huge amounts of data is so crucial.

Traditional Data Collection Techniques

The process of gathering data affects the results of getting, sending, and keeping data while conducting any kind of study. According to Fletcher et al. (2003), there are two popular ways to gather data: using a pencil and paper record or a portable computer with keys. There are advantages and disadvantages to both approaches, so researchers need to decide which is best for their particular study (Gravlee et al., 2006). Traditional methods of data collection for animal research are depicted in Figure 2b.

Paper and Pencil Method

Data recording with paper and pencil is an easy, affordable, and adaptable method (Koster, 2006). Recording the names of the particular behaviors on a lined notebook paper page and then transferring it to a data sheet was the pen and paper technique of data collecting.

Handheld Recorder

Instead of writing down a lot of various sorts of data, the researcher may input information more rapidly when utilizing a handheld recorder to collect data by tapping a few buttons on the device (Novak & Riener, 2015). It would be more effective if users could identify each of the 25 keys instead of having to label five rows and five columns of the program interface. One drawback of electronic portable recorders is that users may inadvertently hit the wrong key by mistake. Consequently, once the data have been input, it might be challenging to annotate, modify, or rectify them if the gathered data are not shown right away. Most electronic recorders arrange data such that it may be transmitted straight to a spreadsheet on a computer, reducing the possibility of human mistake during data transmission.

Advanced Techniques

Numerous academic subjects have grown greatly as a result of technological advancements like artificial intelligence. This also applies to animal studies, where various sensing devices have made data collecting possible. These data may be processed by sophisticated computer systems with artificial intelligence built in. This enables researchers to pinpoint important behaviors associated with sickness detection, assess the animals' mental states, and even identify specific animal identities (Meng et al., 2020).

Table 2: Comparison between traditional and AI-based methods

Traditional Methods	AI-based Methods
Data Collection: Field observations, manual tracking	**Data Collection:** Sensors, Drones, satellite imagery
Data Analysis: Statistical methods, Hypothesis testing	**Data Analysis:** Machine learning, Deep learning
Accuracy: Depends on manual precision	**Accuracy:** Higher precision with large datasets
Efficiency: Time consuming, labor-intensive	**Efficiency:** Automated, faster processing
Scalability: Limited by human resources and time	**Scalability:** Scalable with computing power

Sensor Fusion

A sensor is a device that measures or detects a mechanical, chemical, biological, or combination of these properties. It then gathers and records the data so that a machine or human can interpret it. Sensor technologies can be categorized into types according to the needs of the animal farming market, such as sensors for feeding systems and precision milking robots; applications based on species. Hardware sensors include motion sensors, pedometers, infrared thermal imaging sensors, temperature sensors, RFID (radio frequency IDentification) tags, accelerometers, cameras, vision sensors, microphones, and facial recognition machines. Multimodal sensor fusion is the process of combining data from several sensors to overcome the limits of each sensor alone (Frazier et al., 2023). Data collected from various sensors that are attached with animal bodies is recorded and analyzed by using AI algorithms (Figure 5).

AI as Genome Informant

The term genome informant, in theory, may refer to any program, information, or data source that advances the knowledge of the genome. This could include tools, processes, databases, or scientific discoveries that shed light on the genetic composition, variations, and functional features of genomes in various organisms. For example, genome sequencing technologies and bioinformatic tools may be regarded as "genome informants" because of the enormous volumes of data they produce and evaluate, which contribute to our comprehension of genetic information.

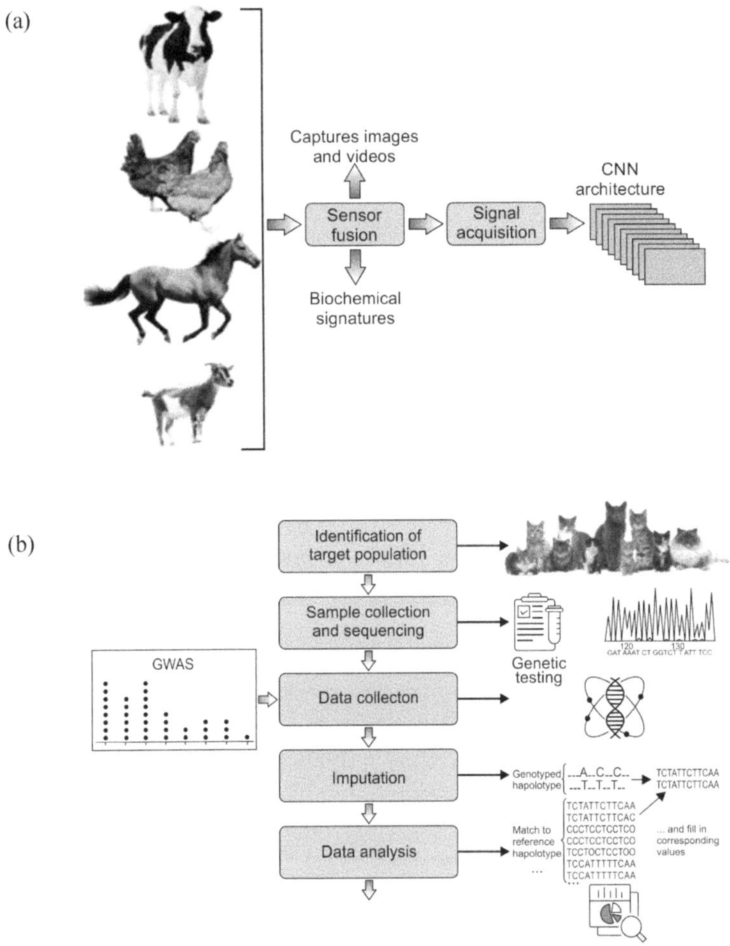

Figure 5: (a) Different types of wearable sensors are integrated with animal bodies in order to monitor their behavior and biochemical markers. The data from these sensors is recorded and AI algorithms are applied on the collected data for advanced analysis and study. The data helps the researchers for timely interventions in the developmental processes. (b) Figure shows the steps involved in GWAS. First of all, the target population is analyzed. Then, the sample is collected from this population. The samples are sequenced using advanced molecular biology techniques. The data from these techniques is collected and imputation is done by sequencing the missing genes. At last, the data is analyzed for predictions and interventions

AI acts as a guide by driving the enhancement of gene sequencing technologies and bioinformatic tools with high power to elucidate associated developmental problems offering increased accuracy, high precision and more in-depth genomics analysis. The databases that collect genetics and annotations of genes serve as resources by the scientists and medical practitioners dealing with the future of genomics and other related fields. In this genetic approach, the connection between genetic and ecosystem data is optimized to have a comprehensive understanding. Besides improving the analytical procedures by multimodal integration, AI algorithms have been used to disclose the genetic roots of developmental disorders.

Use of AI in Decoding Genetic Components for Developmental Problems

An observational investigation of a genome-wide set of genetic variants in various individuals is called a genome-wide association study (GWA study, or GWAS) in genomics to determine whether any variant is connected to a trait. Thousands of loci linked to disease have been found through GWAS, but many of the loci still need to have their molecular mechanisms investigated. Following GWAS, it makes sense to analyze these genetic relationships to comprehend the etiology of the disease (GWAS functional studies) and apply this understanding to animal care (GWAS translational studies). Despite the development of numerous functional genomics-based datasets and methodologies to aid in these investigations, substantial obstacles still exist because of the high dimensionality, multiplicity, and heterogeneity of the data.

While single-nucleotide polymorphisms (SNPs) and characteristics such as significant human diseases are the usual focus of GWA studies, the technique may be for various genetic variants and other organisms. AI is being used in decoding the genetic components for developmental problems by the methods given.

Sequence Assembly and Correction

The accuracy and efficiency of sequence assembly and correction which are essential for comprehending the genetic components that contribute to animal developmental problems are greatly increased by the use of AI technologies. Utilizing the massive amounts of information produced by transcriptomic and genomic projects (RNA sequencing and genome sequencing, for example), functional genomics makes use of these resources.

Instead of concentrating on the static components of hereditary data, like DNA sequence or structures, functional genomics focuses on the dynamic aspects, such as gene transcription, translation, regulation of gene expression, and protein–protein interconnections.

Practical genetic makeup has benefited from the implementation of two significant deep architectures, yielding highly significant scientific results. An entirely autonomous stand-alone programme called DeepBind is used to predict the sequence specificities of DNA and RNA binding proteins. DeepSEA (DL-based sequence analyzer) predicts the chromatin effects of sequence modifications with single-nucleotide resolution by learning regulatory sequences from large amounts of chromatin-profiling data. Accurate genomic sequences are crucial in the context of developmental challenges because they enable the identification of genetic variants and mutations that may cause diseases or abnormalities in animal development (Gouda et al., 2022). Both deep architecture-based approaches have successfully surmounted numerous obstacles, including processing millions of sequences, integrating data from various sources, tolerating noise and absence of data, and achieving end-to-end, fully automatic learning having no requirement for manual tuning. These strategies work more efficiently than other cutting-edge techniques, inspiring numerous scientists to pursue equally fascinating avenues.

Gene Prediction and Annotation

Finding the genomic DNA regions that encode genes is known as gene finding or gene prediction in computational biology. This covers both RNA and protein-coding genes, and it might also include predicting other functional components like regulatory regions. Once a species' genome has been sequenced, one of the first and most crucial steps in understanding it is finding its genes. CNNs can handle genomic sequences as sequences of patterns, recognizing and deciphering the different genetic elements like exons, introns, and regulatory regions, despite the fact that genomic sequences are not visual in nature (Liu et al., 2021).

DL algorithms have transformed the fields of gene annotation and prediction in genomics. DeepGene, a gene annotation tool that makes use of convolutional neural networks (CNNs), a class of DL models, particularly trained at identifying patterns in visual data, is one prominent illustration of this AI-driven approach in genomics. Procedure involved in conducting GWASs is shown in Figure 5.

Introduction of Multi-modality for Advanced Examination

Diagnostic and therapeutic radiography were the only imaging modalities available in veterinary medicine for obtaining an image during the diagnosis process. Unfortunately, radiographs only show an organ's shadow and don't provide much detail about the anatomical structures inside. While endoscopy is a particular method that offers full-color views of body tissues, when it became available, ultrasonography gave veterinarians a better imaging option.

Multi-modality can also help us understand how different species' developmental processes interact in the larger framework of developmental studies (Figure 6) (Oren, 2021).

Integration of Genomic Data and Ecological Data

The integrative field of ecological genetics aims to establish a connection

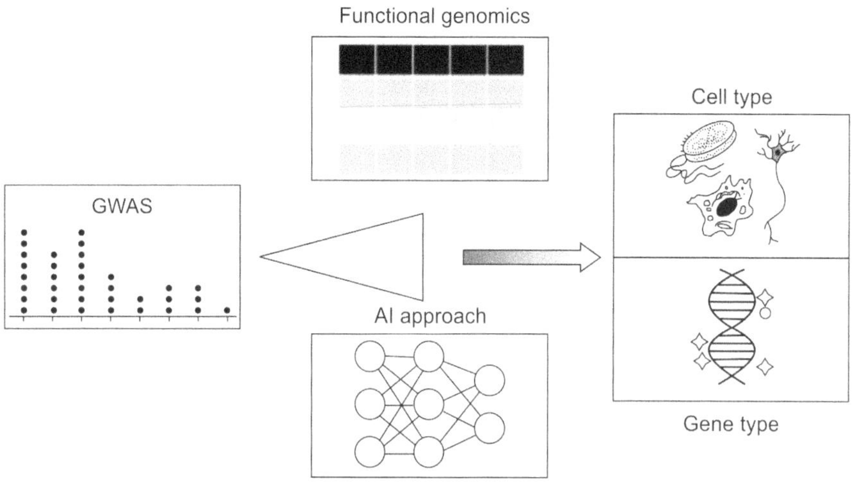

Figure 6: This figure shows the importance of multi-modality through the fusion of AI, ecological genetics, and GWAS in detecting the target gene or cell for each type of disorder, which can be genetic or environmentally induced.

between population hereditary characteristics, such as population contrast, demographic background, and adaptive hereditary changes, and changes in environmental or phenotypic variables. Different types of analyses, including comparative analyses, association studies, and landscape genetics, are based on the same principle, which is to quantify the interaction between an ecological and genetic dataset.

Ecological genetics is evolving from primarily population-level summaries to ecosystem-wide and individual-based analyses as instrumentation and genomic data become more readily available in large numbers of individuals. AI algorithms can identify environmental factors that trigger or exacerbate genetic vulnerabilities leading to developmental problems by analyzing ecological data, including habitat characteristics, exposure to pollutants, and diet, in conjunction with genetic data.

A natural extension of long-standing work in ecological genetics is the integration of genomic data with sophisticated animal instrumentation, and (Shafer et al., 2016) offers a framework for connecting the disparate data streams from these platforms (Zheng et al., 2022).

AI in Habitat Evaluation and Management

AI can improve the efficiency and hence the outcome of habitat evaluation and management which can help to overcome the development challenges faced by the animals. This way it matches the process carried out by the human brain, becoming more exact and comprehensive than all the other techniques.

Predictive Modeling for Habitat Quality

In ecology, the creation of predictive habitat distribution models has surged with the advent of new, potent statistical methods and GIS tools. AI-based predictive modeling is a new approach in environmental science, especially for habitat quality and conservation research. Since these models statistically relate the geographic distribution of species or communities to their current environment, they are static and probabilistic in nature. Numerous models have been created to address a wide range of topics, including habitat or species management, biogeography, conservation biology, and research on climate change. Scientists can process and analyze enormous volumes of environmental data to forecast future changes in habitat conditions by utilizing AI algorithms.

The objectives of the study should be the main factor in selecting an evaluation measure. This could result in assigning varying weights to distinct categories of prediction errors, such as omission, commission, or confusion. Determining the range of applications for which the model predictions are appropriate will be made possible by testing the model in a larger variety of scenarios (both in space and time). Consequently, rather than relying solely on statistics, the model's qualification is largely determined by the study's objectives, which establish the model's usability and the qualification criteria (Li et al., 2018).

Analyzing Animal Behavior and Needs

Understanding the complex behaviors and spatial requirements of different animal species has become easier due to the incorporation of machine learning models into wildlife conservation techniques. Through the analysis of data gathered from GPS collars and other tracking devices, scientists are now able to learn more about the critical behaviors, habitat preferences, and movement patterns of wild animals.

There is an increasing prevalence of wearable devices that utilize the Inertial Measurement Unit (IMU) sensor technology. For instance, it can help with the evaluation of production performance and the automated behavior classification of grazing sheep (Stern et al., 2015).

Intelligent Animal Monitoring System (IAMS)

Tsinghua-Qingdao was the main developer of the Vector Intelligent Monitoring System (VIMS), also known as Intelligent Animal Monitoring System (IAMS).

The recording with various variables are sent distantly and in an instant through the transport layer to the platform layer following collection and preprocessing of the animal image data by the intelligent terminal method (Arshad et al., 2022).

AI-Driven Intervention Strategies

AI can be utilized to develop interventions that are tailor-made for certain needs, modeling through foresight to infer such behaviors and outcomes, and improving delivery of the interventions to increase their efficacy and effectiveness.

AI-based Intervention Strategies

AI-based intervention strategies are especially effective where traditional methods might not be sufficient. These tactics frequently include the following elements:

- **Data Analysis and Pattern Recognition:** AI algorithms leverage current data to create prediction models that assess habitat suitability and species distribution. Planning conservation interventions and identifying high conservation priority areas are made easier with the help of this information (Travers et al., 2019).
- **Predictive Modeling:** By evaluating ecological data and suggesting suitable restoration methods, AI aids in the restoration of ecosystems. The efficiency and efficacy of conservation efforts can be increased by using AI algorithms to optimize interventions like invasive species management and reforestation (Wood et al., 2018).

Cases of Application of the AI Approaches in the Planning and Performing Interventions

- **Conservation of Habitats:** AI has been utilized to map and track habitats, pinpointing those most vulnerable to human activity or climatic shifts. The establishment of protected areas and the rehabilitation of degraded habitats can both benefit from this information. AI analysis of satellite photos, for instance, can spot patterns in deforestation and direct actions to stop the loss of habitat for some animal species (Pedro et al., 2019).
- **Protection of Wildlife:** To stop poaching, AI-driven interventions are being employed. In order to forecast the likely location of the next poaching incident, predictive AI models examine historical data on poaching incidents and patrol routes. In this way, wildlife rangers can maximize the routes they patrol and stop poaching incidents before they happen (Bao & Xie, 2022).
- **Disease Control in Animal Populations:** AI models forecast disease outbreaks in livestock and wildlife, allowing for preventative action. AI has been used, for example, to track and forecast the spread of avian influenza, directing immunization efforts and other countermeasures against large-scale outbreaks.
- **Genetic Conservation:** To maintain the long-term survival of species, AI-driven genetic analysis assists in identifying genetic bottlenecks and other risks

in endangered populations. This information is then used to guide breeding programs and other genetic interventions (Neethirajan, 2024).

Challenges and Limitations of AI in Ecology

There are certain challenges and limitations associated with AI technologies that can impact animals. AI systems used in chicken manufacturing facilities are made to recognize information about the birds and their surroundings, change the birds' surroundings, and ultimately improve their quality of life. Human-targeting lethal autonomous weapons have sparked intense discussion both inside and outside of the AI ethics community. Drones, however, also target animals, particularly those that some people consider to be 'invasive' or 'pests' (Chatelain and Konar, 2015). For instance, an organization in New Zealand named Aeronavics is creating a completely autonomous drone that can locate possums, an Australian native mammal that is protected, but is seen as a feral species that is detrimental to New Zealand's forests, and then release poisons to eliminate them. Although there aren't many animals affected by AI technology intended to eliminate undesired wild creatures right now, these technologies have the potential to expand over the world. To make sure that its progress adheres to moral guidelines, it must be closely examined. For example, if shooting is determined to be essential for killing, the animal's head should be shot, not its body, as this might cause the animal to die slowly (Frankham, 2010).

Conclusion and Future Prospects

AI has transformative possibilities for studying and treating developmental issues in animals in the future. With the use of artificial intelligence technology, livestock farmers can now greatly improve the welfare of their animals. It is essential for their product quality as well as from an ethical and legal perspective. It's now easier than ever to monitor the living conditions of the animals and identify any anomalies that could endanger them, possibly with the help of smart technology and state-of-the-art software. Livestock farmers have the ability to mitigate the adverse environmental impacts of their industry and discontinue dubious, unethical, and non-sustainable practices.

Data entry into farm records, farm activity monitoring, economic performance analysis, animal health improvement, and soil fertility enhancement are all made easier by AI. Through the integration of knowledge from these fields, scientists are able to construct comprehensive AI models that consider the complex aspects of animal development.AI is generally a boon for increasing production and efficiency while reducing the likelihood of human error.

There are a few disadvantages as well, like the cost of development and the possibility that automation will eventually replace them. It's crucial to keep in mind, though, that the AI industry has the potential to create jobs, some of which are still unimagined. Consequently, problems include unemployment, the high cost of technology such as drones which can be costly to develop, assemble, and maintain, as well as the substantial volume of data required to train AI.

References

Ahn, J.C., A. Connell, D.A. Simonetto, C. Hughes, and V.H. Shah. (2021). Application of artificial intelligence for the diagnosis and treatment of liver diseases. *Hepatology*, 73(6): 2546–2563.

Aithal, P.S. (2023). Super-intelligent machines-analysis of developmental challenges and predicted negative consequences. *International Journal of Applied Engineering and Management Letters (IJAEML)*, 7(3): 109–141.

Arshad, J., A.U. Rehman, M.T.B. Othman, M. Ahmad, H.B. Tariq, M.A. Khalid, ... and H. Hamam. (2022). Deployment of wireless sensor network and IoT platform to implement an intelligent animal monitoring system. *Sustainability*, 14(10): 6249.

Bao, J., and Q. Xie. (2022). Artificial intelligence in animal farming: A systematic literature review. *Journal of Cleaner Production*, 331: 129956.

Bass, B.L., K.D. Linney, A.B. Butler, and J.G. Grzywacz. (2007). Evaluating PDAs for data collection in family research with non-professional couples. *Community, Work, and Family*, 10(1): 57–74.

Bendel, O. (2022). Passive, active, and proactive systems and machines for the protection and preservation of animals and animal species. *Frontiers in Animal Science*, 3: 834634.

Bossert, L., and T. Hagendorff. (2021). Animals and AI. The role of animals in AI research and application: An overview and ethical evaluation. *Technology in Society*, 67: 101678.

Brooks, S.K., and N. Greenberg. (2023). The well-being of companion animal caregivers and their companion animals during the COVID-19 pandemic: Scoping review. *Animals*, 13(20): 3294.

Charlier, J., L. Rinaldi, V. Musella, H.W. Ploeger, C. Chartier, H.R. Vineer, ... and E. Claerebout. (2020). Initial assessment of the economic burden of major parasitic helminth infections to the ruminant livestock industry in Europe. *Preventive Veterinary Medicine*, 182: 105103.

Chatelain, E., and N. Konar. (2015). Translational challenges of animal models in Chagas disease drug development: A review. *Drug Design, Development, and Therapy*, 4807–4823.

Dekkers, J.C., and F. Hospital. (2002). The use of molecular genetics in the improvement of agricultural populations. *Nature Reviews Genetics*, 3(1): 22–32.

Ditria, E.M., E.L. Jinks, and R.M. Connolly. (2021). Automating the analysis of fish grazing behavior from videos using image classification and optical flow. *Animal Behaviour*, 177: 31–37.

Driscoll, D.A., D. Armenteras, A.F. Bennett, L. Brotons, M.F. Clarke, T.S. Doherty, ... and T. Wevill. (2021). How fire interacts with habitat loss and fragmentation. *Biological Reviews*, 96(3): 976–998.

Fletcher, L.A., D.J. Erickson, T.L. Toomey, and A.C. Wagenaar. (2003). Handheld computers: A feasible alternative to paper forms for field data collection. *Evaluation Review*, 27(2): 165–178.

Fowler, M.E. (1996). Development and challenges of wildlife husbandry: Concluding remarks. *Revue Scientifique et Technique (International Office of Epizootics)*, 15(1): 337–342.

Frankham, R. (2010). Challenges and opportunities of genetic approaches to biological conservation. *Biological Conservation*, 143(9): 1919–1927.

Frazier, T.W., R.M. Busch, P. Klaas, K. Lachlan, S. Jeste, A. Kolevzon, ... and M. Uljarević. (2023). Development of webcam-collected and artificial-intelligence-derived social and cognitive performance measures for neurodevelopmental genetic syndromes. *American Journal of Medical Genetics Part C: Seminars in Medical Genetics*, 193(3): e32058.

Gonzalez-Rivas, P.A., S.S. Chauhan, M. Ha, N. Fegan, F.R. Dunshea, and R.D. Warner. (2020). Effects of heat stress on animal physiology, metabolism, and meat quality: A review. *Meat Science*, 162: 108025.

Gorjanc, G., and J.M. Hickey. (2018). AlphaMate: A program for optimizing selection, maintenance of diversity and mate allocation in breeding programs. *Bioinformatics*, 34(19): 3408–3411.

Gouda, H.F., F.A. Hassan, E.E. El-Araby, and S.A. Moawed. (2022). Comparison of machine learning models for bluetongue risk prediction: A seroprevalence study on small ruminants. *BMC Veterinary Research*, 18(1): 394.

Gravlee, C.C., S.N. Zenk, S. Woods, Z. Rowe, and A.J. Schulz. (2006). Handheld computers for direct observation of the social and physical environment. *Field Methods*, 18(4): 382–397.

Guo, W., C. Lv, M. Guo, Q. Zhao, X. Yin, and L. Zhang. (2023). Innovative applications of artificial intelligence in zoonotic disease management. *Science in One Health*, 2: 100045.

Harbers, H., C. Zanolli, M. Cazenave, J.C. Theil, K. Ortiz, B. Blanc, ... and T. Cucchi. (2020). Investigating the impact of captivity and domestication on limb bone cortical morphology: An experimental approach using a wild boar model. *Scientific Reports*, 10(1): 19070.

Hoban, S., M. Bruford, J.D.U. Jackson, M. Lopes-Fernandes, M. Heuertz, P.A. Hohenlohe, ... and L. Laikre. (2020). Genetic diversity targets and indicators in the CBD post-2020 Global Biodiversity Framework must be improved. *Biological Conservation*, 248: 108654.

Jeantet, L., V. Vigon, S. Geiger, and D. Chevallier. (2021). Fully convolutional neural network: A solution to infer animal behaviors from multi-sensor data. *Ecological Modelling*, 450: 109555.

Jibrila, I., J. Ten Napel, J. Vandenplas, R.F. Veerkamp, and M.P. Calus. (2020). Investigating the impact of preselection on subsequent single-step genomic BLUP evaluation of preselected animals. *Genetics Selection Evolution*, 52: 1–10.

Kardos, M., E.E. Armstrong, S.W. Fitzpatrick, S. Hauser, P.W. Hedrick, J.M. Miller et al. (2021). The crucial role of genome-wide genetic variation in conservation. *Proceedings of the National Academy of Sciences*, 118(48): e2104642118.

Kelly, S., S.A. Kaye, and O. Oviedo-Trespalacios. (2023). What factors contribute to the acceptance of artificial intelligence? A systematic review. *Telematics and Informatics*, 77: 101925.

Koster, J. (2006). The use of the observer 5.0 and a psion handheld computer in a remote fieldwork setting. *Field Methods*, 18(4): 430–436.

Kumar, D., and S.D. Jakhar. (2022). Artificial intelligence in animal surveillance and conservation. In: S. Balamurugan, S. Pathak, A. Jain, S. Gupta, S. Sharma, and S. Duggal (Eds.), *Impact of Artificial Intelligence on Organizational Transformation*, pp. 73–85. Scrivener Publishing, Beverly, Massachusetts, USA.

Lahoz-Monfort, J.J., and M.J. Magrath. (2021). A comprehensive overview of technologies for species and habitat monitoring and conservation. *BioScience*, 71(10): 1038–1062.

Li, S., X. Hao, B. Sun, J. Bi, Y. Zhang, and W. Du. (2021). Phenotypic consequences of maternally selected nests: A cross-fostering experiment in a desert lizard. *Integrative Zoology*, 16(5): 741–754.

Li, X., B.Y. Cai, and C. Ratti. (2018). Using street-level images and deep learning for urban landscape studies. *Landscape Architecture Frontiers*, 6(2): 20–30.

Li, Y., K.M. Slavik, H.C. Toyoda, B.R. Morehouse, C.C. de Oliveira Mann, A. Elek, ... and P.J. Kranzusch. (2023). cGLRs are a diverse family of pattern recognition receptors in innate immunity. *Cell*, 186(15): 3261–3276.

Liu, F., Y. Li, H. Yu, L. Zhang, J. Hu, Z. Bao, and S. Wang. (2021). MolluscDB: An integrated functional and evolutionary genomics database for the hyper-diverse animal phylum Mollusca. *Nucleic Acids Research*, 49(D1): D988–D997.

March, D., L. Boehme, J. Tintoré, P.J. Vélez-Belchi, and B.J. Godley. (2020). Towards the integration of animal-borne instruments into global ocean observing systems. *Global Change Biology*, 26(2): 586–596.

Marković, B., D. Nedić, and S. Minić. (2018). ICT systems for monitoring and protection of wildlife in their natural environment. *Veterinary Journal of Republic of Srpska*, 18(1): 157–181.

Martinez, B., J.K. Reaser, A. Dehgan, B. Zamft, D. Baisch, C. McCormick, A.J. Giordano et al. (2020). Technology innovation: Advancing capacities for the early detection of and rapid response to invasive species. *Biological Invasions*, 22(1): 75–100.

Masoudzadeh, S.H., M. Mohammadabadi, A. Khezri, R.V. Stavetski, V.P. Oleshko, O.I. Babenko, Z. Yemets et al. (2020). Effects of diets with different levels of fennel (*Foeniculum vulgare*) seed powder on DLK1 gene expression in brain, adipose tissue, femur muscle, and rumen of Kermani lambs. *Small Ruminant Research*, 193: 106276.

Meng, T., X. Jing, Z. Yan, and W. Pedrycz. (2020). A survey on machine learning for data fusion. *Information Fusion*, 57: 115–129.

Muthukrishnan, N., F. Maleki, K. Ovens, C. Reinhold, B. Forghani, and R. Forghani. (2020). Brief history of artificial intelligence. *Neuroimaging Clinics of North America*, 30(4): 393–399.

Neethirajan, S. (2024). Artificial intelligence and sensor innovations: Enhancing livestock welfare with a human-centric approach. *Human-Centric Intelligent Systems*, 4(1): 77–92.

Novak, D., and R. Riener. (2015). A survey of sensor fusion methods in wearable robotics. *Robotics and Autonomous Systems*, 73: 155–170.

Oren, A. (2021). Exploring multi-modality in animal-centered computing. In: *Proceedings of the Eighth International Conference on Animal-Computer Interaction*, pp. 1–3. Association for Computing Machinery, New York, United States.

Pala, A.K.I.N. (2012). A new selection procedure in animal breeding: Classification, error rates, and comparison with BLUP. *The Journal of Animal & Plant Sciences*, 22(3): 537–544.

Pedro, F., M. Subosa, A. Rivas, and P. Valverde. (2019). *Artificial Intelligence in Education: Challenges and Opportunities for Sustainable Developmen*t. UNESCO, Paris.

Remelgado, R., B. Leutner, K. Safi, R. Sonnenschein, C. Kuebert, and M. Wegmann. (2018). Linking animal movement and remote sensing–mapping resource suitability from a remote sensing perspective. *Remote Sensing in Ecology and Conservation*, 4(3): 211–224.

Roh, Y., G. Heo, and S.E. Whang. (2019). A survey on data collection for machine learning: A big data-AI integration perspective. *IEEE Transactions on Knowledge and Data Engineering*, 33(4): 1328–1347.

Samal, K., S. Mahapatra, and M.H. Ali. (2022). Pharmaceutical wastewater as emerging contaminants (EC): Treatment technologies, impact on environment, and human health. *Energy Nexus*, 6: 100076.

Song, W., W. Song, H. Gu, and F. Li. (2020). Progress in the remote sensing monitoring of the ecological environment in mining areas. *International Journal of Environmental Research and Public Health*, 17(6): 1846.

Stern, U., R. He, and C.H. Yang. (2015). Analyzing animal behavior via classifying each video frame using convolutional neural networks. *Scientific Reports*, 5(1): 14351.

Travers, H., M. Selinske, A. Nuno, A. Serban, F. Mancini, T. Barychka, ... and E.J. Milner-Gulland. (2019). A manifesto for predictive conservation. *Biological Conservation*, 237: 12–18.

Ventrella, D., N. Ashkenazi, A. Elmi, K. Allegaert, C. Aniballi, A. DeLise, ... and M.L. Bacci. (2021). Animal models for *in vivo* lactation studies: Anatomy, physiology, and milk compositions in the most used non-clinical species: A contribution from the ConcePTION project. *Animals*, 11(3): 714.

Wood, K.A., R.A. Stillman, and G.M. Hilton. (2018). Conservation in a changing world needs predictive models. *Animal Conservation*, 21(2): 87–88.

Zheng, S., Z. Zhu, Z. Liu, Z. Guo, Y. Liu, Y. Yang, and Y. Zhao. (2022). Multi-modal graph learning for disease prediction. *IEEE Transactions on Medical Imaging*, 41(9): 2207–2216.

Zimring, J.C., S.R. Stowell, J.M. Johnsen, and J.E. Hendrickson. (2012). Effects of genetic, epigenetic, and environmental factors on alloimmunization to transfused antigens: Current paradigms and future considerations. *Transfusion Clinique et Biologique*, 19(3): 125–131.

9 | Importance of AI-Tech in Managing a Protected Area: Considering Forest Health Dynamics

Amlan Jyoti Kalita[1] , Jyotishman Deka[2], Chayanika Devi[3] and Pankaj Kalita[4*]

[1] Department of Electronics and Communication Engineering, School of Engineering, Tezpur University, Tezpur, India

[2] Manash Tiger Reserve (Research Wing), Barpeta Road, Barpeta, Assam, India

[3] Department of Zoology, Pub Kamrup College, BaiahataChariali, Kamrup, Assam, India

[4] Department of Zoology, Eastern Karbi Anglong College, Sarihajan, Karbi Anglong, Assam, India

Forest health can be defined as the integration of ecological indicators of forest condition and function, which are executed at various geographical scales. Humans rely on healthy forests for energy, construction materials, food, as well as functions like carbon storage, biodiversity, climate regulation, and, more importantly, oxygen generation. Although natural forests are acclimated to some degree of disturbances, all forests are currently facing new pressures from climate change. Artificial Intelligence (AI) technologies can play a crucial role in managing a protected area as they facilitate tasks such as image classification, object detection, species identification, counting individuals, species segregation etc. Big data management of forest ecology relies heavily on camera traps for studying animal behaviour and estimating density. Camera traps powered by AI can automatically capture and analyze images of wildlife, aiding researchers in tracking and identifying species in their natural habitats. Satellite imagery, combined with AI, is employed to detect deforestation, illegal logging, and habitat changes, facilitating timely conservation interventions. Predictive modelling, specifically species distribution models, assists conservationists in efficiently allocating resources and planning habitat restoration by predicting the distribution of various species across space and time. The technology also enables behavioural analysis, detecting mating, predation, and social interactions. Population estimation benefits from AI by analyzing the number

*Corresponding author: kalita.pankaj9@gmail.com

of individuals in camera trap images, contributing to population models. Habitat assessment is also improved by image processing for habitat quality and changes over time. This is crucial for maintaining the ecosystem's health and balance, particularly in the presence of key predators like tigers, leopards, wild dogs, and clouded leopards. Further, AI-equipped drones and sensor networks are deployed to detect and track poachers in protected areas, enabling authorities to respond swiftly to potential threats. In addition, AI is utilized for predicting the distribution of amphibians through the analysis of audio data from remote microphones, enabling researchers to monitor species that are challenging to observe visually. This chapter aims to discuss the different AI technologies contributing significantly to the effective management of protected areas, particularly to ensure the health and sustainability of forests.

Introduction

Forest ecosystems sustain their complexity while providing for human needs. Traits like diverse species composition, including both native and non-native plants and animals are the constituent elements of a forest ecosystem. A balanced structure with trees of different ages and sizes provides a habitat for a wide range of wildlife (Smith et al., 2009). A dynamic forest ecosystem has its resilience to natural disturbances, and efficient nutrient cycling and contributes to water regulation and purification, crucial for both ecosystems and human communities. Furthermore, healthy forests serve as a contributor towards different resources including significant carbon sinks, storing and sequestering carbon dioxide, thus mitigating climate change impacts. Lastly, they demonstrate resistance to invasive species, maintaining the balance of ecosystem dynamics and supporting sustainability (U.S. Forest Service, 2012).

"Ecological integrity" provides a useful framework for ecologically based monitoring and can provide valuable information for assessing ecosystem conditions and management effectiveness (Tierney et al., 2009). Monitoring and managing forest health is imperative for sustaining ecological integrity, preserving biodiversity, and ensuring the continued provision of ecosystem services essential for human well-being. By understanding and maintaining forest health, sustainable resource management practices can be promoted that support not only the health of forests themselves but also the interconnected ecosystems and communities that depend on them. In recent years, forest pathology has gained significant importance in the changing world making it necessary to provide an update of recent literature (Pautasso et al., 2015). It provides crucial ecosystem services necessary for human well-being, including clean air and water, carbon sequestration, climate regulation, soil erosion control, and recreational opportunities.

Ensuring forest health is vital for sustaining these services. Moreover, forests play a significant role in mitigating climate change by sequestering carbon dioxide from the atmosphere and storing it in biomass and soils. Healthy forests are more effective at carbon sequestration, helping to offset greenhouse gas emissions and mitigate climate change impacts. Furthermore, forests contribute to economic prosperity through industries like timber production, ecotourism, and non-timber forest products. Healthy forests support sustainable economic activities and livelihoods while also providing valuable ecosystem services.

Lastly, forests hold cultural and spiritual significance for many indigenous and local communities worldwide, providing habitats for traditional practices, ceremonies, and spiritual connections to nature. Furthermore, forests positively impact human health by reducing stress, improving mental well-being, and offering opportunities for recreation and physical activity. Overall, maintaining forest health is essential for sustaining ecosystems, supporting biodiversity, mitigating climate change, and promoting human well-being and socio-economic development.

Planted forests are increasingly threatened by insects and microbial pathogens, which are introduced accidentally and/or have adapted to new host trees (Wingfield et al., 2015). Among both forest practitioners and the general public, "forest health" has become an issue of contention. Whereas the debate over which treatments will best achieve healthy forests has been framed largely by the popular media and politicians as a struggle between industry and environmentalists, the views of the general public remain unexplored (Abrams et al., 2005). Found a higher number of multipurpose and preferred species than in the 'conservation' forest, which responded to the needs of the community in the long term to have more wood products (both firewood and timbers) (Kijtewachakul et al., 2004). Forest health is under threat from various factors such as deforestation, climate change, invasive species, pests, diseases, pollution, overexploitation, and fire suppression. Over the past 50 years, human agents of deforestation have changed in ways that have potentially important implications for conservation efforts (Rudel et al., 2009).

Climate change exacerbates these issues by increasing the frequency and intensity of wildfires, droughts, storms, and pest outbreaks, disrupting ecosystem dynamics and reducing resilience. Invasive species, including plants, animals, insects, and pathogens, can outcompete natives, alter habitats, and exacerbate vulnerability to pests and diseases. Pests and diseases like bark beetles and fungal pathogens also pose significant threats by causing tree mortality and ecosystem changes. Addressing these threats necessitates integrated approaches involving sustainable land management, conservation, policy interventions, community engagement, and scientific research to enhance the resilience and sustainability of forest ecosystems.

Maintaining Forest Health: Shift from Traditional Method to Big Data Methods

The shift from traditional methods to big data techniques is transforming forest health management (Raihan, 2023; Budnik et al., 2023; Sun & Scanlon,

2019; Kourtz, 1990). Big data leverages remote sensing (Peterson et al., 1999), artificial intelligence (AI) (Zulfiqar et al., 2021), and predictive modelling (Chavan et al., 2018) to monitor forests comprehensively and predict threats like disease outbreaks (Trumbore et al., 2015) and habitat loss (Liff et al., 1994). This transition offers more efficient and data-driven solutions for preserving biodiversity and safeguarding the long-term health of forests.

Limitations of Traditional Method

Traditional non-AI methods of forest health management mostly rely on manual inspection and monitoring (Liff et al., 1994). With recent advancements in technologies, monitoring methods evolved towards remote sensing technologies (Barka et al., 2018; Lausch et al., 2017), such as aerial photography (Backsen & Howell, 2013), satellite imagery (Wang et al. 2010), and Light Detection and Ranging (LiDAR) (Wulder et al., 2008). These services can provide valuable data for assessing forest health over large areas; however, there are various limitations associated with traditional methods. Traditional methods rely heavily on foresters and field workers for manual inspections. This approach can be effective in identifying visible signs of stress, such as pest infestations or disease outbreaks. Again, a human intervention always gives a better understanding of a situation in practice. However, it is often labour-intensive (Zhang et al., 2019), time-consuming (Wang et al., 2019), and limited in scale (Rahimi et al., 2021). Covering large forested areas thoroughly is challenging, leading to gaps in monitoring and delayed response to emerging threats. The labour-intensive work can be minimized using smart monitoring systems like remote monitoring (Torres et al., 2021; Kamaruidzaman and Rahmat, 2020; Lechner et al., 2020). However non-AI remote monitoring systems also require technical expertise to interpret remote sensing data to distinguish between various types of vegetation, detect subtle changes in forest conditions, and correlate them with potential stressors (Bravo-Oviedo et al., 2020). This is also required for systemic silviculture (Puettmann et al., 2015; Goodchild, 1991). This means long hours of work, which affects human work efficiency. This issue can be resolved using AI-driven monitoring, where the area of interest can be identified by an AI algorithm. Then an expert can manually monitor the selected portion only, increasing the work efficiency of manpower in forest departments. A commonly used traditional method of keeping records for mapping forest resources is the Geographic Information Systems (GIS) (Wieczorek & Delmerico, 2009; Store and Jokimäki, 2003). This is useful in assessing habitat suitability (Chambers, 2006), and planning animal conservation (González et al., 2011). GIS can provide valuable insights into forest composition, structure, and distribution, they rely on accurate input data and assumptions about ecological processes. Inaccuracies in inventory data, incomplete coverage, and outdated information can undermine the effectiveness of management decisions based on GIS analysis (Sisodiya et al., 2020). Human error can be reduced in the GIS system with an AI-driven automatic update system. While non-AI approaches have proven useful in forest management for ages, they frequently need tremendous human effort, skill, and resources to be effective. Furthermore, they may struggle to manage complex, dynamic, and linked ecological issues like climate change, invasive species, and habitat loss. Integrating AI and machine learning (ML) technology into existing forest management techniques may supplement traditional approaches by offering real-time monitoring, predictive analytics, and data-driven decision support, ultimately improving forest ecosystem resilience and sustainability.

Advantages of AI-powered Solution over Traditional Methods

AI offers several advantages over traditional methods in maintaining forest health. Where non-AI deeply relies on human expertise, AI-driven techniques depend on data and data only. AI-powered systems can analyze vast amounts of data from various sources such as satellite imagery (Reckling et al., 2021), drone images (Serrano et al., 2022), LiDAR (Sakr et al., 2011), weather patterns (Nikhilesh et al., 2023), and sensor networks (Fuentes & Tongson, 2021; Yu et al., 2021) to detect early signs of diseases and pest infestations in forests. This technology is useful for taking early action to prevent widespread damage (Liu et al., 2021). Apart from early detection, predictive action can also be achieved using AI-driven systems. AI algorithms can analyze historical data on forest health, including factors like climate conditions, species composition, and past disease outbreaks, to make accurate predictions about future threats. This predictive capability allows forest managers/workers to implement proactive measures to mitigate risks and preserve ecosystem balance.

AI can enhance remote sensing technologies to monitor forests in real time with high precision (Minakshi et al., 2020). Drones equipped with AI algorithms can identify changes in vegetation health (Haq et al., 2024), detect illegal logging activities (Hernández et al., 2021), and assess the impact of natural disasters (Mühling, 2023) more efficiently than traditional ground surveys. This can reduce significant human effort and enhance efficiency in forest monitoring. The precision forestry techniques by analyzing data at a small level minimize resource wastage and ecological disruption while maximizing the economic yield of forestry operations. AI-based decision support systems can assist forest officials in making informed decisions by synthesizing complex datasets and recommending the most effective management strategies (Tien, 2017). These systems consider multiple variables simultaneously, leading to more holistic and adaptive management practices.

Along with these benefits, AI systems can adapt and learn from ongoing monitoring and feedback, continuously improving their predictive accuracy and decision-making capabilities (Elavarasan & Vincent, 2020). This adaptability is particularly valuable in dynamic environments where forest conditions change rapidly due to factors like climate change and human activities. Reinforcement Learning (RL)-based systems also discover new ways to maintain forest health

(Wu et al., 2021). The use of AI is to assist humans only to improve human efficiency and not replace humans. AI offers a suite of tools and techniques that empower forest managers to monitor, analyze, and respond to forest health challenges more effectively than traditional methods. By using AI, the resilience and sustainability of forest ecosystems can be enhanced for future generations. The role of AI in practice with forest monitoring is described in the subsequent section.

Role of AI in Maintaining Forest Health

The role of AI in maintaining forest health is increasingly significant due to its ability to process large amounts of data, identify patterns, and make predictions. The role of AI in different applications in forest monitoring is detailed in the following points:

(a) **Early Detection of Diseases and Pests:** AI-powered systems play a vital role in analyzing satellite imagery and drone footage to detect signs of diseases, pests, or other stressors in forests (Gonzalez et al., 2016; Paters et al., 2020; Ye et al., 2022; Knebel et al., 2022; Pal et al., 2023). By recognizing subtle changes in vegetation patterns or colour, AI algorithms can identify potential outbreaks much earlier than traditional methods, allowing forest departments to take protective measures to reduce the disease or infestation. ML/DL can analyze historical data on forest health, including factors like weather patterns, soil composition, and previous disease outbreaks, to develop a prediction of disease (Nova, 2023). These models can forecast future changes in forest health, allowing managers to pre-emptively allocate resources and plan interventions to prevent or mitigate potential threats. Thus, AI can assist forest officials in making informed decisions. Further, AI-based decision support systems can consider multiple variables and scenarios to recommend the most effective strategies for maintaining forest health while balancing ecological, economic, and social objectives.

(b) **Monitoring Environmental Conditions:** AI algorithms can process data from various sensors deployed in forests to monitor environmental conditions such as temperature, humidity, soil moisture, and air quality (Rahardja, 2022; Folliot et al., 2022). By continuously analyzing this data, AI can help identify factors contributing to forest stress, such as drought or pollution, enabling forest managers to implement targeted interventions to maintain ecosystem health (Chisom et al., 2024). AI technologies can analyze data from wildlife cameras (Nitoslawski et al., 2021) and acoustic sensors (Kucera & Barrett, 2011) to monitor biodiversity in forests. By identifying species presence, population trends, and habitat preferences, AI can help assess the overall health of ecosystems and identify areas in need of conservation efforts. AI-powered systems equipped with remote sensors and cameras can detect wildfires in their early stages by analyzing smoke patterns, heat signatures, and other relevant data. Additionally, AI algorithms can assist in predicting the spread of wildfires based on factors like wind speed, terrain, and fuel moisture content, enabling more effective deployment of firefighting resources and evacuation efforts.

(c) Forest Inventory and Management: AI-powered systems can streamline forest inventory processes by analyzing remote sensing data to estimate tree species, density, height, and biomass. Software like AID-FOREST (Artificial Intelligence for Digital Forest) (Welbourne et al., 2016) can process point clouds obtained via mobile terrestrial laser scanning (MTLS) and then, deliver a collection of many valuable as well as precise dendrometric and forest stand parameters (Bhatt et al., 2022). This information is crucial for sustainable forest management practices, such as planning timber harvesting operations, monitoring carbon sequestration, and assessing habitat suitability for wildlife.

(d) Adaptive Management: AI-driven monitoring systems can continuously learn from new data and feedback, allowing for adaptive management strategies that evolve in response to changing environmental conditions and emerging threats (Stančić et al., 2022). By incorporating real-time information into decision-making processes, AI helps optimize resource allocation and maximize the resilience of forest ecosystems. By leveraging AI, forest managers may improve their capacity to protect biodiversity, alleviate hazards, and promote the long-term sustainability of forest ecosystems (Liu et al., 2024). In the next section, key techniques and tools used for maintaining forest health are discussed.

Key Techniques and Tools

The key AI tools employed in forests include AI-powered camera traps, AI-equipped drones, forest audio data analysis, and satellite imagery. Figure 1 represents the basic workflow of AI tools, which are discussed in the subsequent sections.

AI-powered Camera Traps

A camera trap is a camera that is automatically activated by motion in its area, such as the presence of an animal or a person. It is frequently equipped with a motion sensor like a passive infrared (PIR) sensor or an active infrared (AIR) sensor employing an infrared light beam (Alzuhair & Alghaihab, 2023; Maharajan et al., 2024). Camera traps have transformed the way ecologists investigate animal species distributions, activity patterns, and interspecific interactions (Bilski et al., 2017). Although video traps are an inexpensive way to monitor species, the time required for data processing might restrict survey efficiency. Thus, the potential of AI, notably DLg, to handle camera-trap data has received a lot of interest. To automatically recognize items and categorize species, DL applications use training techniques such as Convolutional Neural Networks (CNNs) (Sharma et al., 2023). The workflow of AI-powered camera traps is discussed in the following section.

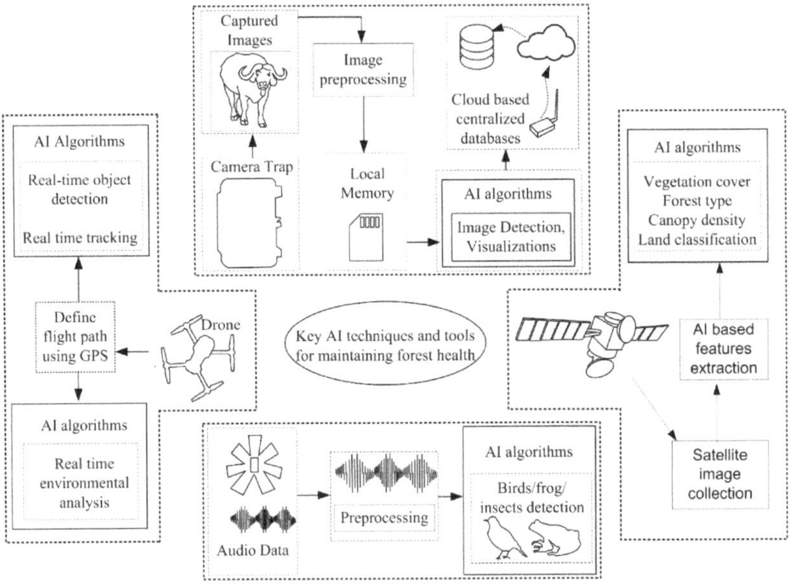

Figure 1: Key AI techniques and tools for maintaining forest health: AI-powered camera traps, AI-equipped drones, audio data, and satellite imagery.

The first stage involves deploying a camera trap. An appropriate location was selected based on ecological qualities, wildlife migration patterns, and conservation goals. Placement of the camera traps needs to ensure that they are securely placed and positioned to record a wide range of views. Camera traps are equipped with motion sensors that trigger image or video capture when movement is detected within the camera's range. A definite parameter is set such as capture frequency, duration, and resolution to optimize data collection while conserving power and storage capacity. Captured images or videos are stored locally on the camera trap device or transmitted wirelessly to a central database or cloud storage. Gather metadata such as timestamps, Global Positioning System (GPS) coordinates, and environmental conditions (e.g., temperature, humidity) to contextualize the captured data. The acquired data then needs preprocessing to remove noise, false positives, and irrelevant images or videos to improve the accuracy of subsequent analysis. To enhance image quality, adjust lighting, and correct distortions to facilitate accurate species identification. This step usually follows an AI model for adaptivity. The improved image is then used to recognize and classify species present. For this, at first AI algorithms need to be trained on vast datasets of animal images. Supervised ML methods such as multilayer perceptron (MLP) (Murariu et al., 2017) are widely used to identify and categorize animals present in the captured images or videos. To distinguish between individual animals based on unique features such as markings, scars, or physical characteristics more efficiently DL methods like CNN (Himeur et al., 2022) can be employed. The animal behaviour patterns and interactions within the captured footage are analysed to gather social dynamics, foraging strategies, and habitat preferences. Usually analyzed data are integrated into centralized databases or management systems for easy access, retrieval, and sharing among scholars and conservationists. Visualizations in heat maps, species distribution maps, and trend graphs are created automatically to present findings and aid decision-making. AI-driven the system to send alerts or notifications when specific species of interest are detected, enabling rapid response to conservation threats or emergencies. The AI-driven system also generates comprehensive reports summarizing key findings, trends, and conservation implications for stakeholders, funding agencies, and policymakers. AI algorithms get refined with time-based feedback and new data to enhance species recognition accuracy, reduce false positives, and improve overall system performance.

AI Equipped Drones

AI-equipped drones combine the capabilities of unmanned aerial vehicles (UAVs) with AI algorithms to perform a variety of tasks ranging from surveillance and monitoring to delivery and disaster response (Villa et al., 2009; Mohapatra & Trinh, 2022). The first step of a drone system, mission planning takes precedence. The objectives of the mission, be it surveillance, mapping, or delivery, are meticulously determined. The area to be covered and any specific points of interest are carefully considered, and a flight path is planned, considering factors such as terrain, obstacles, and regulations. Ensuring the drone's proper working condition, including checking battery levels, propeller integrity, and sensors, is imperative. Verification of all required equipment, such as cameras or sensors, functioning correctly is conducted. Weather conditions and other environmental factors that may impact the flight are assessed. The drone is then launched, either manually or through automated take-off procedures, with confirmation of its airborne and stable status before proceeding with the mission. The predefined flight path is executed while adjusting for real-time obstacles or changes in conditions, utilizing GPS and other navigation systems to maintain position accuracy. Images, videos, or other sensor data are captured as specified by the mission objectives, using onboard cameras, LiDAR, thermal sensors, or other specialized equipment to gather relevant information. The collected data is processed onboard the drone using AI algorithms in real-time, performing tasks such as object detection, tracking, or environmental analysis to extract actionable insights. Decisions are made based on the analyzed data, such as adjusting the flight path or prioritizing specific areas of interest, with relevant information and updates transmitted to ground control stations or remote operators. Data exchange with other drones or systems participating in the mission occurs, if applicable. Evaluation of the collected data and analysis results is undertaken to verify accuracy and completeness, identifying any anomalies or areas requiring further investigation. Insights are extracted, and reports are generated to inform decision-making or planning for future missions. Navigation of the drone to a designated landing area is carried out, followed by a controlled landing, either manually or through automated landing procedures, ensuring the safety of the drone and surrounding personnel or infrastructure. Post-flight maintenance checks and inspections are conducted to ensure the drone is ready for subsequent missions, with collected data securely transferred and stored for further analysis or archival purposes. Debriefing personnel involved in the mission and documenting any lessons learned or recommendations for improvement concludes the process. Through adherence to this workflow, AI-equipped drones can effectively execute a wide range of tasks with efficiency, accuracy, and adaptability, rendering them invaluable tools across various industries and applications.

Audio Data

Audio data are very useful for biodiversity assessment for small life which are hard to detect with camera traps (Mohan et al., 2021; Hino et al., 2018; Rakova & Winter, 2020). Deploying audio recording devices such as microphones or acoustic sensors in the forest area of interest, strategically placing them to cover a wide area and capture a variety of sounds. Metadata collection entails gathering information such as location, date, time, and environmental conditions (e.g., weather, temperature) during recording to contextualize the audio data. Non-biological noises like wind, rain, or anthropogenic sounds (e.g., vehicles) are filtered out from the raw audio recordings to focus on biological sounds. Subsequently, the audio recordings are segmented into smaller segments (e.g., one-minute clips) for easier processing, and relevant features are extracted, including spectral features (e.g., frequency, amplitude), temporal features (e.g., duration, timing), and statistical features (e.g., mean, variance) to characterize different sounds. Segments of audio data are manually labelled with the corresponding species or events (e.g., bird calls, frog calls, animal vocalizations, insect sounds). Data augmentation techniques such as pitch shifting, time stretching, or adding background noise are applied to enhance the training dataset's diversity and robustness. An appropriate AI model architecture for audio analysis tasks, such as CNNs, recurrent neural networks (RNNs), or transformer-based models, is selected (Delaney & Larson, 2023). The selected AI model is trained using the annotated audio data and extracted features to recognize patterns and classify different sounds based on provided labels. The performance of the trained model is assessed using validation datasets not seen during training, with metrics such as accuracy, precision, recall, and F1-score commonly used for evaluation. The trained AI model is deployed to analyze new audio recordings in real-time or batch processing mode, integrated into a forest monitoring system or platform for automated sound analysis and species detection. Model outputs are analyzed to interpret results, such as species presence/absence, abundance estimation, or biodiversity assessment, and visualized using maps, graphs, or other visualization tools to facilitate interpretation and decision-making by stakeholders. Continuous updates and improvements to the AI model are made by incorporating feedback, collecting additional data, and refining the training process to enhance accuracy and generalization ability over time.

Satellite Imagery

Using AI-equipped satellite imagery for forest monitoring plays a crucial role (De Bem et al., 2020; Sabu et al., 2021; Bhatt & Lyngdoh, 2023). The first step is to acquire satellite imagery. Various satellites orbit the Earth, capturing images of the planet's surface at different spatial and temporal resolutions. Depending on the specific monitoring objectives, researchers or organizations may choose satellite imagery with appropriate characteristics such as spatial resolution (detail level) and revisit frequency (how often the satellite passes over the same area). Satellite imagery often requires pre-processing to correct distortions, atmospheric effects, and other artefacts. This step ensures that the imagery is ready for analysis and that the data accurately represents the Earth's surface. AI algorithms are then utilized to extract relevant features from the satellite imagery. For forest monitoring, these features may include vegetation cover, forest type, canopy density, and land use/land cover classification. AI techniques such as CNNs are commonly employed for feature extraction tasks due to their effectiveness in analyzing visual data like satellite imagery. Change detection algorithms are subsequently applied to detect changes in the forest cover over time. This involves comparing multiple satellite images taken at different time points and identifying areas where significant changes have occurred, such as deforestation, reforestation, or forest degradation. AI algorithms can also be utilized for anomaly detection, which involves identifying unusual or unexpected changes in the forest cover that may indicate illegal logging, wildfires, or other disturbances. This aids authorities in prioritizing areas for intervention and conservation efforts. The extracted features and detected changes are integrated with other relevant data sources, such as ground-based observations, historical records, and environmental variables (e.g., climate data, and terrain characteristics). This integrated dataset is then analyzed to gain insights into forest dynamics, trends, and threats. The results of the analysis are visualized in maps, graphs, and other formats to facilitate interpretation and communication. Reports summarizing the findings are generated for stakeholders, policymakers, and conservation organizations to inform decision-making and management actions. The workflow is often iterative, with continuous refinement of AI algorithms, data processing techniques, and analysis methods based on feedback and new data. This iterative approach ensures that forest monitoring efforts remain up-to-date and effective in addressing emerging challenges.

Real-world Applications

Resources Allocation and Planning Habitat Restoration

AI plays a crucial role in resource allocation and planning for habitat restoration efforts by leveraging various technologies and techniques to optimize decision-making processes (Lee et al., 2020; de Araújo et al., 2021). AI algorithms can analyze vast amounts of environmental data, including habitat conditions, species distributions, climate patterns, soil quality, and more. ML models can then predict the potential outcomes of different restoration strategies based on historical data and environmental parameters. This predictive analysis helps in identifying the most effective and efficient allocation of resources. AI-driven optimization algorithms can determine the best allocation of resources such as manpower, equipment, and funding to maximize the impact of habitat restoration projects (Kikon & Deka, 2022). These algorithms consider various constraints and objectives, such as budget limitations, ecological priorities, and regulatory requirements, to generate optimal resource allocation plans (Linkie et al., 2003). AI-powered remote sensing technologies, such as satellite imagery analysis and drone-based monitoring, enable real-time tracking of habitat restoration progress. These technologies can identify changes in vegetation cover, habitat fragmentation, and invasive species encroachment, allowing for timely adjustments to resource allocation plans based on observed outcomes.

AI techniques, including agent-based modelling and simulation, enable the creation of virtual environments to simulate the effects of different restoration interventions on ecosystems and biodiversity (Chen et al., 2021). These simulations help planners assess the long-term ecological impacts of resource allocation decisions and identify optimal strategies for achieving restoration goals. AI-driven adaptive management frameworks continuously learn from ongoing restoration efforts by incorporating feedback from monitoring data and adjusting resource allocation strategies accordingly. Machine learning algorithms can identify patterns in ecosystem dynamics and recommend adaptive management actions to optimize the effectiveness of habitat restoration initiatives over time (Isabelle & Westerlund, 2022). AI algorithms can assess potential risks associated with habitat restoration projects, such as habitat degradation, species displacement, and unforeseen ecological impacts. By identifying and quantifying these risks, planners can proactively implement mitigation measures and allocate resources more effectively to minimize negative outcomes.

Detect Deforestation

AI applications in detecting deforestation have become increasingly prevalent and effective due to the advancement of technology and the availability of high-resolution satellite imagery (Nguyen et al., 2017; Valletta et al., 2017; Kumar & Jakhar, 2022). AI algorithms can analyze satellite images to detect changes in land cover over time. By comparing current images with historical ones, AI can identify areas where deforestation has occurred. ML algorithms can be trained on labelled data to recognize patterns associated with deforested areas, such as changes in colour, texture, and shape. Traditional methods of detecting deforestation often involve manual inspection of satellite images, which can be time-consuming and prone to human error (Hamedianfar et al., 2022). AI algorithms automate this process by rapidly scanning large amounts of imagery to pinpoint areas of interest. This automation enables quicker response times and more efficient monitoring of deforestation hotspots.

AI can classify different types of land cover, including forests, agricultural land, urban areas, and water bodies, using high-resolution remote sensing data (Wei & Cheng, 2022). By accurately distinguishing between forested and non-forested areas, AI algorithms can identify areas undergoing deforestation and track changes in forest cover over time. AI algorithms can analyze satellite images to detect signs of illegal logging activities, such as logging roads, clear-cutting patterns, and logging machinery (Hassija et al., 2024). By flagging suspicious activities, AI can help law enforcement agencies target their monitoring and enforcement efforts more effectively. Deforestation not only leads to the loss of tree cover but also has significant impacts on biodiversity. AI can help monitor changes in biodiversity by analyzing satellite imagery and identifying habitat loss, fragmentation, and changes in species distributions. This information can inform conservation efforts and prioritize areas for protection.

AI can be used to develop early warning systems for deforestation by analyzing real-time satellite data and environmental variables such as temperature, precipitation, and soil moisture (Kochupillai et al., 2022). By detecting conditions conducive to deforestation, such as increased logging activity during dry seasons, AI can alert authorities to potential threats and facilitate proactive intervention. AI algorithms can be integrated with other sources of data, such as ground-based sensors, drones, and field observations, to provide a more comprehensive understanding of deforestation dynamics. By combining different data streams, AI can improve the accuracy and reliability of deforestation monitoring and enable more informed decision-making.

Illegal Logging Detection and Anti-poaching Efforts

Illegal logging and poaching are significant environmental issues that threaten biodiversity and contribute to habitat destruction and species extinction (Mporas et al., 2020). AI applications have been increasingly utilized in efforts to combat these activities due to their ability to process vast amounts of data quickly and accurately (Shoaib et al., 2023). AI algorithms can analyze satellite imagery to detect changes in forest cover, such as deforestation or logging activities. These algorithms can identify patterns associated with illegal logging, such as the presence of logging roads, clear-cut areas, or changes in canopy density. By regularly monitoring large forested areas, authorities can identify and respond to illegal logging activities more effectively. ML algorithms can be trained to recognize patterns associated with illegal logging activities, such as the sound of chainsaws, vehicle movements in remote areas, or the presence of logging equipment. By analyzing audio recordings from sensors deployed in forests or acoustic monitoring devices, AI systems can identify suspicious activities and alert authorities in real-time.

AI-powered camera traps equipped with image recognition capabilities can automatically detect and identify wildlife and human intruders in protected areas (Peters et al., 2020). These camera traps can distinguish between legal activities, such as research or ecotourism, and illegal activities, such as poaching or trespassing. When unauthorized activities are detected, the system can send alerts to park rangers or law enforcement agencies, enabling a rapid response. AI algorithms can analyze historical data on illegal logging and poaching incidents, as well as environmental variables such as weather patterns and terrain features, to predict future hotspots of illegal activity. By identifying areas at high risk of illegal logging or poaching, authorities can prioritize their enforcement efforts and allocate resources more effectively. AI systems can integrate data from various sources, including satellite imagery, sensor networks, law enforcement records, and citizen reports, into a unified platform for better decision-making. Advanced data visualization techniques can help authorities identify trends, track illegal activity patterns over time, and allocate resources strategically.

Drones equipped with AI-powered image recognition software can patrol large areas of forests or wildlife reserves, providing real-time aerial surveillance to detect illegal activities such as logging, poaching, or encroachment (Maharajan et al., 2024). Drones can cover rugged terrain more efficiently than ground patrols and can gather high-resolution imagery for detailed analysis. AI-powered platforms can facilitate collaboration and information sharing among government agencies, law enforcement authorities, conservation organizations, and local communities involved in anti-poaching and anti-illegal logging efforts. These platforms can streamline communication, coordinate joint operations, and share intelligence to enhance the effectiveness of conservation initiatives.

Wildlife Monitoring

AI applications in wildlife monitoring, including species identification, population estimation, and behavioral analysis, have significantly transformed conservation efforts (González et al., 2011; Mühling, 2023; Sabu et al., 2021). AI algorithms can analyze images or videos captured by camera traps or drones to identify species automatically. CNNs are commonly used for this purpose (Bhatt & Lyngdoh, 2023). These models can learn distinctive features of various species, enabling accurate identification even in challenging conditions like low light or dense foliage. Beyond visual identification, AI can also identify species based on their vocalizations. ML algorithms, particularly those employing spectrogram analysis and deep learning techniques, can recognize specific animal calls amidst ambient noise. This is particularly useful for identifying elusive or nocturnal species. AI algorithms can detect and count individual animals in images or videos, aiding in population estimation. This involves techniques like object detection, where bounding boxes are drawn around each animal detected. By analyzing patterns of movement and behaviors, algorithms can distinguish between individual animals and avoid double counting. Traditional methods of population estimation often require invasive techniques like trapping or tagging. AI-powered monitoring offers non-invasive alternatives, reducing stress on wildlife populations while still providing accurate population estimates.

AI can analyze movement patterns and behaviors of animals captured in video footage to understand their behavior better (de Araújo et al., 2021). By tracking individual animals over time, algorithms can identify feeding, mating, or migration patterns. This information is crucial for understanding species ecology and informing conservation strategies. AI algorithms can detect unusual behaviors or events, such as poaching activities or habitat disturbances, by comparing current observations with established behavioral patterns. This enables rapid response to threats and proactive conservation interventions. AI facilitates the integration of diverse data sources, including satellite imagery (Minakshi et al., 2020), GPS tracking data (Lee et al., 2020), and environmental variables (Kucera & Barrett, 2011). By combining multiple data streams, conservationists gain a more comprehensive understanding of ecosystems and species dynamics. AI automates the analysis of large datasets, accelerating the process of extracting meaningful insights. This allows conservationists to focus their efforts on interpretation and decision-making rather than manual data processing. AI can predict species distributions, population trends, and habitat suitability under different scenarios. These predictive models inform conservation planning, helping prioritize areas for protection or restoration efforts. AI enables real-time monitoring of wildlife populations and habitats, providing timely information for adaptive management strategies. Conservationists can respond swiftly to emerging threats or changes in ecological conditions, maximizing conservation effectiveness.

AI applications in wildlife monitoring offer unprecedented opportunities for conservationists to monitor, understand, and protect biodiversity more effectively. By using advanced technologies, conservation efforts can be more targeted, efficient, and sustainable, ultimately contributing to the long-term preservation of species and ecosystems.

A Case Study in Manas National Park

Manas National Park is a national park, tiger reserve, and elephant reserve in Assam, India, situated in the Himalayan foothills. In the context of Manas National Park, the camera trap is a fully AI-based system that helps to concisely figure out the population dynamics. The workflow of the AI-based system is presented in Figure 2. This process is carried out every year without any delay, and because of AI implementation, yearlong processes are summarized in just a month. The process of implementing an AI-based system starts with a carnivore sign survey. During the survey, all the surveyors used an Android-based app named MSTrIPES Ecological, which is capable of automatically recording the travelled path and GPS points with an accuracy of up to one meter. During the survey period whenever that surveyor finds any indirect evidence (footprint, scat, pellet, dung, rake mark, scrap mark, digging, etc.) take a photograph and all the ground information (GPS location, date, time, elevation, etc.) were displayed or stamped on that photograph as presented in Figure 3a. All the MSTrIPES field data were collected at range level and shared with the office of the field director.

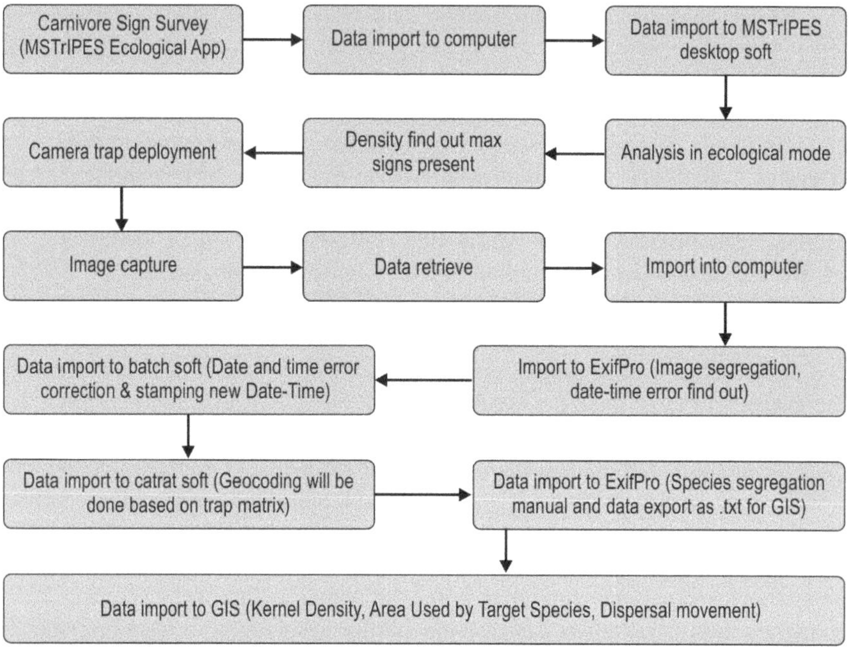

Figure 2: Workflow of the AI-powered system implemented in Manas National Park.

The photographs taken during the survey were brought into a computer which was later imported month-wise to the MSTrIPES software. In the MSTrIPES software, ran the analysis for carnivore occupancy in the ecological part where the AI algorithm automatically figures out the high-density and low-density areas within 1×1 km^2 grid size. In ecological mode, all the necessary data were imported and started the analysis. The densities will be revealed with the help of AI as presented in Figure 3b.

After figuring out the density a team deployed camera traps based on the highest occupancy trails. Based on density the deployment of camera traps in a 1.41×1.41 km^2 grid was accomplished. In each grid, one pair of camera traps were installed. It is prescribed by the National Tiger Conservation Authority (Ministry of Environment and Climate). The installed camera trap was presented zone-wise in Figure 3c. Camera traps were capable of capturing images like Figures 3d, e round the clock as they were equipped with infrared and motion sensors. Also, camera traps can stamp the current date and time which helps in the mass processing of images for tiger population estimation. After the deployment of camera traps, the dataset was retrieved in 10–12 days depending on animal encounter rate. All the secure digital (SD) cards of camera traps were collected and stored in a hard disk drive (HDD) in a particular sequence.

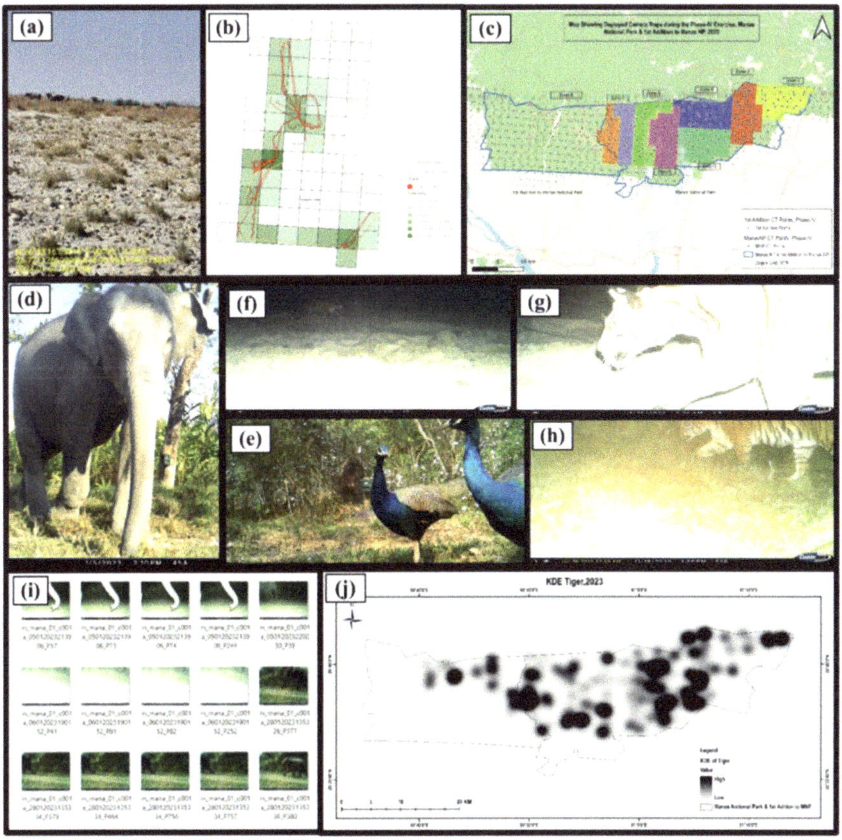

Figure 3: (a) A photograph taken on an Android based app named MSTrIPES Ecological. (b) AI algorithm automatically figures out the high-density and low-density areas within 1×1 km^2 grid size. (c) Deployment of camera traps in a 1.41×1.41 km^2 grid. (d) and (e) Image captured by a camera trap. (f) and (g) Instances of a tiger passing a camera trap. (h) Photograph with new date and time rectifying the error. (i) Each species will have a designated folder as predefined by the software itself before running the geo-tagging process. (j) The dark area indicates the used area by the tigers in Manas National Park.

Four AI-based software were used in the process of determining the population dynamics. ExifPro is an AI-based software that can help to figure out the date and time error of a particular image. Example: If we want to estimate the number of tigers, we need to have both flanks with the same date and time stamp. Only then we will be able to find the complementary image of a particular tiger individual. Figures 3f, g present instances of a tiger passing a camera trap. If there was an error then we must fix it before going to the next step of our analysis part. Another software named batch can stamp the new date and time and can cross the error so that a complimentary image can be found easily. In Figure

3h, the yellow marked date and time is the rectified image and the white colored image is the error date and time. CaTRATt is another software where camera-trapped images can be geo-tagged. A trap sheet needed to be prepared where camera-wise GPS locations and dates were available. Based on that trap sheet CaTRAT software ran the process of geo-tagging. After geo-tagging, camera trap data were imported into ExifPro software. Geo-tagged images at the species level were segregated in a predefined folder as shown in Figure 3i. After completing the segregation process, GPS locations from each image were exported in a .txt file to produce some population dynamics-related maps using a GIS. The dark area in Figure 3j indicates the used area by the tigers in Manas National Park.

This is how a large dataset can be quickly analyzed with the help of AI-based software, which has a negligible error rate. By producing such maps, daily monitoring of animal movements can be enhanced in the region. Also, a continuous study or regular AI-based survey can further help to accurately determine the forest health and population structure.

Advantages, Limitations, and Future Direction

Advantages of AI-driven methods over Traditional Methods

A pictorial representation of a comparative analysis of traditional methods versus AI-driven methods in forest health management, with results from studies is presented in Figure 4. The AI-driven methods have higher accuracy than the traditional methods in terms of disease detection, habitat detection, etc. (Wang et al., 2017; Wiesner-Hanks et al., 2019; Shoaib et al., 2023). This is primarily because of the elimination of human error and fatigue in methods followed in conventional forest health management. The efficiency in forest

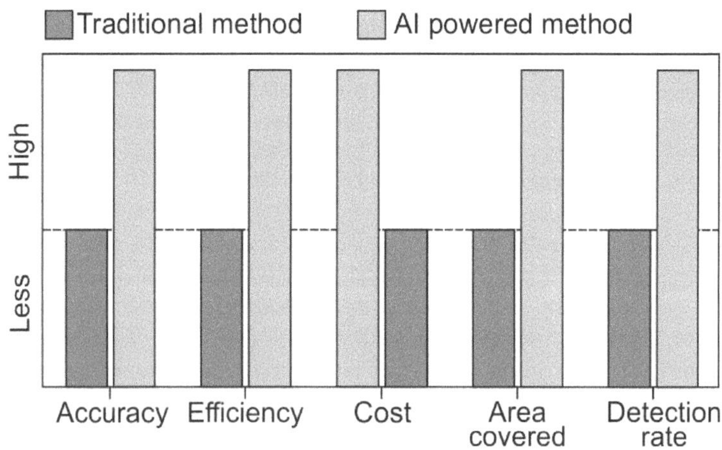

Figure 4: A pictorial representation of a comparative analysis of traditional methods versus AI-driven methods in forest health management.

health management increased a lot due to the introduction of AI. The labor cost has dropped significantly with the increase in AI adoption (Barreto et al., 1998). However, the initial infrastructure cost is higher in terms of AI-driven methods. To instal AI-powered solutions, skilful labor is very much essential. But over time human effort and cost reduced. This is also true because the area covered in forest health inspection and detection rate is higher in case of AI-driven solutions (Sarkar & Chapman, 2021; Raihan, 2023). These facts show the clear benefits of AI-driven solutions in forest health management over traditional methods. However, advancement in AI to date holds certain technical and social disadvantages associated with it.

Technical Limitations

The efficiency of AI in maintaining forest health can indeed be a powerful tool, but it also comes with certain limitations. AI systems rely heavily on data for training and decision-making. In the case of forest health monitoring, data might be sparse or of varying quality, especially in remote areas (Mporas et al., 2020; Kikon & Deka, 2022). Limited or low-quality data can hinder the effectiveness of AI algorithms, leading to inaccurate predictions or recommendations. Forest ecosystems are incredibly complex, with a multitude of interacting factors influencing their health. AI models may struggle to capture the full complexity of these systems, leading to oversimplified representations or missing critical interactions. For example, an AI model might detect a decline in tree cover but fail to understand the underlying causes such as soil nutrient depletion or invasive species encroachment.

Many AI algorithms, especially deep learning models, are often seen as "black boxes" because they provide results without clear explanations of how they reached their conclusions (Linkie et al., 2003). In the context of forest management, this lack of interpretability can make it challenging for forest managers to trust AI recommendations, especially when the decisions could have significant ecological or economic consequences. Forest ecosystems are dynamic and can be subject to rapid changes due to factors like climate change, natural disasters, or human activities. AI models trained on historical data may struggle to adapt to these changing conditions or anticipate emerging threats. Continuous retraining and updating of AI models are necessary to keep pace with evolving environmental conditions. Implementing AI solutions for forest health monitoring requires significant resources, including computational power, data collection infrastructure, and expertise in AI and forestry management. Many forest management agencies or organizations in developing regions may lack these resources, limiting the widespread adoption of AI technologies.

Social and Regulatory Concerns

The use of AI in forest management raises ethical and social concerns, such as data privacy, equity in access to technology, and the potential displacement of human workers (Chen et al., 2021). Additionally, there may be cultural or indigenous knowledge about forests that AI systems cannot incorporate, leading to conflicts or marginalization of local communities. There may be legal and regulatory barriers to the deployment of AI systems in forest management, particularly concerning issues like data ownership, liability for decision-making, and compliance with environmental regulations. Ensuring that AI applications comply with existing laws and standards adds another layer of complexity to their implementation (Sood, 2022). While AI holds great promise for maintaining forest health, its effectiveness is subject to various limitations related to data, complexity of ecosystems, interpretability, adaptability, resource constraints, ethical considerations, and legal challenges. Addressing these limitations requires interdisciplinary collaboration, ongoing research, and careful consideration of societal values and environmental goals.

Future Direction

With the evolution of Artificial Narrow Intelligence to Artificial General Intelligence, the AI methods are getting better in terms of accuracy, efficiency, and analysis (Pei et al., 2019). Along with that, with time the data (image, video, sound, etc.) gathered are huge in number as well as of better quality. Advancements in the communication technologies like 5G also enable to transfer and share big data among different researchers or authorities (Aggarwal et al., 2021). This is a perfect condition for AI to fully show its potential in the coming days. AI will be able to process data from sensors measuring parameters like soil moisture, temperature, and tree growth rates to monitor forest health in real-time (Torresan et al., 2021). This kind of solution is available in the pilot stage. However, in the coming time, this kind of solution can be implemented in a large area like the whole Amazon forest. Furthermore, interconnecting data on a global scale can improve early detection and prevention of disease, improve biodiversity conservation and monitoring of forest conditions (Morris et al., 2022; Singh et al., 2022). The solution offered by AI in forest health management can also extend to climate change adaptation (Filho et al., 2022; Chen et al., 2023). AI can model the impacts of climate change on forests and help develop strategies for adaptation and mitigation. Then, optimize forest management practices to maximize carbon sequestration and contribute to climate change mitigation efforts (Nunes et al., 2020; Al-Sakkari et al., 2024). This solution will impact very positively in the coming ages.

Conclusions

Currently, camera trap projects may face limitations due to the time required for footage classification, resulting in unidentified and unused images, thus compromising the significance and potential of datasets. Both citizen science and AI offer potential solutions to this constraint, but each has its limitations, suggesting that neither alone is a complete solution. We advocate for the integration of AI and citizen science, which has shown promising long-term potential, as demonstrated by recent projects. In this integrated approach, citizen scientists submit classifications of camera trap footage, creating a labelled dataset for training neural networks to classify future footage. A consensus is reached by combining AI and citizen science classifications, with the neural network providing an additional 'vote' to reduce the number of human classifications needed. Alternatively, AI could pre-screen footage, presenting only uncertain classifications to citizen scientists for confirmation, or filter footage for species of interest, allowing human observers to extract additional information such as behavior. Camera traps, monitored by citizen scientists, capture footage classified using AI. These approaches are not mutually exclusive, and projects may transition between them over time. Despite advancements in AI, citizen science remains essential for its engagement benefits and its role beyond classification, including camera placement and servicing. Combining AI with human efforts allows for more comprehensive analysis, particularly in behavioral studies. Integrating citizen science and AI technology maximizes data collection and processing efficiency while engaging and educating people about the natural world. While not suitable for all projects, AI and citizen science should be seriously considered and integrated wherever feasible, as they offer numerous potential benefits.

References

Abrams, J., E. Kelly, B. Shindler, and J. Wilton. (2005). Value orientation and forest management: The forest health debate. *Environmental Management*, 36: 495–505.

Aggarwal, P.K., P. Jain, J. Mehta, R. Garg, K. Makar, and P. Chaudhary. (2021). Machine learning, data mining, and big data analytics for 5G-enabled IoT. In: Sudeep Tanwar (Ed.), *Blockchain for 5G-Enabled IoT: The New Wave for Industrial Automation*, pp. 351–375. Springer.

Al-Sakkari, E.G., A. Ragab, H. Dagdougui, D.C. Boffito, and M. Amazouz. (2024). Carbon capture, utilization and sequestration systems design, and operation optimization: Assessment and perspectives of artificial intelligence opportunities. *Science of The Total Environment*, 170085.

Alzuhair, A., and A. Alghaihab. (2023). The design and optimization of an acoustic and ambient sensing AIoT platform for agricultural applications. *Sensors*, 23(14): 6262.

Backsen, J.C., and B. Howell. (2013). Comparing aerial detection and photo interpretation for conducting forest health surveys. *Western Journal of Applied Forestry*, 28(1): 3–8.

Barka, I., P. Lukeš, T. Bucha, T. Hlásny, R. Strejček, M. Mlčoušek, and Š Křístek. (2018). Remote sensing-based forest health monitoring systems: Case studies from Czechia and Slovakia. *Central European Forestry Journal*, 64(3–4): 259–275.

Barreto, P., P. Amaral, E. Vidal, and C. Uhl. (1998). Costs and benefits of forest management for timber production in eastern Amazonia. *Forest Ecology and Management*, 108(1–2): 9–26.

Bhatt, U., and S. Lyngdoh. (2023). Secrets of the clouded leopard: Abundance, habitat use, and carnivore coexistence in tropical forest of Manas National Park, Assam, India. *Oryx*, 57(6): 757–768.

Bhatt, U., B.S. Adhikari, and S. Lyngdoh. (2022). Monitoring diversity and abundance of mammals with camera-traps: A case study of Manas National Park, Assam, India. *Check List*, 18(5): 1023–1043.

Bilski, P., P. Bobiński, A. Krajewski, and P. Witomski. (2017). Detection of wood boring insects' larvae based on the acoustic signal analysis and the artificial intelligence algorithm. *Archives of Acoustics*, 42(1): 61–70.

Bravo-Oviedo, A., M. Marchi, D. Travaglini, F. Pelleri, M.C. Manetti, P. Corona, F. Cruz et al. (2020). Adoption of new silvicultural methods in Mediterranean forests: The influence of educational background and sociodemographic factors on marker decisions. *Annals of Forest Science*, 77: 1–17.

Budnik, K., J. Byrtek, B. Skrabanek, and J. Wajs. (2023). AI-accelerated decision making in forest management. In: *IOP Conference Series: Earth and Environmental Science*, 1189(1): 012030.

Chambers, R. (2006). Participatory mapping and geographic information systems: Whose map? Who is empowered and who disempowered? Who gains and who loses? *The Electronic Journal of Information Systems in Developing Countries*, 25(1): 1–11.

Chavan, S.B., C.S. Reddy, S.S. Rao, and K.K. Rao. (2018). Assessing and predicting decadal forest cover changes and forest fragmentation in Kinnerasani wildlife sanctuary, Telangana, India. *Journal of the Indian Society of Remote Sensing*, 46: 729–735.

Chen, L., Z. Chen, Y. Zhang, Y. Liu, A.I. Osman, M. Farghali, J. Hua et al. (2023). Artificial intelligence-based solutions for climate change: A review. *Environmental Chemistry Letters*, 21(5): 2525–2557.

Chen, W., W. Zhang, D. Liu, W. Li, X. Shi, and F. Fang. (2021). Data-driven multimodal patrol planning for anti-poaching. In: *Proceedings of the AAAI Conference on Artificial Intelligence*, 35(17): 15270–15277.

Chisom, O.N., P.W. Biu, A.A. Umoh, B.O. Obaedo, A.O. Adegbite, and A. Abatan. (2024). Reviewing the role of AI in environmental monitoring and conservation: A data-driven revolution for our planet. *World Journal of Advanced Research and Reviews*, 21(1): 161–171.

de Araújo, M., E. Andrade, and F. Machida. (2021, November). Performance analysis of machine learning-based systems for detecting deforestation. In: *2021 XI Brazilian Symposium on Computing Systems Engineering* (*SBESC*), pp. 1–8.

De Bem, P.P., O.A. de Carvalho Junior, R. Fontes Guimarães, and R.A. Trancoso Gomes. (2020). Change detection of deforestation in the Brazilian Amazon using landsat data and convolutional neural networks. *Remote Sensing*, 12(6): 901.

Delaney, J.T., and D.M. Larson. (2023). Using explainable machine learning methods to evaluate vulnerability and restoration potential of ecosystem state transitions. *Conservation Biology*, 14203.

Elavarasan, D., and P.D. Vincent. (2020). Crop yield prediction using deep reinforcement learning model for sustainable agrarian applications. *IEEE Access*, 8: 86886–86901.

Filho, W.L., T. Wall, S.A.R. Mucova, G.J. Nagy, A.L. Balogun, J.M. Luetz, A.W. Ng et al. (2022). Deploying artificial intelligence for climate change adaptation. *Technological Forecasting and Social Change*, 180: 121662.

Folliot, A., S. Haupert, M. Ducrettet, F. Sèbe, and J. Sueur. (2022). Using acoustics and artificial intelligence to monitor pollination by insects and tree use by woodpeckers. *Science of the Total Environment*, 838: 155883.

Fuentes, S., and E.J. Tongson. (2021). Special Issue: Implementation of sensors and artificial intelligence for environmental hazards assessment in urban, agriculture and forestry systems. *Sensors*, 21(19): 6383.

González, A., A. Gilmer, R. Foley, J. Sweeney, and J. Fry. (2011). Applying geographic information systems to support strategic environmental assessment: Opportunities and limitations in the context of Irish land-use plans. *Environmental Impact Assessment Review*, 31(3): 368–381.

Gonzalez, L.F., G.A. Montes, E. Puig, S. Johnson, K. Mengersen, and K.J. Gaston. (2016). Unmanned aerial vehicles (UAVs) and artificial intelligence revolutionizing wildlife monitoring and conservation. *Sensors*, 16(1): 97.

Goodchild, M.F. (1991). Geographic information systems. *Progress in Human Geography*, 15(2): 194–200.

Hamedianfar, A., C. Mohamedou, A. Kangas, and J. Vauhkonen. (2022). Deep learning for forest inventory and planning: A critical review on the remote sensing approaches so far and prospects for further applications. *Forestry*, 95(4): 451–465.

Haq, B., M.A. Jamshed, K. Ali, B. Kasi, S. Arshad, M.K. Kasi, I. Ali et al. (2024). Tech-driven forest conservation: Combating deforestation with internet of things, artificial intelligence, and remote sensing. *IEEE Internet of Things Journal*, 11(14): 24551–24568.

Hassija, V., V. Chamola, A. Mahapatra, A. Singal, D. Goel, K. Huang, S. Scardapane et al. (2024). Interpreting black-box models: A review on explainable artificial intelligence. *Cognitive Computatio*n, 16(1): 45–74.

Hernández, D., J.C. Cano, F. Silla, C.T. Calafate, and J.M. Cecilia. (2021). AI-enabled autonomous drones for fast climate change crisis assessment. *IEEE Internet of Things Journal*, 9(10): 7286–7297.

Himeur, Y., B. Rimal, A. Tiwary, and A. Amira. (2022). Using artificial intelligence and data fusion for environmental monitoring: A review and future perspectives. *Information Fusion*, 86: 44–75.

Hino, M., E. Benami, and N. Brooks. (2018). Machine learning for environmental monitoring. *Nature Sustainability*, 1(10): 583–588.

Isabelle, D.A., and M. Westerlund. (2022). A review and categorization of artificial intelligence-based opportunities in wildlife, ocean, and land conservation. *Sustainability*, 14(4): 1979.

Kamaruidzaman, N.S., and S.N. Rahmat. (2020). Water monitoring system embedded with internet of things (IoT) device: A review. In: *IOP Conference Series: Earth and Environmental Science*, 498(1): 012068).

Kijtewachakul, N., G.P. Shivakoti, and E.L. Webb. (2004). Forest health, collective behaviors, and management. *Environmental Management*, 33: 620–636.

Kikon, A., and P.C. Deka. (2022). Artificial intelligence application in drought assessment, monitoring and forecasting: A review. *Stochastic Environmental Research and Risk Assessment*, 36(5): 1197–1214.

Knebel, P., C. Appold, A. Guldner, M. Horbach, Y. Juncker, S. Müller, and A. Matheis. (2022). An artificial intelligence of things based method for early detection of bark beetle infested trees. *EnviroInfo*, 2022.

Kochupillai, M., M. Kahl, M. Schmitt, H. Taubenböck, and X.X. Zhu. (2022). Earth observation and artificial intelligence: Understanding emerging ethical issues and opportunities. *IEEE Geoscience and Remote Sensing Magazine*, 10(4): 90–124.

Kourtz, P. (1990). Artificial intelligence: A new tool for forest management. *Canadian Journal of Forest Research*, 20(4): 428–437.

Kucera, T.E., and R.H. Barrett. (2011). A history of camera trapping. In: O'Connell, A.F., Nichols, J.D., Karanth, K.U. (Eds.), *Camera Traps in Animal Ecology*, pp. 9–26. Springer, Tokyo.

Kumar, D., and S.D. Jakhar. (2022). *Artificial Intelligence in Animal Surveillance and Conservation. Impact of Artificial Intelligence on Organizational Transformation* (pp. 73–85). Wiley online Edn.

Lausch, A., S. Erasmi, D.J. King, P. Magdon, and M. Heurich. (2017). Understanding forest health with remote sensing – Part II: A review of approaches and data models. *Remote Sensing*, 9(2): 129.

Lechner, A.M., G.M. Foody, and D.S. Boyd. (2020). Applications in remote sensing to forest ecology and management. *One Earth*, 2(5): 405–412.

Lee, S.H., K.J. Han, K. Lee, K.J. Lee, K.Y. Oh, and M.J. Lee. (2020). Classification of landscape affected by deforestation using high-resolution remote sensing data and deep-learning techniques. *Remote Sensing*, 12(20): 3372.

Liff, C.I., K.H. Riitters, and K.A. Hermann. (1994). Forest health monitoring case study. In: Michener, W.K., Brunt, J.W., and Stafford, S.G. (Eds.), *Environmental Information Management and Analysis: Ecosystem to Global Scales*, pp. 101–112. Taylor and Francis, New York.

Linkie, M., D.J. Martyr, J. Holden, A. Yanuar, A.T. Hartana, J. Sugardjito, and N. Leader-Williams. (2003). Habitat destruction and poaching threaten the Sumatran tiger in Kerinci Seblat National Park, Sumatra. *Oryx*, 37(1): 41–48.

Liu, L., C. Mou, and F. Xu. (2024). Improved wildlife recognition through fusing camera trap images and temporal metadata. *Diversity*, 16(3): 139.

Liu, T., Y. Sun, C. Wang, Y. Zhang, Z. Qiu, W. Gong, S. Lei et al. (2021). Unmanned aerial vehicle and artificial intelligence revolutionizing efficient and precision sustainable forest management. *Journal of Cleaner Production*, 311: 127546.

Maharajan, M.S., A. Deepa, C.H. Bhargavi, K.G. Kannan, and E. Howard. (2024). Using AI to improve autonomous unmanned aerial vehicle navigation. *International Journal of Intelligent Systems and Applications in Engineering*, 12(14): 368–376.

Minakshi, M., T. Bhuiyan, S. Kariev, M. Kaddumukasa, D. Loum, N.B. Stanley, S. Chellappan et al. (2020). High-accuracy detection of malaria mosquito habitats using drone-based multispectral imagery and Artificial Intelligence (AI) algorithms in an agro-village peri-urban pastureland intervention site (Akonyibedo) in Unyama SubCounty, Gulu District, Northern Uganda. *Journal of Public Health and Epidemiology*, 12(3): 202–217.

Mohan, M., G. Richardson, G. Gopan, M.M. Aghai, S. Bajaj, G.P. Galgamuwa, M. Vastaranta et al. (2021). UAV-supported forest regeneration: Current trends, challenges and implications. *Remote Sensing*, 13(13): 2596.

Mohapatra, A., and T. Trinh. (2022). Early wildfire detection technologies in practice – A review. *Sustainability*, 14(19): 12270.

Morris, C.E., G. Géniaux, C. Nédellec, N. Sauvion, and S. Soubeyrand. (2022). One health concepts and challenges for surveillance, forecasting, and mitigation of plant disease beyond the traditional scope of crop production. *Plant Pathology*, 71(1): 86–97.

Mporas, I., I. Perikos, V. Kelefouras, and M. Paraskevas. (2020). Illegal logging detection based on acoustic surveillance of forest. *Applied Sciences*, 10(20): 7379.

Mühling, Ş.M.K. (2023). Utilizing Artificial Intelligence (AI) for the Identification and

Management of Marine Protected Areas (MPAs): A review. *Journal of Geoscience and Environment Protection*, 11(9): 118–132.

Murariu, G., V. Hahuie, A. Murariu, L. Georgescu, C. Iticescu, M. Calin, C. Preda et al. (2017). Forest monitoring method using combinations of satellite and UAV aerial images. Case study – Bălăbăneşti forest. *International Journal of Conservation Science*, 8(4): 703–714.

Nguyen, H., S.J. Maclagan, T.D. Nguyen, T. Nguyen, P. Flemons, K. Andrews, E.G. Ritchie et al. (2017). Animal recognition and identification with deep convolutional neural networks for automated wildlife monitoring. In: *2017 IEEE International Conference on Data Science and Advanced Analytics* (*DSAA*) pp. 40–49. Tokyo, Japan.

Nikhilesh Krishna, C., A. Rauniyar, N. Kireeti Sai Bharadwaj, S.B. Raj, V. Valsan, K. Suresh, V. Ravikumar Pandi et al. (2023, December). ECO-Guard: An integrated AI sensor system for monitoring wildlife and sustainable forest management. In: *International Conference on Information and Communication Technology for Competitive Strategies*, pp. 409–419. Jaipur, India.

Nitoslawski, S.A., K. Wong-Stevens, J.W. Steenberg, K. Witherspoon, L. Nesbitt, and C.C. Konijnendijk van den Bosch. (2021). The digital forest: Mapping a decade of knowledge on technological applications for forest ecosystems. *Earth's Future*, 9(8): e2021EF002123.

Nova, K. (2023). AI-enabled water management systems: An analysis of system components and interdependencies for water conservation. *Eigenpub Review of Science and Technology*, 7(1): 105–124.

Nunes, L.J., C.I. Meireles, C.J. Pinto Gomes, and N.M. Almeida Ribeiro. (2020). Forest contribution to climate change mitigation: Management oriented to carbon capture and storage. *Climate*, 8(2): 21.

Pal, O.K., M.S.H. Shovon, M.F. Mridha, and J. Shin. (2023). A comprehensive review of AI-enabled unmanned aerial vehicle: Trends, vision, and challenges. arXiv preprint arXiv: 2310.16360.

Pautasso, M., M. Schlegel, and O. Holdenrieder. (2015). Forest health in a changing world. *Microbial Ecology*, 69: 826–842.

Pei, J., L. Deng, S. Song, M. Zhao, Y. Zhang, S. Wu, G. Wang et al. (2019). Towards artificial general intelligence with hybrid Tianjic chip architecture. *Nature*, 572(7767): 106–111.

Peters, D.P., D.S. McVey, E.H. Elias, A.M. Pelzel-McCluskey, J.D. Derner, N.D. Burruss, T.S. Schrader. et al. (2020). Big data-model integration and AI for vector-borne disease prediction. *Ecosphere*, 11(6): e03157.

Peterson, D.J., S. Resetar, J. Brower, and R. Diver. (1999). *Forest Monitoring and Remote Sensing*. White House Office of Science and Technology Policy.

Puettmann, K.J., S.M. Wilson, S.C. Baker, P.J. Donoso, L. Drössler, G. Amente, B.D. Harvey et al. (2015). Silvicultural alternatives to conventional even-aged forest management: What limits global adoption? *Forest Ecosystems*, 2: 1–16.

Rahardja, U. (2022). Camera trap approaches using artificial intelligence and citizen science. *International Transactions on Artificial Intelligence*, 1(1): 71–83.

Rahimi, A., M. Anvaripour, and K. Hayat. (2021). Object detection using deep learning in a manufacturing plant to improve manual inspection. In: *2021 IEEE International Conference on Prognostics and Health Management* (*ICPHM*), pp. 1–7. (Online)

Raihan, A. (2023). Artificial intelligence and machine learning applications in forest management and biodiversity conservation. *Natural Resources Conservation and Research*, 6(2): 3825.

Rakova, B., and A. Winter. (2020). Leveraging traditional ecological knowledge in ecosystem restoration projects utilizing machine learning. arXiv preprint arXiv: 2006.12387.

Reckling, W., H. Mitasova, K. Wegmann, G. Kauffman, and R. Reid. (2021). Efficient drone-based rare plant monitoring using a species distribution model and AI-based object detection. *Drones*, 5(4): 110.

Rudel, T.K., R. Defries, G.P. Asner, and W.F. Laurance. (2009). Changing drivers of deforestation and new opportunities for conservation. *Conservation Biology*, 23(6): 1396–1400.

Sabu, M.M., S.V. Pasha, C.S. Reddy, R. Singh, and R. Jaishanker. (2021). The effectiveness of Tiger Conservation Landscapes in decreasing deforestation in South Asia: A remote sensing-based study. *Spatial Information Research*, 1–13.

Sakr, G.E., I.H. Elhajj, and G. Mitri. (2011). Efficient forest fire occurrence prediction for developing countries using two weather parameters. *Engineering Applications of Artificial Intelligence*, 24(5): 888–894.

Sarkar, D., and C.A. Chapman. (2021). The smart forest Conundrum: Contextualizing pitfalls of sensors and AI in conservation science for tropical forests. *Tropical Conservation Science*, 14: 19400829211014740.

Serrano, F.L., E. Rubio, F.G. Morote, M.A. Abellán, M.P. Córdoba, F.G. Saucedo, E.M. García et al. (2022). Artificial intelligence-based software (AID-FOREST) for tree detection: A new framework for fast and accurate forest inventorying using LiDAR point clouds. *International Journal of Applied Earth Observation and Geoinformation*, 113: 103014.

Sharma, S., K. Sato, and B.P. Gautam. (2023). A methodological literature review of acoustic wildlife monitoring using artificial intelligence tools and techniques. *Sustainability*, 15(9): 7128.

Shoaib, M., B. Shah, S. Ei-Sappagh, A. Ali, A. Ullah, F. Alenezi, T. Gechev et al. (2023). An advanced deep learning models-based plant disease detection: A review of recent research. *Frontiers in Plant Science*, 14: 1158933.

Singh, R., A. Gehlot, S.V. Akram, A.K. Thakur, D. Buddhi, and P.K. Das. (2022). Forest 4.0: Digitalization of forest using the Internet of Things (IoT). *Journal of King Saud University – Computer and Information Sciences*, 34(8): 5587–5601.

Sisodiya, N., N. Dube, and P. Thakkar. (2020). Next-generation artificial intelligence techniques for satellite data processing. *Artificial Intelligence Techniques for Satellite Image Analysis*, 235–254.

Smith, W.B., P.D. Miles, C.H. Perry, and S.A. Pugh. (2009). *Forest Resources of the United States, 2007*. U.S. Department of Agriculture, Forest Service, General Technical Report WO-78.

Sood, R. (2022). *Interconnection between Environmental Law, Policy, and Technology vis-a-vis Forest Management*. 1-15. Available at SSRN: https://ssrn.com/abstract=4638771 or http://dx.doi.org/10.2139/ssrn.4638771.

Stančić, A., V. Vyroubal, and V. Slijepčević. (2022). Classification efficiency of pre-trained deep CNN models on camera trap images. *Journal of Imaging*, 8(2): 20.

Store, R., and J. Jokimäki. (2003). A GIS-based multi-scale approach to habitat suitability modeling. *Ecological Modelling*, 169(1): 1–15.

Sun, A.Y., and B.R. Scanlon. (2019). How can Big Data and machine learning benefit environment and water management: A survey of methods, applications, and future directions. *Environmental Research Letters*, 14(7): 073001.

Tien, J.M. (2017). Internet of things, real-time decision-making, and artificial intelligence. *Annals of Data Science*, 4: 149–178.

Tierney, G.L., D. Faber-Langendoen, B.R. Mitchell, W.G. Shriver, and J.P. Gibbs. (2009). Monitoring and evaluating the ecological integrity of forest ecosystems. *Frontiers in Ecology and the Environment*, 7(6): 308–316.

Torres, P., M. Rodes-Blanco, A. Viana-Soto, H. Nieto, and M. García. (2021). The role of remote sensing for the assessment and monitoring of forest health: A systematic evidence synthesis. *Forests*, 12(8): 1134.

Torresan, C., M. Benito Garzón, M. O'grady, T.M. Robson, G. Picchi, P. Panzacchi, E. Tomelleri et al. (2021). A new generation of sensors and monitoring tools to support climate-smart forestry practices. *Canadian Journal of Forest Research*, 51(12): 1751–1765.

Trumbore, S., P. Brando, and H. Hartmann. (2015). Forest health and global change. *Science*, 349(6250): 814–818.

U.S. Forest Service. (2012). *Forest Health Conditions in the United States, 2009*. Forest Service, Washington Office, Forest Health Protection, State and Private Forestry.

Valletta, J.J., C. Torney, M. Kings, A. Thornton, and J. Madden. (2017). Applications of machine learning in animal behaviour studies. *Animal Behaviour*, 124: 203–220.

Villa, F., M. Ceroni, K. Bagstad, G. Johnson, and S. Krivov. (2009, September). ARIES (Artificial Intelligence for Ecosystem Services): A new tool for ecosystem services assessment, planning, and valuation. In: *Proceedings of the 11th Annual BIOECON Conference on Economic Instruments to Enhance the Conservation and Sustainable Use of Biodiversity*, pp. 21–22. Venice, Italy.

Wang, G., Y. Sun, and J. Wang. (2017). Automatic image-based plant disease severity estimation using deep learning. *Computational Intelligence and Neuroscience*, 2017(1): 2917536.

Wang, J., T.W. Sammis, V.P. Gutschick, M. Gebremichael, S.O. Dennis, and R.E. Harrison. (2010). Review of satellite remote sensing use in forest health studies. *The Open Geography Journal*, 3(1).

Wang, S., S.A. Zargar, C. Xu, and F.G. Yuan. (2019). An efficient augmented reality (AR) system for enhanced visual inspection. In: *12th International Workshop on Structural Health Monitoring: Enabling Intelligent Life-Cycle Health Management for Industry Internet of Things (IIOT), IWSHM*, pp. 1543–1550.

Wei, S., and S.T. Cheng. (2022). An artificial intelligence approach for identifying efficient urban forest indicators on ecosystem service assessment. *Frontiers in Environmental Science*, 10: 994389.

Welbourne, D.J., A.W. Claridge, D.J. Paull, and A. Lambert. (2016). How do passive infrared triggered camera traps operate and why does it matter? Breaking down common misconceptions. *Remote Sensing in Ecology and Conservation*, 2(2): 77–83.

Wieczorek, W.F., and A.M. Delmerico. (2009). Geographic information systems. *Wiley Interdisciplinary Reviews: Computational Statistics*, 1(2): 167–186.

Wiesner-Hanks, T., H. Wu, E. Stewart, C. DeChant, N. Kaczmar, H. Lipson, M.A. Gore et al. (2019). Millimeter-level plant disease detection from aerial photographs via deep learning and crowdsourced data. *Frontiers in Plant Science*, 10: 1550.

Wingfield, M.J., E.G. Brockerhoff, B.D. Wingfield, and B. Slippers. (2015). Planted forest health: The need for a global strategy. *Science*, 349(6250): 832–836.

Wu, B., A. Liang, H. Zhang, T. Zhu, Z. Zou, D. Yang, W. Tang et al. (2021). Application of conventional UAV-based high-throughput object detection to the early diagnosis of pine wilt disease by deep learning. *Forest Ecology and Management*, 486: 118986.

Wulder, M.A., C.W. Bater, N.C. Coops, T. Hilker, and J.C. White. (2008). The role of LiDAR in sustainable forest management. *The Forestry Chronicle*, 84(6): 807–826.

Ye, W., J. Lao, Y. Liu, C.C. Chang, Z. Zhang, H. Li, and H. Zhou. (2022). Pine pest detection using remote sensing satellite images combined with a multi-scale attention-UNet model. *Ecological Informatics*, 72: 101906.

Yu, R., Y. Luo, Q. Zhou, X. Zhang, D. Wu, and L. Ren. (2021). Early detection of pine wilt disease using deep learning algorithms and UAV-based multispectral imagery. *Forest Ecology and Management*, 497: 119493.

Zhang, C., J. Valente, L. Kooistra, L. Guo, and W. Wang. (2019). Opportunities of UAVs in orchard management. *The International Archives of the Photogrammetry, Remote Sensing and Spatial Information Sciences*, 42: 673–680.

Zulfiqar, A., M.M. Ghaffar, M. Shahzad, C. Weis, M.I. Malik, F. Shafait, and N. Wehn. (2021). AI-ForestWatch: Semantic segmentation based end-to-end framework for forest estimation and change detection using multi-spectral remote sensing imagery. *Journal of Applied Remote Sensing*, 15(2): 024518–024518.

10 | Exploring the Role of Artificial Intelligence in Microbial Identification, Environmental Distribution, and Ecological Interactions

Abrar Hussain[1*], Areej Alvi[1], Shahana Rasheed Qureshi[2], Sara Parveen[3], Um-E-Habiba-U-Nisa[4] and Syed Abid Ali[1]

[1] Third World Center for Science and Technology, HEJ Research Institute of Chemistry, International Center for Chemical and Biological Sciences, University of Karachi, Karachi – 75270, Pakistan

[2] Department of Biochemistry, Jinnah University for Women, Karachi, Pakistan

[3] Department of Zoology, University of Education, Lahore, Pakistan

[4] Pure and Applied Biology Department, Bahauddin Zakariya University, Multan, Pakistan

Microbiology is a complex and dynamic field that addresses fundamental questions about the interactions of microorganisms with their environments and other organisms. The ubiquitous distribution of microorganisms and their intricate interactions with animals, plants, and ecosystems make them an important area of current research. Microorganisms play essential roles in the biotechnological and agricultural processes, such as nutrient recycling and biodegradation, while also contributing to the spread and transmission of diseases. To explore this dual nature, proper identification and characterization remain crucial for understanding their ecological significance and mitigating their negative impacts. Traditional microbial identification methods, including culture-based techniques and microscopy, have provided invaluable insights but are often limited by challenges related to time, cost, sensitivity, and specificity. The introduction of artificial intelligence (AI) has started a paradigm shift in microbial research. AI's ability to analyze complex datasets, recognize patterns, and make predictions has opened new possibilities for identifying

*Corresponding author: abrarhussain0307@gmail.com

and understanding microorganisms with precision. AI and its subfields, such as machine learning (ML) and deep learning, have proven successful in analyzing high-throughput microbial genomic data and enhancing the detection and classification of microbial communities. AI also facilitates predicting microbial behavior and their interactions with other organisms and environments. AI further supports the study of the relationships between microorganisms and their hosts, including the effects of microbial communities on animal behavior, health, and disease transmission. The synergistic approach of AI and microbiology explores the pioneering discoveries, transforming our understanding of the microbial world and its critical contributions to life on Earth.

Introduction

Microorganisms, which are extremely important creatures, help humans and other animals in multifactorial and vast disciplines. These microorganisms, being ubiquitous in nature, can have both positive and negative impacts on the ecological distribution and activities of animals. The unique characteristics of these organisms are crucial for various biotechnological and industrial processes, making their identification and quantification a significant area of interest for researchers and industrialists (Xu et al., 2023; Abrar and Abid, 2024). On the other hand, their prevalence in environments, such as indoor household settings, can lead to various diseases, posing risks to human health (Hussain et al., 2024).

The rapid identification of microorganisms, particularly in diagnosis, aids in this appeal, as it is the matter of someone's life. Hence, the field is open every time to develop the existing methods or elucidate new approaches. Although the various available methods have their own value, the current advancements in science and technology compel people to switch to the new methods. The present world has the hold of AI, and everyday new applications, domains, and usage are identified, which include their potential role in microorganisms' world (Rhoads, 2020).

Artificial Intelligence (AI)

The term AI first appeared in a scientific paper in the 1950s when a tortoise robot was described. Currently, more than 16,000 papers are dedicated to various aspects of AI, with the number continuing to grow. This large number of publications highlights the potential applications of AI in a wide range of areas, including life sciences (Liu et al., 2019). AI is an umbrella term that describes the intelligent criteria of a system that usually require a human mind, i.e., decision-making and critical thinking (Holzinger et al., 2023). In addition to its other potential applications in life sciences, AI has also established a strong presence in environmental microbiology and the study of microorganisms (Sarrafzadeh et al., 2022). The fact that microorganisms are found everywhere and are associated with different aspects shouting to find an easy, cheap, and potential method of their identification.

Machine Learning (ML)

Machine learning (ML), a subfield of AI, has the potential to develop microbial domains with minimal or no human involvement (Goodswen et al., 2021; Xu et al., 2023). It involves training computers to perform tasks using patterns and data without explicit instructions (Holzinger et al., 2023). It has been extensively used in computational problems like drug target predictions, microorganisms' diagnosis, antimicrobial drug classification, showcasing disease outbreaks, vaccine candidate development, and exploring microbial interactions. It is used in almost all aspects of microbial research, including bacteriology, virology, environmental microbiology, mycology, parasitology, etc. (Goodswen et al., 2021).

The widely used ML approaches include artificial neural networks, genetic algorithms, and vector machines (Undru et al., 2022). Particularly in microbial diagnosis, the ML can enhance the speed and accuracy of analyzing complex and large data patterns (Peiffer-Smadja et al., 2020; Ali et al., 2023). Its potential in molecular biology is dedicated to the analysis and predictions of viral and bacterial genomes to explore the binding targets for drugs and to show mutations, thus helping the clinician to prescribe effective treatments (Májek et al., 2021). Besides the technical applications of ML, studies were also conducted to use it in real sample analysis. For instance, researchers used ML to identify genotypic-to-phenotypic predictions in *M. tuberculosis* (Peiffer-Smadja et al., 2020).

Deep Learning (DL)

Deep learning (DL) is a subfield of ML that consists of multiple layers and uses artificial neural network (ANN) algorithms. It has powerful features such as regression capabilities and automatic feature extraction. DL is divided into different types like Recurrent Neural Network and Convolutional Neural Networks, etc. (Liu et al., 2019; Wang et al., 2022; Xu et al., 2023). DL is mostly used and dedicated to the analysis of larger data (Holzinger et al., 2023).

ANNs is a prestigious aspect of AI development and has various applications in different fields. An artificial neuronal network is like the artificial version of a natural neuron, comprised of an input, processing area, and output layer (Undru et al., 2022; Wang et al., 2022). Due to limited computer capacity, the ANN went through a hibernate period, but again arose and got more attention in the current period and is extensively used in microbial image analysis (Zhang et al., 2023).

Hence, the aforementioned subject matter was kept in mind, we documented the recent, updated, and emerging role of AI in microbial identification to correctly measure their potential, particularly their role in pathogenesis and biotechnological applications. For the better understanding of the subject, we first documented the traditional methods, their pros and cons, and the limitations associated, and then we described the emerging role of AI, ML, and DL in microbial identification, ecological distribution, its advancement, accuracy, usefulness, and applications. The traditional methods and the current genomic methods have their pros and cons but have certain challenges that appeal to another potential method. Thus, AI, ML, and DL have the capacity to solve these challenges. The problems present in conventional methods, their challenges and how AI can help to mitigate these problems are illustrated in Figure 1.

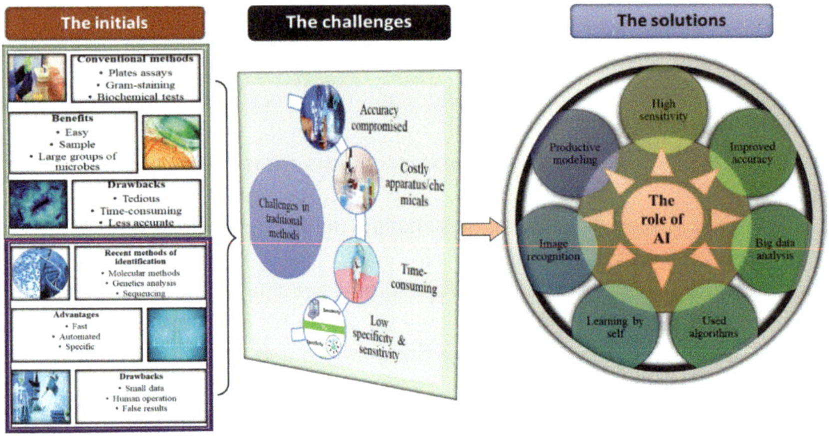

Figure 1: Illustration showing the potential of AI in microorganisms. The initials indicated the different traditional methods used in microbe identification. The challenges examine the existing drawbacks in microbial identification. The solutions provide the role of AI to overcome these limitations.

Ecological Distribution of Microorganisms

Microorganisms appeared approximately 3.5 billion years ago and became one of the earliest living things on earth. Microorganisms, which include bacteria, algae, viruses, fungi, etc., are found ubiquitously, having approximately 159,000 species (Madsen, 2005; Hussain et al., 2024). Their ecological distribution in the environment depends on multiple factors, which makes them more prevalent in one place but lower elsewhere. The potential ecological niches of these microbes include dumps and warm places. Hussain et al. (2024) recently reviewed their distribution in the indoor household environment and concluded their prevalence in all indoor places, including sleeping beds and kitchens (Hussain et al., 2024). Microorganisms have both positive and negative faces, like biotechnological, pharmaceutical, food, and industrial applications, while pathogenic nature, destructive potential, disease transmissions, etc. are their pros and cons, respectively (Hussain, A., 2024). These microorganisms are studied in detail, like their structures, functions, relationships, communications, and associations with the host, particularly humans (Qu et al., 2019). Microorganisms within a community are affected by nearby cells, whether they belong to the same species (intraspecies) or different (interspecies) interactions. Their ecological interactions are categorized based on their overall effect on each interacting species as positive, negative, or having no impact. In natural environments, various eco-interactions occur among microorganisms, involving both beneficial and detrimental effects (Martinez-Rabert et al., 2022).

The microbiome, or microbiota, is the collection of all types of microorganisms in a given environment; e.g., the gut microbiota consists of all types of microbes in the gut. The ecological distribution of microorganisms enables them to have associations with each other and with the environment. Hence, understanding potential sources of microbial finding, identification, and knowing their beneficial effects is an important area of the modern world. Through cultivation, researchers can further explore the interactions between microorganisms and their environment, revealing the vast scope and diversity of microbial distribution. Microbial communities, comprising diverse microorganisms inhabiting different environments or hosts, engage in intricate interactions with their surroundings and hosts, giving rise to various ecosystem types (Xie et al., 2019). The prominent microorganisms present in different environments are summarized in Figure 2.

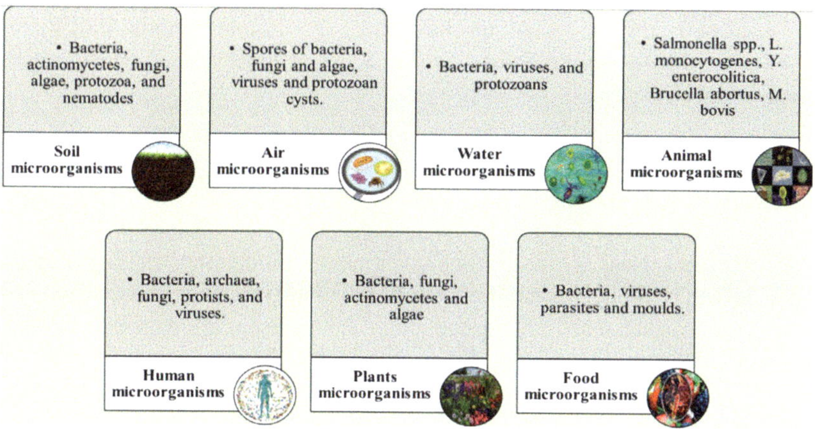

Figure 2: The ecological distribution of microorganisms in different environments.

Potential Ways to Identify the Ecological Distribution of Microorganisms

Due to their minute size, microorganisms necessitate the use of microscopes as a vital tool for investigation. However, microscopy solely facilitates observation, necessitating supplementation with culture techniques to delve into the physiological, biological, pathogenic, metabolic, and other characteristics (Waldron, 2018). The collective dynamics observed within microbial ecosystems across various biomes arise from numerous interactions among community members. These interactions encompass metabolite exchange, signaling mechanisms (quorum sensing), inhibition of growth, and even lethal interactions. Grasping interspecies relationships within microbial communities is crucial for comprehending the functionality of natural ecosystems and crafting synthetic consortia. Various organisms present within a microbial community can profoundly influence each other's growth patterns, thereby greatly impacting community dynamics, with significant ramifications for both human and environmental well-being (DiMucci et al., 2018; Goodswen et al., 2021).

The earliest identification methods comprised culture enrichment, in which an individual culture with certain distinct properties like fixation, specific growth conditions, etc,. is identified. This method is sample-based and gives good results, but due to the growth of different microbes' in culturing media, a more precise method is needed (Madsen, 2005). Identification with physiological characteristics has its own pros and cons. For instance, every microbe's response differs differentially to different substances, i.e., light, oxygen, chemicals, etc., and hence is used to make them separate from each other (Madsen, 2005). In the advanced technological period, identification was shifted to non-culturing methods, and genomic analysis showed its fruits. PCR analysis, microscopy, and genome sequencing are adopted for rapid and accurate microbial identification. However, both of these techniques have their own limitations, and thus there is a need for a highly accurate, fast, cost- and time-effective, and reliable method of microbial detection.

Challenges in Traditional Microbial Diagnosis in Different Settings

Microbial diagnosis, a cornerstone of modern medicine and microbiology, grapples with challenges spanning the entire diagnostic pathway. One such hurdle arises at the outset with sample collection and transportation, which represent critical initial stages (Rodrigues & Groves, 2018). The process commences with sample collection, ensuring their timely analysis in the laboratories presents a significant obstacle. Mishandling, inappropriate storage, and transportation delays can jeopardize the viability and integrity of microorganisms within the sample, thereby compromising diagnostic accuracy (Rodrigues & Groves, 2018; Shrestha & Pokharel, 2020). Consequently, the isolation and culture of microorganisms stand out as the primary traditional diagnostic approach. However, this method is time-consuming, often resulting in delays in obtaining culture results. Improper microbial cultivation increases the risk of false negative outcomes. Moreover, the rampant use of antibiotics in clinical settings can further impede the isolation and growth of specific organisms (Zeeshan, 2019).

Accurately identifying and classifying microbial species presents a significant hurdle. Traditional techniques like biochemical testing and microscopy often fall short of consistently providing accurate identifications due to ambiguous morphological features or unusual behaviors displayed by certain species (Zhang et al., 2022). Additionally, determination of bacteria isolates susceptibility to antibiotics for selection of treatment options presents its own set of challenges.

Disk diffusion method, a traditional susceptibility testing method, is time-consuming and sometimes fails in the prediction of infecting strain reaction to antibiotics, resulting in potential therapeutic consequence and suboptimal treatment decisions. Moreover, traditional diagnostic methods mostly need a team of highly skilled persons, specialized instruments, and substantial resources to diagnose the medical situations. Thus, due to limited resources, there is an elevated risk of human error due to reliance on skilled professionals. Therefore, there is a pressing need to develop new and user-friendly diagnostic procedures.

The Potential Role of AI and ML

The incorporation of AI into microbial identification and investigation has transmuted the field, offering faster and more reliable results than conventional methods (Davenport and Kalakota, 2019). AI algorithms excel at genomic data analysis and assist scientists and clinicians in the identification of pathogens, the prediction of antibiotic resistance, and even the discovery of novel microbial species. Additionally, ML models enhance the speed and precision of microbial diagnosis by swiftly analyzing intricate data patterns (Ali et al., 2023).

The integration of AI with microbe diagnosis has evolved into a multifaceted and crucial tool that helps in the recognition of different patterns and prediction modeling, along with the enhancement of efficiency in microbe analysis. AI-based algorithms have transformed the field by analyzing large datasets containing microbial information and identification of patterns and abnormalities that may be difficult to detect accurately and quickly by human analysts (Peiffer-Smadja et al., 2020). This ability of recognition of various patterns proves particularly helpful in the quick exposure of infectious diseases, where quick identification of different species of viruses and their patterns of transmission can provide essential containment strategies (Ali et al., 2023). Moreover, AI plays an important role in predictive modeling, by utilizing historical data to forecast the behavior of microbes and thus helps in future decision-making. This predictive ability of AI proves valuable in anticipating disease outbreaks, understanding trends in antibiotic resistance, and optimization of treatment protocols. Thus AI-based algorithms accelerate tasks such as sample processing, analysis of different images, and interpretation of data. This not only reduces the time needed for diagnosis but also enables healthcare professionals to focus on the most complex aspects of patient care (Davenport and Kalakota, 2019).

An exemplary instance lies in ML models, such as DNA sequencers, which analyze the genomic sequences of bacteria and viruses to forecast their likelihood of mutations and resistance to particular drugs, aiding clinicians in pinpointing the most suitable treatment options (Májek et al., 2021). AI-driven automation expedites processes like sample processing, analysis of images, and data interpretation. This not only reduces the diagnostic turnaround time but also enables healthcare professionals to focus on the most intricate facets of patient care (Davenport and Kalakota, 2019).

ML is a multidisciplinary field that draws from various disciplines such as statistics, probability theory, approximation theory, algorithm complexity theory, and convex analysis (Zitnik et al., 2019). ML methods can be categorized into two main types including supervised and unsupervised learning. For supervised learning, a model is set by utilizing a labeled dataset, consisting of features and corresponding outcomes. Some of the commonly utilized supervised learning algorithms include statistical classification and regression analysis. While, on the other hand, unsupervised learning, also called clustering, utilizes methods like k-means to identify patterns in data without labeled outcomes. It iteratively establishes centroid and minimizes error to acquire classification. With the advancement of ML, numerous fields have embraced this technique for research purposes. Examples include identification of disease-related microRNAs, drug repositioning (Qu et al., 2019), and identification of disease-related long non-coding RNAs (Chen and Huang, 2017).

Applications of AI and ML for the Ecological Perspective of Microorganisms

Early Pathogen Detection

Due to the diverse characteristics of microorganisms, it is crucial to ensure their accurate identification. Fiannaca et al. (2018) introduced a method for identification of 16S short-read sequences using a combination of k-mer and deep learning techniques. Their results demonstrate the efficacy of the method in accurately classifying both 16S shotgun (SG) and amplicon (AMP) data (Fiannaca et al., 2018). The accurate identification of specific microbial sequences within mixed metagenomics samples is crucial. While gene-based similarity methods are commonly employed for classifying prokaryotic and host organisms from mixed samples, these methods have notable limitations. Consequently, numerous studies have been undertaken to explore improved methods for identifying specific microorganisms. Amgarten et al. (2018) developed a tool called MARVEL, designed for predicting double-stranded DNA bacteriophage sequences in metagenomics (Amgarten et al., 2018).

Ren et al. (2017) introduced VirFinder, an ML method based on k-mer analysis for identifying virus overlap groups without relying on gene-based similarity searches (Ren et al., 2017). These methods cater to specific needs in microbial classification. Additionally, MARVEL, VirSort, and VirFinder excel at identifying specific types of microorganisms. According to Amgarten et al. (2018), these three methods exhibit comparable specificity performance, but MARVEL demonstrates superior recall (sensitivity) performance. AI-powered algorithms have indeed emerged as invaluable tools for swiftly and accurately detecting microbial pathogens within clinical samples. Trained to identify specific patterns or genetic markers associated with various pathogens, these algorithms facilitate rapid and precise pathogen detection. This capability enables immediate therapeutic interventions, thereby substantially mitigating the risk of infection progression (Shelke et al., 2023).

Prediction of Environmental Microorganisms by AI

Microorganisms have emerged as valuable indicators for ecological assessment across various environments. Advancements in sequencing technologies enable the expansion and application of omics-based ML for more ambitious environmental monitoring and mitigation efforts. These indicators can unveil significant insights for land management, especially when conventional field measurements prove inadequate. Significantly, the ML analysis of microbial 16S rRNA abundances can directly forecast soil productivity on arable land and assess risks for agriculture. In cases where conventional analyses struggle to establish clear relationships, machine learning methods may still identify community subpopulations that serve as predictors for relevant environmental parameters and processes. Moreover, machine learning applied to meta-barcoded environmental DNA (eDNA) demonstrates superior performance in environmental quality monitoring compared to traditional bioindicator values, particularly in marine aquaculture monitoring. Thus, ML serves as a valuable tool for enhancing the monitoring of environmental programs (McElhinney et al., 2022).

Microorganisms that have potential similarities with each other make it difficult to identify accurately via conventional methods; thus, ML plays a crucial role in this regard as well (Rani et al., 2022). Currently, images of four types of microorganisms, i.e., bacteria, algae, protozoa, and fungi, are identified. Highly used in bacteria (38.4%), followed by algae image recognition (28.3%). Different neuronal networks and algorithms with minimal human intervention are employed in microbial image recognition, thus making this technique a potential source of microbial identification (Rani et al., 2022). It was documented that DL can be used to train models for substrate secretion mechanisms in bacteria, mostly Gram-negative bacteria. Following this approach with sequence-based non-RTX-motif features in combination with a triple-layer stacking model, the RTX proteins were predicted accurately. It also enables the identification of the proteins secreted by the bacterial membrane into their exterior environment (Sarrafzadeh et al., 2022). Different ML algorithms in supervised form, i.e., Random Forest (RF), Naïve Bayes (NB), Decision Trees (DT), etc., are used to show the enhancement in accuracy of bacterial classification (Oudah & Henschel, 2018).

The Role of AI in Microorganisms' Identification

The potential role of AI in microbial identification is studied exclusively, particularly in the diagnosis and pathogenic detection of clinical settings. Identification and characterization of microbes using AI tools provide an efficient, easy, and time-saving approach with accurate and precise results. These AI algorithms and ML techniques effectively overcome the challenges present in traditional identification methods. Hence, in the following section, we will briefly describe the conventional methods of microbial identification, followed by their shortcomings, and how AI tools can overcome these in more detail. As described, the identification of microorganisms is crucial in any possible application, and these have been used since ancient times in different processes. Therefore, recognized, accurate, and potential methods of identification were available. These conventional techniques and methods mostly rely on the physical, chemical, and biochemical properties of microorganisms. These traditional techniques include staining, culturing, and biochemical assays for phenotypic identification, followed by more potent immunological, biochemical, and molecular analytical techniques (Oudah and Henschel, 2018; Buszewski et al., 2017).

It is also important that these methods can be used alone or in combinations, or that the newly developed methods reinforce the traditional methods in multiple ways. The traditional methods consist of sequential analysis, i.e., sample collections, examination, preservations, morphological identifications, and characterizations. Figure 3 summarizes the traditional methods in a sequential way, from sample collection to strain identification. Each step in this sequential analysis has its own characteristics, advantages and disadvantages, and applications. The next step showed further advancement from the previous and, hence, increased the specificity and identification of the species (Ferone et al., 2020). The importance of identifying bacteria lies in every field of research, e.g., in diagnostics, the physician prescribes medication and treatment after the proper identification of

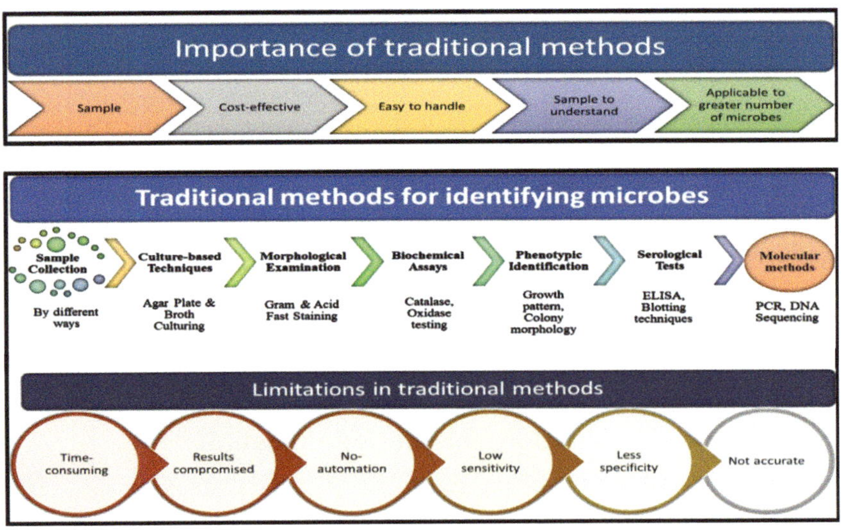

Figure 3: The different spectrum of traditional methods of microbial identification. The types of identification methods, their importance, and associated limitations are described.

bacteria; in animal science, the caregiver is interested in knowing about pathogens that cause diseases in the animals; in plants, identification is important to properly address their potential; and in food science, their exact nature can show the quality status of the products. Likewise, in microbial ecology, identification leads to their activities, while in marine ecology, the detection of pathogens is critical for the healthy and pure status of water.

Currently, AI is extensively used in different domains of life, including drug development and discovery, safety of drugs, genomic analysis, metabolomics, pharmacology, and others, thus enabling researchers to understand AI and use it effectively. Moreover, the microbial biotechnological applications and probiotics potential have recently been raised and become a billion-dollar industry, which totally relies on the proper identification and, thus, characterization of the microorganisms. The potential role of AI in life and in biotechnological applications has been recently reviewed (Holzinger et al., 2023). The AI applications in the fields of microbiology and microbial identification can be viewed by their following potentials and usages: The AI used four types of algorithms in microbiology, like growth with colony counting, discrimination of no growth, phenotypic recognition, and chromogenic detection (Wang et al., 2023).

AI in Bacterial Differentiation, Detection, and Image Recognition

The traditional ways of identification require human intervention to identify and differentiate bacteria (for instance, the Gram-staining), but the ML algorithm does this by analyzing a huge set of images of Gram-staining and thus helps in rapid identification (Peiffer-Smadja et al., 2020). The potential role of AI in the identification of food bacteria is elucidated and documented. Three hours after inoculation, bacteria can be detected at the micro-colony stage using the combinatorial approach of AI and optical imaging. To be more precise, the researchers identified *E. coli* in food samples using the You Only Look Once version 4 (YOLOv4) real-time object recognition and classification system. With this method, bacterial identification may be done without the requirement for resource-intensive molecular approaches and time-consuming culture-based colony isolation. In the food industry, it might be used as a quick and simple method of bacterial detection (Ma et al., 2023).

ANNs have been utilized in the analysis of microorganism images. In research, it was found that ANNs fared better at recognizing dinoflagellates than human specialists. Moreover, food production, illness prevention, medicine discovery, environmental pollution management, etc. all depend on the current AI and associated automated techniques (Goodswen et al., 2021). The detection of antimicrobial resistance is generally considered a time-consuming and laborious process that can be made more effective by using ML, which predicts AMR in different bacterial strains (Májek et al., 2021; Wang et al., 2023). AI algorithms can identify illnesses, including novel and drug-resistant types, with speed and

Table 1: The different types of traditional methods used for the identification of microorganisms

Methods	Description	References
Culture-based Techniques	• Microbial samples are cultured on nutrient agar plates or in broth media under specific conditions (e.g., temperature, pH) to encourage their growth. • The resulting colonies are then examined for morphological characteristics such as size, shape, color, and texture.	Sutherland and Rafii, 2006;
Microscopy and Gram-staining	• Microbial cells are visualized under a microscope to observe their morphological features, such as cell shape, size, arrangement, etc. • Common staining techniques are gram staining, acid-fast staining, and fluorescent staining.	(Lane, 2015)
Biochemical Assays	• Various biochemical tests are performed to assess the metabolic activities of microorganisms. • These tests may include the utilization of specific substrates, the production of enzymes, or other biochemical reactions. • Commercial kits like API strips provide a standardized method for biochemical testing and identification.	Pradhan and Tamang, 2019; Sutherland and Rafii, 2006
Phenotypic Identification	• This identification is based on observable traits such as growth patterns on different media, colony morphology, motility, pigment production, and biochemical characteristics (e.g., fermentation of sugars, production of specific enzymes).	Pradhan and Tamang, 2019
Serological Tests	• Detection of specific antigens or antibodies associated with microbial species using serological techniques such as enzyme-linked immunosorbent assay (ELISA), Western blotting, or agglutination tests. • Serological tests are often used for the diagnosis of infectious diseases and the identification of pathogens.	Sutherland and Rafii, 2006; Eldin et al., 2019
Molecular Techniques	• Molecular methods like PCR, DNA sequencing, and Restriction Fragment Length Polymorphism (RFLP) analyze the genetic material (DNA or RNA) of microorganisms for species identification. • These techniques offer high specificity and can detect microbes even at low concentrations.	Pradhan and Tamang, 2019; Ferone et al., 2020

accuracy. They make better diagnostic methods and possible early antibiotic resistance detection. Pathogen identification, speed, accuracy, and antibiotic resistance prediction are the main focuses of AI in microorganisms (Zhang et al., 2023).

To understand the advanced identification methods of microorganisms, it is important to briefly highlight the conventional methods. In Table 1, an overview of traditional methods of microbial identification is summarized. To correctly identify microorganisms, a mixture of these conventional approaches is frequently employed, each with its own potential. To quantitatively analyze microorganisms and determine their biological activities, biologists and other researchers working with microorganisms need to know how to count them. However, because they are time-consuming, subjective, and require complicated processes, traditional manual methods for counting microorganisms such as hemocytometry, turbidimetry, and plate counting are challenging to use in large-scale applications (Zhang et al., 2022). There are several drawbacks to the conventional method of identification. First of all, they take a lot of time, as cultures usually need to be incubated for several hours or days at a time. The application of these technologies is limited to bacteria that can be cultured in present laboratory settings due to the lack of adequate conditions and mediums. The low resolution of microscopy methods might make it difficult to differentiate between tiny or morphologically identical bacteria, calling for the expertise of trained specialists for precise sample preparation and analysis. Furthermore, for closely related species, differentiation based only on morphological traits may not be adequate. Serological tests may produce false-positive findings because of cross-reactivity with comparable antigens, whereas biochemical assays may have specificity problems that might result in misidentification (Pradhan and Tamang, 2019; Eldin et al., 2019; Ma et al., 2023).

Currently Developed Methods for Microorganism Identification

The aforementioned methods describe the microbial world in terms of their physical appearance, biochemical processes, and other methods in specific domains. Although the existing shortcomings of these methods make them of less use and cannot provide the minute details of microorganisms, the advanced technologies are devoted to proper microorganism identification and thus enhance their biotechnological applications and human use (Buszewski et al., 2017). Likewise, the routinely used biochemical tests for identification are equipped with automated and semi-automated devices, which greatly reduce the time of analysis and enhance the identification qualities. For instance, Vitek 2 Compact, BD Phoenix, Analytical Profile Index (API), etc. are using advanced functions (Buszewski et al., 2017).

The spectrometric methods also give a flavor to the microbial identification, as well-documented. Fourier transform infrared (FTIR) spectroscopy has been used for microbial identification for a long time due to its high quality, accuracy, short analysis time, cheapness, easy sample preparation, and covering of whole bacterial cells, etc. (Hussain, 2023). The identification is mostly based on its common properties, like the fact that the same species has a unique characteristic peak. The probiotic identification was recently performed by Hussain et al. (2023) using FTIR analysis, which indicated a probiotic-specific peak located at 2845 and 1929 cm^{-1} (Hussain et al., 2023).

MALDI-TOF MS is the most recent tool used for the fast identification of microorganisms. The identification is obtained due to the comparison of the spectral profile of the analyzed microorganism with the available automated database. Currently, this method is extensively used in clinical laboratories for the rapid identification of microbes. Identification based on MALDI TOF MS greatly reduced the cost of operation and storage (Buszewski et al., 2017; Han et al., 2021; Wang et al., 2023).

Still, there are many limitations to these modern methods of identification, and we are again shouting for a more reliable identification process. For instance, the sophisticated machine needs to be operated by experts; there must be a proper place of placement; and it must have additional analytical shortcomings. Thus, it's the right time to understand, explore, and take advantage of modern AI tools and develop identification in detail. The spectrum of AI, i.e., its advantages, applications, shortcomings, and future perspectives, is summarized in Figure 4.

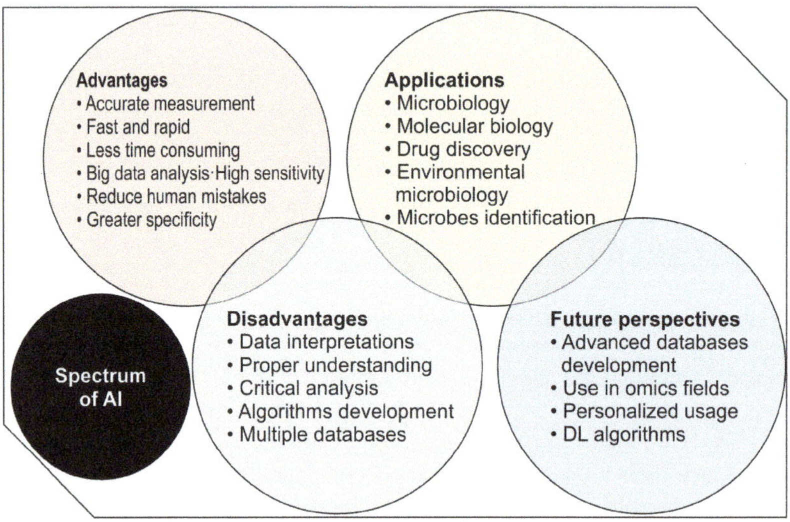

Figure 4: Illustration representing the multiple spectrums of AI in microbial identification, the applications, advantages, disadvantages, and future recommendations.

Effects of Microbial Ecology on Animal's Behavior and Health

The preceding 20 years have seen amazing technology breakthroughs that have raised living standards and improved security for all individuals. Using AI, in various sectors, including medicine, is one of these developments. AI is important for diagnosing and identifying pathogens (Mishra et al., 2023). Most of the conducted studies are based on observations. Recently, Barwant et al. (2024) documented the use of different nano-biosensors for animal health. The effects of AI in different ecosystems are summarized in Table 2.

Table 2: Overview of AI methodologies in different spectrums of environmental microbiology

AI Methodology	Spectrums	References
Metagenomics sequencing analysis	Used to study all genes in environmental microbial samples	Zhang et al., 2021
Satellite-based remote sensing and imaging	Understanding the biology, ecology, and growth dynamics of microorganisms	Grimes et al., 2014
Machine learning model	Helpful for predicting microorganisms	Qu et al., 2019

Interaction with Ecosystem

Eukaryotes likely evolved from prokaryotes and have maintained a close relationship with them ever since. It's therefore unsurprising that animal and plant surfaces host a wide variety and abundance of microorganisms. Additionally, some microorganisms can grow within animal or plant cells as endosymbionts. Thus the number of microbial cells and their collective genetic material often far surpasses that of their hosts. Microorganisms can associate with hosts in various ways. Some associations are temporary and have little impact on the adaptation or evolution of the holobiont. While others are well-established, long-term interactions—such as the rumen system—where host and microorganism become entirely dependent on each other. Between these extremes, there exists a range of interactions of varying strengths, including pathogenesis. It's important to note that studying these host-microorganism interactions is complex because mostly these associated microorganisms have not been cultured, and many interactions involve multiple microorganisms interacting with the host, as seen in the human gut microbiota and the co-aggregating bacteria in the human mouth (Zilber-Rosenberg & Rosenberg, 2008). The bacterial communities of animals are influenced by various factors such as the environment, genetics, physiology, and social interactions of their hosts. In turn, these bacterial communities can convey information about the animals they inhabit by altering the hosts' odor or physical appearance (Archie & Theis, 2011).

Effects on Animal's Ecology

Animals are essential to our ecosystem, coexisting with us and creating a complex web of life. The integration of AI into animal welfare presents an exciting frontier, leveraging advanced technologies such as DL, ML, and neural networks to significantly improve the monitoring, management, and overall well-being of animals. AI-driven solutions can facilitate better understanding of animal behavior, health, and environmental interactions, enabling proactive and preventive measures that enhance the quality of life for various animal species (Zhang et al., 2024). Bacteria are a crucial part of animal bodies, residing on their skin, fur, feathers, scales, and exoskeletons, as well as present in their different systems. In humans, bacterial cells outnumber somatic cells by ten to one, and the collective human-associated microbial communities have about a hundred times more genes than the human genome (Yang et al., 2009). This is likely true for most animals as well. While bacteria in vertebrate systems are often seen as disease-causing agents, many actually have beneficial relationships with their hosts. For example, bacterial communities in animals assist in extracting energy and nutrients from food, producing vitamins, combating pathogens, and modulating immune system function (Ruiz-Rodríguez et al., 2009; Archie and Theis, 2011).

Role of Microorganisms in Animal Gut Health and Disease Transmission

Symbiotic bacteria can influence host health and convey information about their hosts, understanding the microbial communities of animals—such as the species present, their origins, and their functions—can provide insights into long-standing mysteries in animal behavior (Archie & Theis, 2011). Zoonotic diseases, or zoonosis, represent those infections that occur because of the natural transmission of pathogens between humans and animals. These diseases can spread through direct or indirect contact between humans and animals, or via food-borne, vector-borne, and water-borne routes involving bacteria, viruses, parasites, and fungi. Over 70% of infectious diseases are from animal sources, making zoonosis a significant public health issue with an estimated annual mortality of 2.7 million. Besides their impact on human health, zoonosis also affects livestock production and security, leading to economic losses. Recently, AI models have been utilized to study zoonotic pathogens and the factors influencing their spread (Pillai et al., 2022).

Logistic Regression and RF models are commonly employed to analyze and draw insights about different zoonotic diseases and their transmission (Ntampaka et al., 2020; Kiambi et al., 2020). Likewise, ANNs have also shown effectiveness in modeling zoonotic diseases and their causes in various studies (Boleratz & Oscar, 2022; ZareBidaki et al., 2022). These AI-driven approaches hold promise for enhancing our ability to predict, prevent, and control zoonotic disease outbreaks, ultimately improving both human and animal health. Disease prediction models are classified into two main categories: traditional ML models, which require only modest computing power, and DL models.

Using partial least squares regression and hybrid support vector machine (SVM) model, trap counts of the female mosquitoes, Culex Tarsalis, that is responsible for transmitting West Nile Virus (WNV) is accurately predicted by analyzing meteorological data, dead birds, WNV cases, and human fatalities. The study demonstrated that the SVM model, which operates on decision boundaries, performs better when the classes are clearly separable, achieving a mean absolute error of 3.01, thus outperforming other ML models.

American trypanosomiasis, also known as Chagas disease, is a neglected tropical illness caused by *Trypanosoma cruzi*, a flagellated protozoan. It is transmitted by blood-feeding triatomines from the family Reduviidae, subfamily Triatominae. In order to identify differences in the intestinal metabolome of the triatomine Rhodnius prolixus and to predict exposure to *T. cruzi*, RF classifiers, logistic regression, and gradient boosting algorithms were employed by Eberhard et al. (2021). Their findings indicate that the ensemble methods were more effective than logistic regression in detecting the complex interactions between triatomine vectors and the parasites.

Ebola virus disease (EVD) is an uncommon and lethal illness that affects both humans and animals. Price et al. (2020) investigated how host responses to Ebola virus infection in mice differ between tolerant and fatal outcomes, using clinical, virologic, and transcriptomic data. Their analysis revealed that the RF model was highly effective in accurately predicting the outcomes of the disease.

Crimean-Congo hemorrhagic fever (CCHF) is a highly virulent disease in humans caused by a negative-sense single-stranded RNA virus from the genus Nairovirus. Structured Gaussian approach is used to identify high-risk geographic areas for CCHF, incorporating data on climate, land use, and populations of both animals and humans to capture spatiotemporal transmission patterns. Their analysis indicated that CCHF is mainly influenced by geographical factors and climate impacts on ticks. The Gaussian process, which relies on a Gaussian probability distribution, proved effective for reliable classification in uncertain conditions related to climate and spatiotemporal variables (Pillai et al., 2022).

Limitations of Using AI and ML in Microbial Identification and Distribution

Although AI, ML, and DL have a substantial effect in all domains of life, i.e., from identification to confirmation and from validation to analysis, there are still some shortcomings present that raise questions about their supremeness and fastness. The potential shortcomings found in these technologies include the following:

- The robustness of their performance required validation via multiple datasets, particularly in clinical settings, which makes it difficult.
- The construction and development of AI models and their interpretations is not an easy task, and any looseness can lead to negative or incorrect results.

- The AI algorithm is not static, and improvements occur rapidly; hence, regulatory authorities, etc., must be updated and understand their mechanisms and effects.
- Data interpretation is considered a major challenge in ML as most of the users have less understanding of these advanced technologies.
- Microbial morphological similarities make it a hurdle for ML to recognize their images.

Future Perspectives and Potentials

With rising trends like the combination of omics data, personalized microbial analysis, and the creation of ethical and regulatory structures, the field of microbial science has a bright future ahead (Rani et al., 2022). It is anticipated that cooperative study and data exchange will deepen our comprehension of the microbial realm and provide answers to some of the most pressing problems of our day (Goodswen et al., 2021). These days, AI is the table-talk topic for everyone due to its widespread accomplishments. When biotechnology and AI progress together, hitherto unattainable new possibilities open up. This can support significant Sustainable Development Goals (SDGs) and assist with some global issues. Today, AI is pervasive throughout the biological sciences. A wide range of topics are covered, including, biomedical, reasoning, natural language, ML, big data analytics, etc., with applications in biotechnology and related fields (Holzinger et al., 2023). The development of large language models (LLMs) like ChatGPT is considered an advancement in the field because it can summarize and generate different contents, suggesting the robustness and potential of AI technologies (Liu et al., 2023). AI combined with optical imaging provides automated bacterial detection and reduces the human workload in different settings (Ma et al., 2023). Development and automation in the CRISPR-Case system for microorganism identification and functional characterization is also an important application (Hussain and Ali, 2023).

Conclusion

Artificial intelligence (AI) has emerged as a transformative tool for microbial detection and the efficiency of diagnostic processes across various settings. Its influence is particularly significant in resource-limited settings, where traditional methods have limitations of time, cost, and accuracy. AI and its allied technologies like machine learning (ML) and deep learning (DL) enable fast, precise, and cost-effective methods for microbial detection, identification, and surveillance. In a hospital setting, the AI-based protocol provides accurate and efficient pathogen detection, helping in the prevention, early diagnosis, and control of microbial infections. Outside the clinical sites, AI also plays a critical role in environmental microbiology while facilitating the identification of various microbial groups, like bacteria, algae, protozoa, and fungi, of different ecological niches. Owing to the practical applications of AI, these tools greatly help in managing and analyzing large-scale biological datasets with speed, accuracy, and high sensitivity. The integration of AI, ML, and DL algorithms has enabled researchers to identify microbial prevalence, detect pathogens, and explore microbial dynamics in a complex and dynamic environment. Besides ecological microbial applications, these cutting-edge technologies have broader applications, covering forensic microbiology, clinical diagnostics, bioinformatics, and imaging. In conclusion, the proper implementation of these advanced technologies helps researchers to explore the world of microorganisms across various disciplines and delve deep into their applications while deciphering their potential pathogenic nature.

References

Abrar H., and A.A. Syed. (2024). Probiotics and immunity: Exploring the development of probiotics concept. *EC Pharmacology and Toxicology,* 12: 01–06.

Ali, T., S. Ahmed, and M. Aslam. (2023). Artificial Intelligence for Antimicrobial Resistance Prediction: Challenges and opportunities towards practical implementation. *Antibiotics,* 12: 523.

Alowais, S.A., S.S. Alghamdi, N. Alsuhebany, T. Alqahtani, A.I. Alshaya, S.N. Almohareb, A. Aldairem et al. (2023). Revolutionizing healthcare: The role of artificial intelligence in clinical practice. *BMC Medical Education,* 23: 1–15.

Amgarten, D., L.P.P. Braga, A.M. da Silva, and J.C. Setubal. (2018). MARVEL, a tool for prediction of bacteriophage sequences in metagenomic bins. *Frontiers in Genetics,* 9: 1–8.

Archie, E.A., and K.R. Theis. (2011). Animal behaviour meets microbial ecology. *Animal Behaviour,* 82: 425–436.

Barwant, M.M., A. Rasool, V.K. Mannam, A. Hussain, M. Abdullah, Z. Memon, M. Shirani et al. (2024). Nanotechnology applications in livestock farming and animal health. In: Said-Al-Tawaha, Alexious Ghazaryan, Rajput Singh (Eds.), *Harnessing NanoOmics and Nanozymes for Sustainable Agriculture*, pp. 413–442. IGI Global.

Boleratz, B.L., and T.P. Oscar. (2022). Use of ComBase data to develop an artificial neural network model for nonthermal inactivation of Campylobacter jejuni in milk and beef and evaluation of model performance and data completeness using the acceptable prediction zones method. *Journal of Food Safety,* 42: e12983.

Buszewski, B., A. Rogowska, P. Pomastowski, M. Złoch, and V. Railean-Plugaru. (2017). Identification of microorganisms by modern analytical techniques. *Journal of AOAC International,* 100: 1607–1623.

Chen, X., and L. Huang. (2017). LRSSLMDA: Laplacian regularized sparse subspace learning for MiRNA-disease association prediction. *PLoS Computational Biology,* 13: 1–28.

Chen, X.X., H. Tang, W.C. Li, H. Wu, W. Chen, H. Ding, and H. Lin. (2016). Identification of bacterial cell wall lyases via pseudo amino acid composition. *BioMed. Research International,* 1: 1654623.

Davenport, T., and R. Kalakota. (2019). The potential for artificial intelligence in healthcare. *Future Healthcare Journal,* 6: 94–98.

DiMucci, D., M. Kon, and D. Segrè. (2018). Machine learning reveals missing edges and putative interaction mechanisms in microbial ecosystem networks. *MSystems,* 3: 101–128.

Eberhard, F.E., S. Klimpel, A.A. Guarneri, and N.J. Tobias. (2021). Metabolites as predictive biomarkers for *Trypanosoma cruzi* exposure in triatomine bugs. *Computational and Structural Biotechnology Journal,* 19: 3051–3057.

Eldin, C., P. Parola, and D. Raoult. (2019). Limitations of diagnostic tests for bacterial infections. *Medecine et Maladies Infectieuses,* 49: 98–101.

Ferone, M., A. Gowen, S. Fanning, and A.G.M. Scannell. (2020). Microbial detection and identification methods: Bench top assays to omics approaches. *Comprehensive Reviews in Food Science and Food Safety,* 19: 3106–3129.

Fiannaca, A., L. La Paglia, M. La Rosa, G. Lo Bosco, G. Renda, R. Rizzo et al. (2018). Deep learning models for bacteria taxonomic classification of metagenomic data. *BMC Bioinformatics,* 19: 61–76.

Goodswen, S.J., J.L.N. Barratt, P.J. Kennedy, A. Kaufer, L. Calarco, and J.T. Ellis. (2021). Machine learning and applications in microbiology. *FEMS Microbiology Reviews,* 45: 1–19.

Grimes, D.J., T.E. Ford, R.R. Colwell, C. Baker-Austin, J. Martinez-Urtaza, A. Subramaniam, and D.G. Capone. (2014). Viewing marine bacteria, their activity and response to environmental drivers from orbit: Satellite remote sensing of bacteria. *Microbial Ecology,* 67: 489–500.

Han, S.S., Y.S. Jeong, and S.K. Choi. 2021. Current scenario and challenges in the direct identification of microorganisms using MALDI-TOF MS. *Microorganisms,* 9: 1–11.

Holzinger, A., K. Keiblinger, P. Holub, K. Zatloukal, and H. Müller. (2023). AI for life: Trends in artificial intelligence for biotechnology. *New Biotechnology,* 74: 16–24.

Hussain, A. 2023. *Molecular Assessment and Validation of the Selected Enterococcal Strains as Probiotics.* M.Phil. Thesis, University of Karachi, Karachi, 1–130.

Hussain, A., S. Akram, D. Ahmad, M. Rehman, A. Ahmed, and S.A. Ali. (2023). Molecular assessment and validation of the selected enterococcal strains as probiotics. *Probiotics and Antimicrobial Proteins,* 0123456789.

Hussain, A., and S.A. Ali. (2023). Probiotics development and CRISPR-Cas system. *Conference Presentation, Science Innovation Showcase,* 1–18.

Hussain, A., M. Rahman, T. Bibi, R. Fatima, I. Arif, M.M. Barwant, and S.A. Ali. (2024). Prevalence of microorganisms in indoor household environments and their pathogenesis. *Journal of Antimicrobial Agents,* 10(1): 1–5.

Hussain A. (2024). Microorganisms and climate change: The smallest doing the bigger. *The Earth Needs Love.* green blog [online].

Kiambi, S.G., E.M. Fèvre, J. Omolo, J. Oundo, and W.A. De Glanville. (2020). Risk factors for acute human brucellosis in Ijara, north-eastern Kenya. *PLoS Neglected Tropical Diseases,* 14: e0008108.

Lane, N. (2015). The unseen world: Reflections on Leeuwenhoek (1677), concerning little animals. *Philosophical Transactions of the Royal Society B: Biological Sciences,* 370: 1666.

Liu, B., F. Liu, X. Wang, J. Chen, L. Fang, and K.C. Chou. (2015). Pse-in-One: A web server for generating various modes of pseudo components of DNA, RNA, and protein sequences. *Nucleic Acids Research,* 43: W65–W71.

Liu, X., L. Faes, A.U. Kale, S.K. Wagner, D.J. Fu, A. Bruynseels, T. Mahendiran, et al. (2019). A comparison of deep learning performance against healthcare professionals in detecting diseases from medical imaging: A systematic review and meta-analysis. *The Lancet Digital Health,* 1: e271–e297.

Liu, Y., Y. Chen, and L. Han. (2023). Bioinformatics: Advancing biomedical discovery and innovation in the era of big data and artificial intelligence. *The Innovation Medicine,* 1: 100012.

Ma, L., J. Yi, N. Wisuthiphaet, M. Earles, and N. Nitin. (2023). Accelerating the detection of bacteria in food using artificial intelligence and optical imaging. *Applied and Environmental Microbiology,* 89(1): e01828.

Madsen, E.L. (2005). Identifying microorganisms responsible for ecologically significant biogeochemical processes. *Nature Reviews Microbiology,* 3: 439–446.

Májek, P., L. Lüftinger, S. Beisken, T. Rattei, and A. Materna. (2021). Genome-wide mutation scoring for machine-learning-based antimicrobial resistance prediction. *International Journal of Molecular Sciences,* 22: 1304922.

Martinez-Rabert, E., C. van Amstel, C. Smith, W.T. Sloan, and R. Gonzalez-Cabaleiro. (2022). Environmental and ecological controls of the spatial distribution of microbial populations in aggregates. *PLoS Computational Biology,* 18: 1–23.

McElhinney, J.M.W.R., M.K. Catacutan, A. Mawart, A. Hasan, and J. Dias. (2022). Interfacing machine learning and microbial omics: A promising means to address environmental challenges. *Frontiers in Microbiology,* 13: 851450.

Mishra, A., S. Khan, A. Das, and B.C. Das. (2023). Evolution of diagnostic and forensic microbiology in the era of artificial intelligence. *Cureus,* 15: 9.

Ntampaka, P., F. Niragire, P.N. Nyaga, and G. Habarugira. (2021). Canine gastrointestinal nematodiasis and associated risk factors in Kigali city, Rwanda. *Journal of Parasitology Research,* 2021: 9956256.

Oudah, M., and A. Henschel. (2018). Taxonomy-aware feature engineering for microbiome classification. *BMC Bioinformatics,* 19: 1–13.

Peiffer-Smadja, N., S. Dellière, C. Rodriguez, G. Birgand, F.X. Lescure, S. Fourati, and E. Ruppé. (2020). Machine learning in the clinical microbiology laboratory: Has the time come for routine practice? *Clinical Microbiology and Infection,* 26: 1300–1309.

Pillai, N., M. Ramkumar, and B. Nanduri. (2022). Artificial intelligence models for zoonotic pathogens: A survey. *Microorganisms,* 10: 1911.

Pradhan, P., and J.P. Tamang. (2019). Phenotypic and genotypic identification of bacteria isolated from traditionally prepared dry starters of the Eastern Himalayas. *Frontiers in Microbiology,* 10: 1–15.

Price, A., A. Okumura, E., Haddock, F. Feldmann, K. Meade-White, P. Sharma, M. Artami et al. (2020). Transcriptional correlates of tolerance and lethality in mice predict Ebola virus disease patient outcomes. *Cell Reports,* 30: 1702–1713.

Puaschunder, J.M. (2020). The potential for artificial intelligence in healthcare. *SSRN Electronic Journal,* 6: 94–98.

Qu, K., F. Guo, X. Liu, Y. Lin, and Q. Zou. (2019). Application of machine learning in microbiology. *Frontiers in Microbiology,* 10: 1–10.

Rani, P., S. Kotwal, J. Manhas, V. Sharma, and S. Sharma (2022). Machine learning and deep learning based computational approaches in automatic microorganisms image recognition: Methodologies, challenges, and developments. *Archives of Computational Methods in Engineering,* 29: 1801–1837.

Ren, J., N.A. Ahlgren, Y.Y. Lu, J.A. Fuhrman, and F. Sun. (2017). VirFinder: A novel k-mer based tool for identifying viral sequences from assembled metagenomic data. *Microbiome,* 5: 69.

Rhoads, D.D. 2020. Computer vision and artificial intelligence are emerging diagnostic tools for the clinical microbiologist. *Journal of Clinical Microbiology,* 1–3.

Rodrigues, C.M.C., and H. Groves. (2018). Community-acquired pneumonia in children: The challenges of microbiological diagnosis. *Journal of Clinical Microbiology,* 56: 1–9.

Ruiz-Rodríguez, M., E. Valdivia, J.J. Soler, M. Martín-Vivaldi, A.M. Martín-Platero, and M. Martínez-Bueno. (2009). Symbiotic bacteria living in the hoopoe's uropygial gland prevent feather degradation. *Journal of Experimental Biology,* 212: 3621–3626.

Sarrafzadeh, M.H., S.S. Mansouri, J. Zahiri, S.I. Mussatto, and H. Asgharnejad. (2022). Editorial: Artificial Intelligence in Environmental Microbiology. *Frontiers in Microbiology,* 13: 944242.

Shelke, Y.P., A.K. Badge, and N.J. Bankar. (2023). Applications of artificial intelligence in microbial diagnosis. *Cureus,* 15: 1–8.

Shrestha, L.B., and K. Pokharel. (2020). Standard operating procedure for specimen collection, packaging, and transport for diagnosis of SARS-CoV-2. *Journal of the Nepal Medical Association,* 58: 627–629.

Sutherland, J.B., and F. Rafii. (2006). Genetic methods for identification of bacteria. In: *Identification of Microorganisms by Mass Spectrometry,* pp. 1–21. Wiley.

Undru, T.R., U. Uday, J.T. Lakshmi, A. Kaliappan, S. Mallamgunta, S.S. Nikhat, V. Sakthivadivel, and A. Gaur. (2022). Integrating artificial intelligence for clinical and laboratory diagnosis: A review. *Maedica,* 17: 420–426.

Waldron, L. (2018). Data and statistical methods to analyze the human. *Clinical Science and Epidemiology,* 2: 1–4.

Wang, D., M. Zhang, A.S. Mujumdar, and D. Yu. (2022). Advanced detection techniques using artificial intelligence in processing of berries. *Food Engineering Reviews,* 14: 176–199.

Wang, H., C. Chung, W. Lin, S. Kalpana, and J. Horng. (2023). Artificial intelligence in MALDI-TOF MS: Microbial identification, strain typing, and antimicrobial resistance detection. DOI: 10.20944/preeprints202312.1149. ResearchGate.

Xie, K., L. Guo, Y. Bai, W. Liu, J. Yan, and M. Bucher. (2019). Microbiomics and plant health: An interdisciplinary and international workshop on the plant microbiome. *Molecular Plant,* 12: 1–3.

Xu, G., X. Teng, X.H. Gao, L. Zhang, H. Yan, and R.Q. Qi. (2023). Advances in machine learning-based bacteria analysis for forensic identification: Identity, ethnicity, and site of occurrence. *Frontiers in Microbiology,* 14: 1–6.

Yang, X., L. Xie, Y. Li, and C. Wei. (2009). More than 9,000,000 unique genes in the human gut bacterial community: Estimating gene numbers inside a human body. *PloS One,* 4: e6074.

Yang, H., H. Tang, X.X. Chen, C.J., Zhang, P.P. Zhu, H. Ding, W. Chen, and H. Lin. (2016). Identification of secretory proteins in mycobacterium tuberculosis using pseudo amino acid composition. *BioMed Research International,* 1: 5413903.

Zare Bidaki, M., E. Allahyari, T. Zeinali, and M. Asgharzadeh. (2022). Occurrence and risk factors of brucellosis among domestic animals: An artificial neural network approach. *Tropical Animal Health and Production,* 54: 1–13.

Zeeshan A. Khan, M.F. Siddiqui and S. Park. (2019). Current and Emerging Methods of Antibiotic Susceptibility Testing. *Diagnostics,* 49: 1–17.

Zhang, J., C. Li, M.M. Rahaman, Y. Yao, P. Ma, J. Zhang, X. Zhao et al. (2022). A comprehensive review of image analysis methods for microorganism counting: From classical image processing to deep learning approaches. *Artificial Intelligence Review,* 55: 2875–2944.

Zhang, J., C. Li, Y. Yin, J. Zhang, and M. Grzegorzek. (2023). Applications of artificial neural networks in microorganism image analysis: A comprehensive review from conventional multilayer perceptron to popular convolutional neural network and potential visual transformer. *Artificial Intelligence Review,* 2: 1013–1070.

Zhang, L., W. Guo, C. Lv, M. Guo, M. Yang, Q. Fu, and X. Liu. (2024). Advancements in artificial intelligence technology for improving animal welfare: Current applications and research progress. *Animal Research and One Health,* 2: 93–109.

Zhang, L., F.X. Chen, Z. Zeng, M. Xu, F. Sun, L. Yang, X. Bi et al. (2021). Advances in metagenomics and its application in environmental microorganisms. *Frontiers in Microbiology,* 12: 766364.

Zilber-Rosenberg, I., and E. Rosenberg. (2008). Role of microorganisms in the evolution of animals and plants: The hologenome theory of evolution. *FEMS Microbiology Reviews,* 32: 723–735.

Zitnik, M., F. Nguyen, B. Wang, J. Leskovec, A. Goldenberg, and M.M. Hoffman. (2019). Machine learning for integrating data in biology and medicine: Principles, practice, and opportunities. *Information Fusion,* 50: 71–91.

11 | Future Directions and Emerging Technologies in AI for Animal Ecology

Nik Fadzly* **and Zi-Han Yeo**
School of Biological Sciences, Universiti Sains Malaysia, 11800, Penang, Malaysia

Artificial Intelligence (AI) is transforming the landscape of animal ecology, offering innovative tools to enhance ecological research and conservation efforts. This chapter explores the integration of emerging AI technologies, including machine learning, computer vision, and real-time monitoring systems, to improve our understanding of animal behavior, habitat use, and ecosystem dynamics. The discussion emphasizes the convergence of traditional ecological methods with AI-driven predictive modeling, facilitating more accurate predictions and informed decision-making for conservation strategies. A central theme is the potential of AI to decode animal communication, with advances in bioacoustic analysis and neural interfaces enabling deeper insights into interspecies interaction. Examples such as the Cetacean Translation Initiative showcase the possibilities of using AI to understand complex communication systems. Moreover, the application of AI-enhanced environmental DNA (eDNA) analysis provides non-invasive methods to detect elusive species, monitor biodiversity, and track invasive organisms in challenging environments. Further technological innovations, such as blockchain-integrated AI, present solutions for addressing illegal wildlife trade by ensuring traceability and data integrity across supply chains. Multifunctional sensors, coupled with AI algorithms, enable real-time data collection and adaptive responses to environmental changes. Autonomous drones and IoT (internet of things) devices also play a crucial role in enhancing wildlife monitoring while minimizing human interference. However, the chapter also highlights challenges, including power consumption, data security, and the ethical implications of AI in animal ecology. With the rapid development of AI, balancing accessibility, accuracy, and ethical considerations becomes imperative.

*Corresponding author: nfadzly @usm.my

Ultimately, this chapter envisions a future where AI not only aids conservation efforts but also fosters meaningful interactions between humans and animals, promoting a deeper understanding and stewardship of our natural world.

Introduction

In the last few years, the term Artificial Intelligence (AI) has become so ubiquitous that it is now a fundamental part of various fields, including animal ecology (Dhyani et al., 2023). Researchers are increasingly leveraging AI to analyze complex ecological data, enabling more accurate predictions of animal behavior and habitat use. As AI technologies continue to evolve, new methodologies such as machine learning and computer vision are being integrated into wildlife monitoring systems, allowing for real-time data collection and analysis.

Although we can define AI as something new for the study of animal ecology, the fundamentals or the mathematics behind AI is something that most theoretical ecologists have been using for the past few decades. Predictive modeling on population growth, animal movement, and habitat selection has long relied on statistical techniques that share similarities with modern AI approaches (Saarenmaa et al., 1988). This convergence of traditional ecological methods and cutting-edge technology is paving the way for more nuanced insights into species interactions and ecosystem dynamics. As researchers continue to refine these models, the potential for AI to enhance conservation efforts becomes increasingly evident, enabling targeted interventions and more effective resource management.

In this chapter, we will discuss future directions, potential use cases, and emerging technologies that hold promise for integrating AI into ecological research, including machine learning algorithms, remote sensing data, and real-time monitoring systems.

Future Communication with Animals

With the advancement of Machine Learning and Deep Learning models, we are looking into a future where basic communication with animals may become common, allowing us to decode their behaviors and vocalizations in ways previously thought unattainable (Ma, 2015; Jahns, 2013). The Earth Species Project is one of the many initiatives aiming to bridge the gap between human understanding and animal communication, utilizing AI to analyze vast amounts of data collected from various species (Project Earth Species, n.d.). McCowan et al. (2023) have demonstrated that a rudimentary conversation was established between an adult female humpback whale named Twain. Twain showed a remarkable ability to respond to specific sounds made by researchers, indicating a level of comprehension and interaction that opens up new avenues for studying interspecies communication. Additionally, the Cetacean Translation Initiative emphasizes the potential of machine learning in understanding complex communication systems in sperm whales, advocating for a multidisciplinary approach to data collection and analysis (Andreas et al., 2021). This ground-breaking work not only enhances our knowledge of marine life but also raises ethical questions about the implications of such interactions and the responsibilities humans have towards these intelligent beings.

AI-powered machine learning algorithms have shown significant promise in deciphering animal vocalizations, facilitating more effective communication between species. In primates, machine-learning techniques have been applied to analyze vocal communication, leveraging large datasets from passive acoustic monitoring (Cauzinille et al., 2024). Furthermore, mobile applications utilizing AI for real-time animal sound recognition demonstrate practical applications of these technologies in biodiversity monitoring (Lin & Fernando, 2023). Overall, while the advancements in AI and machine learning present exciting opportunities for interspecies communication, challenges such as data availability and model validation remain critical for future research and application in this field (Rutz et al., 2023).

What does this mean for the future of animal ecology and conservation efforts? As researchers continue to refine these technologies, we may see a paradigm shift in how we understand animal behavior, leading to more informed conservation strategies and enhanced protection of endangered species. Observation strategies can incorporate new recognition phrases that would enable scientists to study the hierarchical social structures within animal populations, revealing intricate relationships and interactions that were previously difficult to discern. More importantly, this brings us to another deeper fundamental perspective. Who is watching who? As observers, scientists often adopt a no-interrupt rule, in which the observer will not under any circumstances interfere with the daily activities of the subjects being studied (Bateson & Martin, 2021). This approach, while essential for maintaining natural behaviors, raises questions about the impact of human presence on these interactions and whether the mere act of observation alters the dynamics within these communities. As we might understand these animals in the future, perhaps the animals will be aware of the observers and have their own ways of interpreting our presence, leading to a complex interplay of awareness and behavior that challenges our understanding of their social structures.

More importantly, future communications with complex animals such as elephants, whales, and dolphins may reveal insights into their emotional lives and cognitive abilities, potentially reshaping our perceptions of intelligence in the animal kingdom. A good example comes from the Apple TV series drama, "Extrapolations" in which one of the episodes showcased how future scientists managed to communicate with the last whale by using AI to decode the intricate songs of the whale, unveiling a rich tapestry of emotions and stories that had long been hidden beneath the ocean's surface. While the premise of the story centers on a bleak outlook on climate change, it does offer a glimpse of the possible future. To simplify the concept, future AI would enable us to interview the animals. Rather than tracking and tagging the animals, in the future, we can simply ask them, "How was your day?"

AI-enhanced Environmental DNA (eDNA) Analysis

Environmental DNA (eDNA) technology has emerged as a revolutionary tool for monitoring biodiversity by detecting traces of genetic material left behind by organisms in their surroundings (Mathieu et al., 2020; Suren et al., 2024). With advancements in AI, the analysis of eDNA data has become more efficient, offering unprecedented opportunities for conservationists to monitor ecosystems without direct observation. AI-enhanced eDNA analysis allows researchers to detect species presence, monitor biodiversity, and assess the health of ecosystems with greater speed and accuracy.

One of the most promising applications of AI-enhanced eDNA analysis is in the detection of elusive or endangered species (Mauvisseau et al., 2020). Traditional monitoring methods, such as camera traps or manual tracking, are often limited by terrain or visibility. In contrast, eDNA sampling can detect organisms through genetic traces in water, soil, or air, even when they are not directly observable. AI algorithms can rapidly process these samples, identifying species from large genomic databases, which helps conservationists prioritize their efforts in protecting endangered populations.

AI-enhanced eDNA analysis also offers significant advantages in monitoring biodiversity in remote or difficult-to-access areas. Autonomous devices equipped with AI algorithms could process samples on-site, transmitting data in real-time to researchers. This approach allows for continuous monitoring of ecosystems and the early detection of changes in biodiversity. For example, AI-powered eDNA sensors deployed in marine environments could detect shifts in fish populations, providing timely information to fisheries managers and policymakers (Mauvisseau et al., 2020).

Another critical application of AI-enhanced eDNA analysis is in monitoring invasive species. Early detection of invasive species is essential for preventing their establishment and spread. AI algorithms can analyze eDNA data to identify invasive species and predict their movement patterns, enabling conservationists to take pre-emptive action. This capability is particularly valuable in ecosystems where invasive species threaten native biodiversity and ecological balance.

The use of AI in eDNA analysis also contributes to understanding the impacts of climate change on biodiversity. By comparing eDNA samples collected over time, AI algorithms can detect shifts in species distributions and assess how ecosystems are responding to changing environmental conditions. These insights are essential for developing adaptive management strategies that ensure the long-term sustainability of biodiversity in the face of climate change.

Blockchain-integrated AI for Wildlife Trade Monitoring

The illegal wildlife trade represents one of the most significant threats to global biodiversity, driving species toward extinction and destabilizing ecosystems.

While regulatory frameworks exist to curb this illicit trade, enforcement is often limited by the opacity and fragmentation of supply chains. Blockchain technology, integrated with AI algorithms, offers a novel solution to these challenges by improving transparency, traceability, and data integrity across wildlife trade networks (Dryga et al., 2019; Busse et al., 2019). This integration holds immense potential for conservationists, enabling more effective monitoring of wildlife products and reducing opportunities for exploitation.

Blockchain technology provides a decentralized ledger that records transactions transparently and securely. Every transaction in a blockchain system is timestamped, immutable, and visible to all authorized participants, minimizing the risk of fraud. In the context of wildlife trade, blockchain can track the provenance of products from origin to sale. This ensures that wildlife-derived goods are sourced legally and sustainably. AI complements blockchain by automating the detection of suspicious patterns within trade data, enabling authorities to identify and respond to illegal activities more swiftly.

One significant application of blockchain-integrated AI is in the monitoring of supply chains involving endangered species. AI algorithms can analyze blockchain-logged transactions to detect anomalies, such as sudden surges in the trade of specific species or products. This capability is particularly valuable in wildlife markets, where illegal goods are often laundered into legal supply chains. For example, blockchain systems could record each step in the ivory trade. We can track products from poaching hotspots to end markets. Using AI, we can identify patterns that indicate illegal sourcing. This data-driven approach enhances enforcement by providing real-time evidence that authorities can act upon. Here, we need to address that this concept is different from the Bitcoin concept preferred by criminals. Whilst the transactions are recorded in the blockchain ledgers, the identities of the parties involved can be anonymized, allowing for transparency in legal trades while protecting sensitive information. Or, in certain cases, allows criminals to use this system. Perhaps identifying these criminals might take an effort; tracing the illegal sales activities of wildlife trade can at least be tracked and analyzed through advanced algorithms that monitor unusual patterns and flag suspicious transactions for further investigation.

Another important area where blockchain-integrated AI can make an impact is in improving compliance with international wildlife trade agreements, such as the Convention on International Trade in Endangered Species (CITES). Blockchain can serve as a digital ledger for permits, ensuring that only authorized transactions occur, while AI systems verify the legitimacy of permits by cross-referencing data with government databases. This dual-layered system reduces the likelihood of permit forgery, streamlining compliance checks and enhancing enforcement efforts.

Additionally, blockchain-integrated AI can foster transparency and accountability within conservation programs. Communities and organizations involved in wildlife conservation often rely on donor funding to support their efforts (Stuit et al., 2022). Blockchain technology can track the allocation of these funds, ensuring that they are used effectively and as intended. AI analytics can provide further insights by assessing the impact of conservation interventions, optimizing resource allocation, and identifying areas requiring additional support.

Multifunctional LLM Sensors

Sensor technology has evolved over the past decade by becoming cheaper, smaller, and smarter. We have sensors to record the environment, movement, trajectory, and even the physiology of the animals themselves. Previously, any field equipment would need an operator to either manually record the measurements or to actually retrieve the data. Now, with the integration of AI, these sensors can autonomously gather and analyze information, providing real-time insights into animal behavior and environmental changes.

Multi-function sensors play a crucial role in the conservation of endangered species by enabling precise tracking of their populations and habitats. These sensors, which include IoT (internet of things) devices, GPS trackers, and remote sensing technologies, facilitate real-time monitoring and data collection, enhancing conservation efforts. Key contributions include:

Enhanced Monitoring Capabilities

- **Sensor Fusion:** Combining various sensors (e.g., infrasound, cameras, seismic) allows for comprehensive coverage and detection of poaching activities (Siewert et al., 2024).
- **Enhanced Algorithmic-based Sensing:** Modern motion magnification algorithms like the Eulerian Video Magnification algorithm created by Wu et al. (2012) from MIT are able to reveal subtle changes in videography and are able to allow heart rate or breathing rate monitoring from a distance. The incorporation of such algorithms with AI technologies like computer vision and their deployment in wildlife monitoring sensors would allow such features to be implemented for wildlife, allowing never-before-monitored wildlife data to be collected and potential new insights to be inferred.
- **IoT Integration:** Real-time data from motion sensors and cameras aids in understanding animal behavior and habitat use (Roy et al., 2023). It can also allow for real-time monitoring of wildfires and provide immediate alerts for quick and effective damage control, potentially saving precious endangered ecosystems and resources.
- **Automated Wildlife Monitoring Camera:** Real-time image capture can now detect different types of animals captured by the camera. The camera, equipped with a 4G cellular system, then transmits the data to a server equipped with a data processing capability, enabling researchers to not only monitor wildlife but also analyze patterns in real time. This integration of AI with sensor technology allows for the deployment of biologgers that can intelligently manage their data collection based on specific behavioral triggers, optimizing battery life and storage capacity while providing critical insights into animal behavior (Korpela et al., 2019). Moreover, as these technologies advance, they may facilitate a more holistic understanding of ecosystem interactions by

correlating environmental changes with animal responses, thereby contributing to predictive conservation models that address both immediate threats and long-term sustainability goals. Such advancements could ultimately lead to innovative strategies for habitat restoration and species reintroduction efforts, reinforcing the necessity for ongoing research and collaboration across disciplines in the field of animal ecology (Kissling et al., 2024). This enables researchers to analyze population dynamics and track movements, leading to more informed conservation strategies.

- **Integrated TinyML using Cheap and Small Microcontrollers:** These cameras can facilitate on-device processing, allowing for immediate analysis of data at the source, reducing latency and bandwidth usage while enhancing the efficiency of wildlife monitoring efforts (Panda et al., 2022). This technology not only streamlines data collection but also empowers conservationists to respond swiftly to emerging threats, ensuring that interventions are timely and effective.

The next step is to integrate these technologies into a cohesive framework that enables real-time monitoring and adaptive responses to environmental changes. We expect the development of a multifunctional sensor system that is integrated with a Large Language Model (LMM), which would enable real-time data interpretation and predictive analytics, enhancing our ability to respond swiftly to ecological shifts. Users can simply give commands using voice commands, and the sensor will react and adapt accordingly. A basic example of this technology is the SeeedStudio Watcher (Seeedstudio Watcher, n.d.). This ESP32 is equipped with a camera and a development kit that integrates ChatGPT within its ecosystem. The device has extra connectivity pins that would allow users to add multiple sensors and customize the functionality to suit specific environmental monitoring needs, such as air quality, temperature, and humidity levels. The commands to initiate the changes and trigger specific actions can be tailored through a user-friendly interface, making it accessible even for those with minimal technical expertise. Furthermore, with the integration of ChatGPT, users can use voice commands to change the device's directives. This is similar to the Universal Scanner (Tricoder) that was depicted in the Star Trek series.

While the development of these LLM sensors is still in its infancy, the potential applications for animal ecology science are vast, ranging from tracking wildlife movements to monitoring habitat conditions in real time. This would bring us to the next possible future direction in AI and animal ecology: The deployment phase. This phase will involve field testing these technologies in various ecosystems, allowing researchers to gather data and refine the systems based on real-world feedback.

Automated AI Drone for Data Collection

Autonomous drones are revolutionizing animal ecology by enhancing wildlife monitoring and conservation efforts. These unmanned vehicles (UVs) facilitate data collection in remote areas, minimize human disturbance, and integrate advanced technologies like machine learning for real-time analysis. Key aspects of their application include:

Enhanced Data Collection

- UAVs can cover vast and rugged terrains, collecting data that is often inaccessible by traditional methods (Mazumdar, 2022). As we embrace the deployment of autonomous drones in wildlife monitoring, it is essential to consider how these technologies can be integrated with AI-driven data analytics to enhance our understanding of animal behavior and ecosystem health. For instance, combining drone-collected imagery with machine learning algorithms enables more sophisticated analyses of population dynamics and habitat changes over time, allowing researchers to identify critical areas for conservation efforts (Raghuwanshi et al., 2023). This integration between aerial surveillance and ground-level data collection not only improves our ability to monitor species at risk but also fosters a deeper appreciation for the complexities of ecological networks, emphasizing the importance of protecting biodiversity in an ever-changing landscape. They enable in situ imageomics, allowing researchers to infer biological traits from images, thus providing valuable insights into animal behavior (Kline et al., 2024).

Moreover, the integration of AI with autonomous drones opens up exciting avenues for collaborative research initiatives that transcend geographical boundaries. For instance, combined efforts across different regions can utilize UAVs equipped with machine learning algorithms to collect and analyze data on migratory patterns of species such as birds or marine mammals, thereby enhancing our understanding of their ecological needs and conservation strategies. Such collaborative frameworks could also incorporate animal-borne data loggers or biologgers, which allow researchers to gather real-time physiological and behavioral data from individual animals in their natural habitats (Korpela et al., 2019). By employing these technologies together, scientists can create a more holistic view of ecosystem dynamics, improving predictive models and ultimately leading to more effective conservation interventions tailored to specific species and environments. This synergy not only amplifies the scope of ecological research but also fosters an adaptive management approach that is responsive to changing environmental conditions and anthropogenic pressures.

Reduced Disturbance to Wildlife

- Innovative navigation methods, such as motion camouflage, allow drones to observe animals with minimal visual disturbance (Li et al., 2022).
- This capability is crucial for studying sensitive species without altering their natural behavior (Raffik et al., 2024).

There are two main types of drones that are currently used in animal ecology. The aerial drones and the submersible drones each serve unique purposes in monitoring and studying different species in their respective habitats. Current technology and law regulations require that drone operations be controlled and monitored by humans. The new advancements in artificial intelligence are paving the way for more autonomous systems, potentially allowing drones to operate with greater independence while still adhering to safety protocols. Currently, autonomous flight drones are still bound by their pre-programmed routes, but future developments may enable them to adapt in real-time to changing environmental conditions and animal behaviors. Similarly, underwater drones are mostly remote-controlled due to various restrictions such as power distribution, signal telemetry, and navigation challenges. However, innovations in battery technology and communication systems could soon allow for more autonomous underwater exploration, enhancing our understanding of marine ecosystems.

To illustrate the possible future scenario, there was a famous scene from the movie Minority Report where police officers in the future deployed 'spider' bots with the ability to navigate complex environments and identify potential threats. This concept, while fictional, highlights the potential for advanced robotics to revolutionize surveillance and security measures in our own world. Robotic fish are being designed to assess the swimming formation and behavior of real fish, providing insights into their social structures and responses to environmental changes (Butail et al., 2015).

However, as of now, the movements are still limited to pre-programmed movements and basic reactive behaviors. Researchers are now exploring ways to incorporate machine learning algorithms that would allow these robotic fish to adapt their movements in real time, mimicking the fluid dynamics of their biological counterparts.

In terms of the possible future, we expect to see the development of these drones, both aerial, nautical, and land, that are equipped with General Artificial Intelligence that would enable them to learn from the environment and interact more autonomously with their surroundings, potentially revolutionizing fields such as environmental monitoring and search-and-rescue operations. We actually do have these smart rovers, but they are currently not on this earth, and with a total development cost of 1.08 billion USD (including launch and deployment), they are hardly cost-effective for our ecological survey needs. Perhaps the technology could be adapted for a cheaper solution, allowing for smaller, more efficient units that could be deployed in various ecosystems on Earth, thus making ecological monitoring more accessible and widespread.

Big Data in the Field of Ecology

With the increasing amount of sensors being used for ecological monitoring, massive amounts of new data are being collected as we speak. Citing our previously mentioned examples, sensors are being installed in forests and in the air in the form of UAVs. Furthermore, locations like oceans, wetlands, volcanic regions, and concrete jungles could also benefit from sensor installations. All this provides an abundance of real-time information, from species behavior to climate changes, all contributing to an enormous treasure trove of ecological Big Data. Applying advanced data analysis in the form of machine learning and AI models could shine a light on previously undiscovered insights into the complex relationships that govern ecosystems, enabling informed decision-making and novel approaches for ecological efforts like conservation, resource management, and species preservation.

Resource Optimizations for Sustainable Ecosystem Management

The most immediate impact we can derive from data-driven insights would be the optimization of natural resource use. With real-time data being easily acquired and a slew of historical data within our grasp, new protocols and policies regarding natural resources can be crafted in such a way that long-term ecosystem viability is sustained alongside responsible and sustainable resource use.

By tracking critical natural resources in real-time while applying insights for optimized natural resource harvest, we can prevent exerting ecosystem stress that leads to ecological disruption. For instance, sensors monitoring soil moisture, tree growth rates, and biodiversity can feed into AI models that help predict optimal logging intervals or identify areas at risk of deforestation. This data-driven approach ensures that timber harvesting is done in a methodological way that preserves forest health and biodiversity, reducing the chances of habitat loss and species extinction. Moreover, in marine ecosystems, big data from sensor networks that monitor fish populations, ocean temperatures, and nutrient levels can help implement sustainable harvesting practices for fisheries. By integrating this collected big data into AI models, governing bodies can predict population declines or overfishing risks, adjusting quotas and fishing policies accordingly. This ensures that fish stocks are harvested at sustainable levels, preserving marine biodiversity while supporting the livelihoods of fishing communities. Besides, an adaptive resource management system could also be implemented, where environmental management strategies can be updated continuously based on real-time environmental changes and feedback. For example, in water resource management, real-time data on rainfall, river flow, and groundwater levels, combined with temperature metrics, can be fed into an AI model that predicts optimal water allocation for agriculture, energy production, and urban use, ensuring that usages do not exceed natural replenishment rates.

By applying these data-driven insights, AI and big data enable a balance between human exploitation of natural resources and the long-term health of ecosystems, promoting sustainability and reducing negative ecological impacts. The ability to optimize resource use with precision not only supports conservation efforts but also supports the growth of developing economies that rely heavily on natural resource use.

Taxonomical and Evolutionary Insights

By applying big data concepts to our growing database of wildlife genomics, we can also unveil previously unknown taxonomical and evolutionary links and connections that would otherwise remain invisible to the naked eye. Taxonomy is the scientific study of naming and categorizing groups of organisms based on shared characteristics. Traditionally, taxonomy is an old field of science that relied more on morphological similarities than genetic similarities. However, over the past few decades, there has been a rise in the use of molecular methods in the field of taxonomy, accompanying the advancement of molecular biology. By using molecular methods like DNA sequencing, genetic data can be cataloged for all sorts of organisms, and these vast amounts of genetic data benefit greatly from the use of powerful AI predictive models (López-Rubio et al., 2021). AI-driven algorithms can be used to analyze and group species more efficiently, allowing the taxonomic tree to be reshaped in an unprecedented manner. This will deepen our understanding of global biodiversity and will bring to light previously unknown species that may be at risk of extinction, boosting conservation efforts.

Similarly, with big data and data science introduced to this field, genetic variation in the gene pool can easily be tracked over time, uncovering truths about evolution and even witnessing microevolution occurring in real-time. It can also reveal how species adapt to environmental pressures like climate change, habitat loss, and diseases genetically. For example, in a paper by Sunday et al. (2011), an attempt was made to quantify evolutionary adaptations in response to ocean acidification using a full-factorial breeding design that evaluates genetic variation by assessing different breeding combinations. By leveraging IoT sensors and big genomic data, a clearer link could be established between the two while using fewer resources.

Disease Ecology Insights

Especially after the COVID-19 pandemic, disease ecology is seeing more importance with the impact it can bring. Big disease data have become readily available, and data science methods are being increasingly used in the field (Doherty et al., 2021). Zoonotic diseases, like COVID-19 or the avian flu, can be predicted and tracked by feeding environmental and animal migration data derived from sensors, as well as genetic data, into AI prediction models. By using big data analytics, disease ecologists can effectively identify patterns of disease transmission, model potential outbreaks, and develop mitigation strategies to prevent the spread of pathogens to new populations or even to humans. This kind of disease-predictive capability is invaluable in today's world, where habitat encroachment and climate change are increasing the likelihood of wildlife diseases spilling over into human populations.

Challenges for the Future Direction

In this particular section, we will explore the technical and logistical hurdles that must be overcome to achieve this goal, including miniaturization of components, energy efficiency, and the integration of advanced AI for real-time data analysis.

Software

We can divide the source of these challenges based on the two basic fundamentals of AI. The software and hardware. In the first portion, we address the rapid advancement of these AI tools that are readily available and how they can be tailored to process vast amounts of ecological data efficiently. This includes developing algorithms that can learn from diverse datasets (Korzeniowski & Goczyła, 2019), improving their predictive capabilities, and ensuring they can operate in real-time under varying environmental conditions. AI development has become more streamlined and geared towards the common people compared to 10 years ago when it was more geared towards scientists and researchers. The democratization of AI technology has led to a surge in citizen science initiatives, where everyday individuals can contribute to ecological monitoring and data collection, further enhancing the breadth of information available for analysis. This shift not only empowers communities but also fosters a deeper connection between people and their local environments, encouraging stewardship and awareness of ecological issues.

Nowadays, platforms with graphical user interfaces (GUIs) have made AI accessible to a broader audience, eliminating the need for extensive coding knowledge. This has empowered individuals from various backgrounds, including young students and older adults, to engage with AI, fostering a more inclusive and innovative community. These tools often include drag-and-drop functionalities, pre-built algorithms, and visualizations, making AI learning and application more intuitive and engaging for everyone. This collaborative approach has the potential to drive innovative solutions to pressing environmental challenges as diverse perspectives and local knowledge come together to inform decision-making and policy development.

Whilst AI might empower citizen science participation, it does bring into another set of ethical considerations that must be addressed to ensure responsible use and equitable access (Ceccaroni et al., 2019). The open sharing of ecological data, while beneficial for scientific progress, presents a risk of misuse. For instance, poachers could exploit publicly available datasets to reverse-engineer AI models initially designed for wildlife protection. This could enable them to track the movements of wildlife or even locate park rangers, undermining conservation efforts. To mitigate this risk, it's crucial to implement data-sharing protocols that balance accessibility with security and develop AI systems with robust safeguards against exploitation.

Furthermore, without proper validation of the dataset gathered, there is a high chance of the AI developing a biased model that may misinterpret species' behavior or habitat needs, leading to ineffective conservation strategies. For example, an AI trained on incomplete or skewed data might prioritize certain species over others, neglecting those that are equally endangered but less visible in the dataset. This could result in a misallocation of resources, ultimately harming biodiversity and the ecosystems that depend on it. A person might report a sighting of a whale in the harbor; this would then be promoted on social media, which would invariably create a viral trend of reporting sightings, potentially overwhelming researchers with unverified information and diverting attention from critical conservation efforts that require immediate action. This influx of data, while seemingly beneficial, can lead to confusion and misinformation, making it challenging for scientists to discern genuine threats to marine life from mere social media buzz. Another disparaging situation might be the over-reporting of certain species, which could skew public perception and funding towards those animals while neglecting others that are equally or more endangered. This imbalance in attention can ultimately hinder comprehensive conservation strategies, as resources become misallocated and critical habitats remain unprotected.

Nowadays, there exists a number of MLOps, such as Roboflow and EdgeImpulse (Motta et al., 2024). These two platforms curate hundreds and thousands of datasets, enabling researchers to train machine learning models that can analyze and interpret data more effectively. Whilst essentially a free platform, these MLOps work by utilizing a tiered subscription model that offers advanced features and support for users seeking to enhance their projects. This is a good feature for developers, although it does put a huge restraint on the common free-tiered users. Free-tiered user models are obligated to be shared on the platform, which can limit the potential for proprietary research and innovation. As a result, many developers are exploring alternative solutions that offer more flexibility while still providing robust tools for model training and deployment. This shift has led to a growing interest in open-source frameworks that prioritize user autonomy and data privacy, allowing developers to maintain control over their intellectual property while still benefiting from community-driven advancements. On the other hand, this raises the problem of data security and ethical use of sensitive information, as the open-source nature can sometimes lead to vulnerabilities if not managed properly. Some of the data shared on the public platform of Roboflow Universe contains images of actual people, places, and even properties that might be protected by privacy laws. This necessitates a careful approach to data governance, ensuring that contributors are aware of the implications of sharing such content and that robust measures are in place to anonymize or secure sensitive information before it is made publicly accessible.

Another future trend that might emerge is the creation of specialized GPTs for animal ecology projects. Currently, OpenAI encourages the creation of custom-made GPTs that can be trained via uploading datasets specific to various fields, allowing researchers and conservationists to develop tailored models that can more effectively analyze animal behavior, habitat usage, and ecological impacts. This could lead to more precise conservation strategies and an improved understanding of species interactions within their ecosystems, ultimately fostering better decision-making in wildlife management and preservation efforts. As these specialized models evolve, they may also incorporate real-time data from field studies, enhancing their predictive capabilities and enabling proactive measures to protect endangered species. Ultimately, we might be seeing the end of the role of human project managers and the rise of AI-driven systems that can autonomously manage projects, analyze data, and implement strategies with minimal human intervention.

Currently, the advancement in the software portion of AI seems to be heading in the right direction, with the current advancement and also the availability of a virtual Graphical Processing Unit (GPU) that would enable training the AI model in the cloud, therefore reducing the cost and time associated with local hardware limitations. The most applied system is Google Colab, which graciously allows users to leverage powerful GPUs for their machine-learning projects without the need for expensive infrastructure. This democratization of access is paving the way for more innovative solutions and accelerating the pace of research and development across various fields.

Hardware

The second part of future AI in animal ecology deals with the hardware portion. As discussed earlier, the future trend shows promising advancements in LLM that enable sensors to collect vast amounts of data in real time, allowing researchers to monitor animal behavior and environmental changes more effectively. One of the first hurdles is the price of these sensors. First-generation technology tends to oversell and often falls short of expectations. Also, they tend to be expensive due to the relatively new market and perhaps less demand. This is intricately tied with the development of the rare earth materials and components that are essential for manufacturing these advanced devices. As the market matures and production scales up, we can anticipate a decrease in costs, making these tools more accessible to researchers worldwide.

Portability for Deployment

When creating these multifunctional LLM sensors, the design and portability of the device must be taken into account. This includes ensuring that they are lightweight and user-friendly, allowing for easy deployment in various environments. Referring back to the fictional tricorders, they must fit into your hand. Unsurprisingly, our current smartphones are full of sensors that can be turned into useful ecological data loggers. However, they are not as hardy nor dispensable as the dedicated devices we envision.

Power Source

This would bring us to another challenge: the power source. The power source needs to be both efficient and sustainable, ideally utilizing renewable energy options to ensure longevity in the field. For any smart sensors, there must be at least a microcontroller that can handle basic processing powers. While we usually think of Intel or AMD for larger-size CPUs, the most popular choice for processing prowess for small devices is often ARM architecture, which offers a balance of performance and energy efficiency suitable for compact applications. Other options include the Raspberry Pi ecosystem, ESP32, ESP8266, RISVC, and various microcontroller units (MCUs) that can be tailored to specific tasks, allowing for flexibility in design and functionality. These platforms not only support a wide range of programming languages but also come with extensive community support, making it easier for developers to innovate and troubleshoot as they create interconnected systems. These cheap microcontrollers are quite powerful enough to run as Edge devices, capable of running local TinyML. This brings us back to our original hurdle: the battery. To ensure longevity and efficiency, selecting the right battery technology is crucial, as it directly impacts the performance and operational lifespan of these devices in the field.

If the sensors are to be portable, the batteries must be light and small while still providing sufficient energy to support continuous operation. Lithium polymer batteries, for instance, offer a good balance between weight and capacity, making them a popular choice for such applications. However, running an AI model, even on small Edge devices, consumes a lot of computing power. This translates to more energy needed. A tethered device can mitigate this issue by drawing power from a stable source, allowing for more robust processing capabilities without the constraints of battery life. If deployed in the field, solar panels can provide a sustainable energy solution, enabling longer operational periods and reducing reliance on traditional power sources. Additionally, integrating energy-efficient algorithms can further optimize power consumption, ensuring that the device operates effectively while minimizing energy use. For example, the ESP32 and ESP8266 both have deep sleep functions that essentially allow the device to enter a low-power state when not in use, significantly extending battery life and making them ideal for IoT applications where energy conservation is crucial.

Heat Dissipation

Moreover, utilizing energy harvesting techniques, such as kinetic or thermal energy conversion, can complement these strategies, providing additional power sources that enhance the device's autonomy and functionality in various environments. While on the subject of the power challenges for AI systems, this is inevitably linked to another challenge. Heat dissipation. All electronic systems release heat; the more complex or rigorous the system, the more heat is generated, which can lead to performance degradation or even failure if not managed properly. To address the challenge of heat dissipation in advanced AI systems deployed for ecological monitoring, innovative cooling solutions are emerging that leverage passive and active thermal management techniques. For instance, integrating phase change materials (PCMs) within device enclosures can absorb excess heat generated during operation, thereby maintaining optimal temperatures without relying heavily on additional power sources. This approach not only enhances system reliability but also contributes to energy efficiency, an essential aspect given the constraints often faced by remote devices operating in challenging environments (Molina-Molina et al., 2021). Moreover, the use of advanced materials such as graphene and aerogels is being explored to further improve thermal conductivity and insulation, allowing for even more effective heat management strategies.

Data Retrieval

The third other possible challenge is the data retrieval portion from the sensors. As envisioned for the future, we would ideally have robots and drones that could initiate the sampling process autonomously, collecting and transmitting data in real time to a central hub for analysis. This would not only streamline operations but also enhance the accuracy of the data collected, enabling quicker decision-making and response times in critical situations. However, as most projects involving tracking animals occur in parts of the world where the environment can be unpredictable, ensuring reliable connectivity and power sources for these autonomous systems presents another significant hurdle that must be addressed. To further enhance the effectiveness of autonomous ecological monitoring systems, integrating advanced communication technologies such as 5G and satellite networks could play a pivotal role in ensuring real-time data transmission from remote locations. These technologies would not only facilitate seamless connectivity for drones and robots but also support larger-scale deployments across diverse ecosystems, addressing the challenge of reliable data retrieval amidst unpredictable environmental conditions. Moreover, leveraging AI-driven predictive analytics can optimize the timing and location of data collection efforts by analyzing historical patterns and current environmental factors, thus increasing the likelihood of capturing critical data on species behavior and habitat changes (Hegde & Bargavi, 2024). This holistic approach to incorporating cutting-edge communication infrastructure with intelligent data processing capabilities promises to revolutionize conservation strategies, ultimately enabling more responsive and adaptive management practices that are essential for preserving biodiversity in an era of rapid ecological change.

Lora Technology

One of the cheapest and easiest setups for data retrieval and information sharing is through the LoRa network. LoRa (Long Range) technology is a low-power, wide-area networking protocol designed for long-range communication with minimal energy consumption. One notable application of this technology is Meshtastic, an open-source project that enables users to create a mesh network for text messaging and data transfer using LoRa-enabled devices. This capability allows for reliable communication in remote areas where traditional cellular networks may be unavailable or unreliable.

In the context of animal ecology, LoRa technology can be a game-changer for wildlife monitoring and conservation efforts. By deploying LoRa-enabled sensors and tracking devices, researchers can create a network that collects and transmits data on animal movements, habitat use, and environmental conditions in real-time (Bandari et al., 2022). This real-time data collection can facilitate more informed decision-making and adaptive management strategies, especially in challenging terrains where accessibility is limited. Additionally, the low power requirements of LoRa devices mean they can operate for extended periods without the need for frequent battery replacements, making them ideal for long-term ecological studies. Overall, the integration of LoRa technology in animal ecology holds great promise for enhancing data collection, improving conservation strategies, and fostering a deeper understanding of wildlife behavior. Moreover, as the reliance on technologies like LoRa for wildlife monitoring grows, it becomes essential to consider the role of data integration platforms that can combine diverse datasets collected from various sources. Such platforms could facilitate a more comprehensive understanding of ecological dynamics by merging real-time sensor data with historical records and citizen science contributions, thus enhancing the granularity and accuracy of analyses. For instance, integrating data from camera traps, acoustic sensors, and satellite imagery could create a multifaceted view of species interactions and habitat changes, enabling researchers to identify emerging trends or threats in biodiversity conservation effectively. Furthermore, this approach aligns with the principles of FAIR (Findable, Accessible, Interoperable, Reusable) data management, which encourages collaborative efforts among ecologists and technologists to ensure that valuable insights are not only generated but also shared responsibly across communities engaged in conservation work (Mergen et al., 2023). Ultimately, fostering such integrative frameworks will be vital in addressing complex challenges in animal ecology while ensuring that data-driven decisions support both scientific advancement and ethical stewardship of natural resources. By leveraging these technologies, conservationists can develop targeted strategies that not only protect endangered species but also promote ecosystem resilience in the face of climate change and habitat degradation. This technology allows for long-range communication with minimal power consumption, making it ideal for remote areas where traditional networks may be unreliable or non-existent.

Brain-Computer Interface (BCI)

Earlier in this chapter, we discussed the future communication with animals. Using AI, we hope to decipher the animal's language. There is, however, a new possible future AI trend that combines both the software and the hardware portion of AI and animal communications. Brain-computer interfaces (BCIs), initially developed for medical applications in humans, are now being explored for their potential use in animal studies. By directly monitoring neural activity, these interfaces could enable a deeper understanding of animal mental states, emotions, and behavioral patterns, facilitating improved conservation and management strategies (Mercier-Ganady et al., 2013). This innovative approach could lead to breakthroughs in how we interact with wildlife, allowing researchers to gather real-time data on stress responses or social dynamics within animal groups.

One of the primary applications of BCIs in animal ecology lies in wildlife rehabilitation. Animals under rehabilitation often exhibit stress or discomfort that may not be immediately apparent through external behaviors (Bamdad et al., 2015). Neural interfaces, integrated with non-invasive sensors, could monitor neural responses in real-time, alerting caretakers to signs of distress. Such insights would enable more responsive and effective care, reducing recovery time and improving animal welfare. This form of monitoring could also play a critical role in conservation breeding programs by tracking animals' physiological responses during reproduction efforts, thereby increasing the success rate of these programs.

Another critical application of neural interfaces is in the management of animal translocation programs. Relocating animals to new habitats, especially in human-wildlife conflict areas, requires careful planning and monitoring. BCIs could provide valuable insights into how animals adapt to unfamiliar environments, revealing their emotional and behavioral states. For instance, elephants, known for their complex social structures, might experience heightened anxiety when separated from their group. Monitoring their neural activity could help conservationists tailor relocation strategies, minimizing stress and ensuring smoother transitions.

BCIs also hold the potential to transform the way researchers study animal cognition. Neural data can provide unprecedented insights into how animals perceive their environment, make decisions, or communicate within their species. Understanding these cognitive processes will allow scientists to build more accurate models of animal behavior, leading to enhanced conservation strategies. Additionally, insights into emotional states, such as fear or contentment, can improve our understanding of social dynamics in species like primates or cetaceans.

However, the application of BCIs in animal ecology is not without challenges. Ethical considerations regarding the use of neural technology in animals must be addressed to ensure that the well-being of animals remains a priority. Developing non-invasive or minimally invasive interfaces will be critical to achieving this balance. Furthermore, the integration of interdisciplinary knowledge, combining neuroscience, ecology, and conservation biology, will be essential to unlocking the full potential of BCIs in animal studies.

Dangers of AI in the Future

Suleyman (2023) wrote an interesting excerpt in his book about the pitfalls and dangers of AI, emphasizing the need for careful consideration of ethical implications and the potential for unintended consequences in its application. In his book, the author describes a tech-sharing event that the stakeholders and leaders of AI attended. This was an event several years before the COVID-19 outbreak. During one of the breakout sessions, there was a presentation that stood out. This excerpt is rephrased from his book: "The presenter illustrated how the cost of DNA synthesizers, which can generate custom DNA sequences, has been declining rapidly. These devices, priced in the range of several tens of thousands of dollars, are compact enough to fit on a workbench at home, enabling individuals to produce DNA. This capability is now accessible to anyone with a graduate-level understanding of biology or a strong interest in self-directed online learning. The presenter offered a grim warning: it may soon be possible to engineer synthetic pathogens that are more contagious and deadly than any naturally occurring ones. Such pathogens could potentially bypass current defences, spread without noticeable symptoms, or even be designed to resist existing treatments. Additionally, individuals could enhance home experiments by ordering DNA sequences online and assembling them on their own, making the scenario of catastrophic biothreats alarmingly feasible through mail-order resources."

And there was the chilling remark that a single person today can kill millions of people. All they need is motivation. The implications of this reality are staggering, raising urgent questions about bioethics, regulation, and the responsibility of scientists in an age where knowledge can be weaponized.

Conclusion

Although the previous paragraph is quite ominous, it does not suggest a negative concluding outcome for AI in the future. AI is a tool. As it is historically depicted, any tool can be used either to create or to destroy. From the perspective of science, especially in animal ecology, there is more good that can be gained from AI than bad. Future trends remain a prediction; future readers reading this chapter in the next 10 or 50 years might be smiling and amused at what might be viewed as naivety by our generation. What remains important is the role of humans in utilizing, creating, and carefully managing AI for a brighter future and co-existence with the animals.

References

Andreas, J., G. Beguš, M.M. Bronstein, R. Diamant, D. Delaney, S. Gero, S. Goldwasser et al. (2021). Cetacean translation initiative: A roadmap to deciphering the communication of sperm whales. arXiv preprint arXiv:2104.08614.

Bamdad, M., H. Zarshenas, and M.A. Auais. (2015). Application of BCI systems in neurorehabilitation: A scoping review. *Disability and Rehabilitation: Assistive Technology*, 10(5): 355–364.

Bandari, G., L.N. Devi, and P. Srividya. (2022, December). Wild animal detection using a machine learning approach and alerting using LoRa communication. In: *2022 International Conference on Smart Generation Computing, Communication and Networking (SMART GENCON)*, pp. 1–5. IEEE.

Bateson, M., and P. Martin. (2021).*Measuring Behaviour: An Introductory Guide,* 4th Edn. Cambridge University Press.

Busse, A., J. Eberhardt, S. Frost, D.-H. Kim, T. Weilbier, L. Renner, M. Roth et al. (2019, May 14). *A Response to the United Nations CITES Blockchain Challenge: Incremental and Integrative PoA-based Permit Exchange.* https://doi.org/10.1109/BLOC.2019.8751373.

Butail, S., N. Abaid, S. Macrì, and M. Porfiri. (2015). Fish–robot interactions: Robot fish in animal behavioral studies. In: *Robot Fish: Bio-Inspired Fishlike Underwater Robots*, pp. 359–377. ResearchGate. https://www. researchgate.

Cauzinille, J., B. Favre, A. Rey, and R. Marxer. (2024). *Applying Machine Learning to Primate Bioacoustics: Review and Perspectives.* https://doi.org/10.1002/ajp.23666.

Ceccaroni, L., J. Bibby, E. Roger, P. Flemons, K. Michael, L. Fagan, and J.L. Oliver. (2019). *Opportunities and Risks for Citizen Science in the Age of Artificial Intelligence.* https://doi.org/10.5334/CSTP.241.

Dhyani, S., S. Kathuria, S. Pandey, and G. Chhabra. (2023, March 14). *AI Technologies, Innovations, and Possibilities in Wildlife Conservation.* https://doi.org/10.1109/ICIDCA56705.2023.10099721.

Doherty, J.F., X. Chai, L.E. Cope, D. de Angeli Dutra, M. Milotic, S. Ni, E. Park, and A. Filion. (2021). The rise of big data in disease ecology. *Trends in Parasitology*, 37(12): 1034–1037. https://doi.org/10.1016/J.PT.2021.09.003.

Dryga, A., S. Tsiulin, M. Valiavko, Y. Qing, and K.H. Reinau. (2019, August 28). Blockchain-based wildlife data-management framework for the WWF Bison Rewilding project. In: *International Conference on Big Data.* https://doi.org/10.1145/3358528.3358530.

Hegde, N.R., and Dr. S.M. Bargavi. (2024). Smart Conservation: Integrating AI for enhanced wildlife monitoring. *International Research Journal of Computer Science.* https://doi.org/10.26562/irjcs.2024.v1101.10.

Jahns, G. (2013). *Computational Intelligence to Recognize Animal Vocalization and Diagnose Animal Health Status.* https://doi.org/10.1007/978-3-642-32378-2_16.

Kissling, W.D., J.C. Evans, R. Zilber, T.D. Breeze, S. Shinneman, L.C. Schneider, and L.H. Geelen. (2024). Development of a cost-efficient automated wildlife camera network in a European Natura 2000 site. *Basic and Applied Ecology*, 79: 141–152.

Kline, J., M. Kholiavchenko, O. Brookes, T.Y. Berger-Wolf, C.V. Stewart, and C. Stewart. (2024). Integrating biological data into autonomous remote sensing systems for *in situ* imageomics: A case study for Kenyan animal behavior sensing with unmanned aerial vehicles (UAVs). arXiv.Org. https://doi.org/10.48550/arxiv.2407.16864.

Korpela, J., H. Suzuki, S. Matsumoto, Y. Mizutani, M. Samejima, T. Maekawa, J. Nakai. (2019). AI on animals: AI-assisted animal-borne logger never misses the moments that biologists want. *bioRxiv*. https://doi.org/10.1101/630053.

Korzeniowski, Ł., and K. Goczyła. (2019). Artificial intelligence for software development: The present and the challenges for the future. *Biuletyn Wojskowej Akademii Technicznej*. https://doi.org/10.5604/01.3001.0013.1464.

Li, X., H. Huang, and A.V. Savkin. (2022). Autonomous navigation of an aerial drone to observe a group of wild animals with reduced visual disturbance. *IEEE Systems Journal*. https://doi.org/10.1109/jsyst.2021.3135982.

Lin, Y.-H., and O.N.N. Fernando. (2023). *Animal Hunt: AI-Based Animal Sound Recognition Application*. https://doi.org/10.1007/978-3-031-36004-6_64.

López-Rubio, E., and E. Ratti. (2021). Data science and molecular biology: Prediction and mechanistic explanation. *Synthese*, 198(4): 3131–3156. https://doi.org/10.1007/S11229-019-02271-0.

Ma, Z. (Sam). (2015). Towards computational models of animal communications, an introduction for computer scientists. *Cognitive Systems Research*. https://doi.org/10.1016/J.COGSYS.2014.08.001.

Mathieu, C., S.M. Hermans, G. Lear, T.R. Buckley, K.C. Lee, and H.L. Buckley. (2020). A systematic review of sources of variability and uncertainty in eDNA data for environmental monitoring. *Frontiers in Ecology and Evolution*, 8: 135.

Mauvisseau, Q., E. Kalogianni, B. Zimmerman, M. Bulling, R. Brys, and M. Sweet. (2020). eDNA-based monitoring: Advancement in management and conservation of critically endangered killifish species. *Environmental DNA*, 2(4): 601–613.

Mazumdar, S. (2022). *Drone Applications in Wildlife Research: A Synoptic Review*. https://doi.org/10.1007/978-981-19-2083-7_14.

McCowan, B., J. Hubbard, L. Walker, F. Sharpe, J. Frediani, and L. Doyle. (2023). Interactive bioacoustic playback as a tool for detecting and exploring nonhuman intelligence: "Conversing" with an Alaskan humpback whale. *PeerJ.*, 11(e16349). https://doi.org/https://doi.org/10.7717/peerj.16349.

Mercier-Ganady, J., E. Loup-Escande, L. George, C. Busson, M. Marchal, and A. Lécuyer. (2013, October). Can we use a brain-computer interface and manipulate a mouse at the same time? *Virtual Reality Software and Technology*. https://doi.org/10.1145/2503713.2503744.

Mergen, P., S. Meeus, and F. Leliaert. (2023). FAIR but not necessarily open: Sensitive data in the domain of biodiversity. *Biodiversity Information Science and Standards*. https://doi.org/10.3897/biss.7.112296.

Molina-Molina, J.C., M. Salhaoui, M. Salhaoui, A. Guerrero-González, and M. Arioua. (2021). Autonomous marine robot based on AI recognition for permanent surveillance in marine protected areas. *Sensors*. https://doi.org/10.3390/S21082664.

Motta, I.V.D.C., F.A.P. de Figueiredo, and S.B. Mafra. (2024). *Tiny Machine Learning for Classifying Specialty Coffees*. Authorea Preprints.

Panda, P.K., C.S. Kumar, B.S. Vivek, M. Balachandra, and S.K. Dargar. (2022). Implementation of a wild animal intrusion detection model based on internet of things. In: *2022 Second International Conference on Artificial Intelligence and Smart Energy (ICAIS)*, pp. 1256–1261.

Project Earth Species. (n.d.). *Project Earth Species*. https://www.earthspecies.org/

Raffik, R., M. Swetha, R.R. Sathya, V. Vaishali, B. Adithya, and S. Balavedhaa. (2024). *Integration of Unmanned Aerial Vehicle Systems with Machine Learning Algorithms for Wildlife Monitoring and Conservation*. https://doi.org/10.4018/979-8-3693-0578-2.ch006.

Raghuwanshi, S.S., S. Sharma, S. Singh, and S. Kumar. (2023). Application of artificial intelligence in wildlife disease surveillance. *Journal of Nonlinear Analysis and Optimization: Theory & Applications*. https://doi.org/10.36893/jnao.2022.v13i02.054-059.

Roy, Dr. R.R., D.V.K. Pandey, Dr. R. Raman, Dr. E.N. Ganesh, and Dr. M. Amanullah. (2023). *IoT Applications in Wildlife Conservation: Tracking and Protecting Endangered Species*. https://doi.org/10.1109/iceca58529.2023.10395145.

Rutz, C., S.C. Vernes, K. Zacarian, and D.E. Blasi. (2023). Using machine learning to decode animal communication. *Science*. https://doi.org/10.1126/science.adg7314.

Saarenmaa, H., N.D. Stone, L.J. Folse, J.M. Packard, W.E. Grant, M.E. Makela, and R.N. Coulson. (1988). An artificial intelligence modeling approach to simulating animal/habitat interactions. *Ecological Modelling*. https://doi.org/10.1016/0304-3800(88)90085-3.

Seeedstudio Watcher. (n.d.). Watcher. Retrieved October 20, 2024. https://www.seeedstudio.com/watcher?gad_source=1&gclid=CjwKCAjw1NK4BhAwEiwAVUHPUC7HJk8EtdCbyBltl81u4K2EHLVP4Hq3UMMOXzHCjs9_KvE__gQwMBoCtEcQAvD_BwE.

Siewert, S., L.F. Zapata-Rivera, C. Aranzazu-Suescun, G. Waldron, R. Mangar, D. Raval, P. Vaddkkepurakkal et al. (2024). *Sensor Fusion with Multi-modal Ground Sensor Network for Endangered Animal Protection in Large Areas*. https://doi.org/10.1117/12.3012684.

Stuit, A., D. Brockington, and E. Corbera. (2022). Smart, commodified, and encoded: Blockchain technology for environmental sustainability and nature conservation. *Conservation and Society*, 20(1): 12–23.

Suleyman, M. (2023). *The Coming Wave: Technology, Power, and the Twenty-first Century's Greatest Dilemma*. Crown Publisher.

Sunday, J.M., R.N. Crim, C.D.G. Harley, and M.W. Hart. (2011). Quantifying rates of evolutionary adaptation in response to ocean acidification. *PLoS One*, 6(8): e22881. https://doi.org/10.1371/JOURNAL.PONE.0022881.

Suren, A.M., F.J. Burdon, and S.P. Wilkinson. (2024). eDNA is a useful environmental monitoring tool for assessing stream ecological health. *Environmental DNA*, 6(4): e596.

Wu, H.Y., M. Rubinstein, E. Shih, J. Guttag, F. Durand, and W. Freeman. (2012). Eulerian video magnification for revealing subtle changes in the world. *ACM Transactions on Graphics*, 31(4). https://doi.org/10.1145/2185520.2185561.

Index

A

Acoustics, 71, 75
AI applications, 168, 173, 176
AI genome informant, 181
AI limitations, 188
Animal behavior, 44-47, 58, 60
Animal ecology, 69, 140, 144-152, 156, 163, 164, 169
Animal kingdom, 65-67, 71
Artificial Intelligence, 44, 45, 60, 65, 67, 83, 107, 122, 136-138, 146, 168, 169, 171, 172, 181, 188, 193, 196, 199, 221, 222, 238

B

Behavioral strategies, 145
Big data management, 193
Bio-inspired optimization algorithm, 1, 13
Brain-computer interface, 260, 261

C

Camera trap, 193, 194, 199, 200, 205
Causal inference, 106
Communications, 122, 129
Convolutional neural networks, 44, 45

D

Deep learning, 222, 223, 228, 238
Developmental challenges, 169-171, 173, 176, 179
Differential evolution, 1, 4
Dolphin echolocation algorithm, 18, 19, 23, 24
Dragonfly algorithm, 18, 20, 33

E

Ecological modeling, 107
Ecological sustainability, 164
Elephant herding optimization, 18, 19, 26
Environmental DNA (eDNA), 244, 247
Environmental stimuli, 145, 153

G

Genetic algorithm, 1, 2

H

Habitat evaluation, 185

I

Illegal logging and poaching, 205
Interactions, 122-127

L

Large Language Model (LMM), 250
Lion optimization algorithm, 18, 19, 27
Lora technology, 259, 260

M

Machine learning, 44-47, 49, 55-57, 60, 82, 83, 87, 88, 93, 222, 223, 229, 238

Maintaining forest health, 195, 197-199, 211
Microbiology, 221-223, 226, 231
Microorganisms, 221-226, 228-236, 239
Moth-flame optimization, 18, 20

P

Pathogens, 227, 228, 231, 235, 236, 239
Population dynamics, 144, 145, 158, 161

S

Satellite imagery, 203
Species interaction model, 82, 84
Species, 122-130
Swarm optimization, 1, 2, 5

V

Vocal communication, 65-69, 71, 74

W

Whale optimization algorithm, 18-20